Neem: Today and in the New Millennium

Neem: Today and in the New Millennium

Edited by

Opender Koul
Director, Insect Biopesticide Research Centre,
Jalandhar, India

and

Seema Wahab
Advisor, Department of Biotechnology,
CGO Complex, New Delhi, India

KLUWER ACADEMIC PUBLISHERS
DORDRECHT / BOSTON / LONDON

A C.I.P. Catalogue record for this book is available from the Library of Congress.

ISBN 1-4020-1229-2

Published by Kluwer Academic Publishers,
P.O. Box 17, 3300 AA Dordrecht, The Netherlands.

Sold and distributed in North, Central and South America
by Kluwer Academic Publishers,
101 Philip Drive, Norwell, MA 02061, U.S.A.

In all other countries, sold and distributed
by Kluwer Academic Publishers,
P.O. Box 322, 3300 AH Dordrecht, The Netherlands.

Printed on acid-free paper

Printed in the Netherlands.

Contents

Preface

The chemicals from plant sources, generally termed as phytochemicals, play an important role in acceptance or rejection of the plant by the pests as they could be distasteful or toxic on one hand or on the other hand specialist herbivores have the capability to feed on many such chemicals, as they are able to process these natural products in a manner that is beneficial to them. In the wake of increasing environmental degradation due to burgeoning synthetic chemicals, there has been a process going on to rediscover the usefulness of plants and herbs and a continued effort for more than 2 decades has been to study the green products for cures for several ailments and pest management. In fact, according to Indian Medicinal Plants: A Sectoral Study, the global trade for medicinal plants amounts to about US $ 60 billion and the world demand continues to grow at the rate of 7 per cent per annum. Although many such plants are known in literature, neem has been one of trees with mani-fold virtues.

Indian neem tree, *Azadirachta indica* A. Juss, which is a large evergreen tree, is an outstanding example among plants that has been subject matter of numerous scientific studies concerning its utilization in medicine, industry and agriculture. So far neem preparations have been evaluated against more than 500 species of insects and more than 400 hundred are reported to be susceptible at different concentrations. Many commercial formulations of neem are now available in various countries; therefore, neem has come to the fore as the model botanical biopesticide for the new millennium, which has been successfully used as pesticide in practically all crops. It is also effective in animal husbandry for the control of ectoparasites. This implies that neem has numerous attributes that should ensure its adoption and success, i.e. efficacy against a broad spectrum of pests, minimal mammalian toxicity, minimal impact on pollinators and natural enemies, and rapid disappearance from the environment. In spite of these attributes and enormous scientific effort spawning more than two decades, neem has made little impact in the US market, and has yet to be approved for use in many other industrialized countries (though it has been approved recently in Sweden and Switzerland, and registration is pending in Germany and Canada). Part of the problem has been that these countries require highly refined and standardized products. The main problem hindering greater acceptance of neem by farmers includes poor dissemination of neem related knowledge and the fact in those regions in which neem could be used successfully, there are not enough trees or none at all. Another important reason is the lack of professional marketing strategies for neem.

In terms of the active allelochemicals from neem, more than 200 active components occurring in different parts of the tree, have variety of effects on pests. The most active among these are azadirachtins (tetranortriterpenoids) that occur in the seed core in the range of 0.1 to 0.9 per cent and 30 to 60 g of azadirachtin per ha suffice, in order to combat and repel the key biting and sucking pests. However, many concepts in azadirachtin mode-of-action have been documented, but lot is required to be done in this direction.

If we look at overall scenario of neem research, scores of articles and some books have been published which are based on general evaluation of neem preparations against various pests. The question, therefore, is that why another volume on neem? The answer is simple that the present volume has different approach and answers some pertinent questions of applied nature. Although the first chapter gives a general global perspective of neem status, second chapter discusses the place of neem among modern natural pesticides. As mentioned above that neem has little impact on US market, the factors limiting commercial success of neem insecticides in North America and Western Europe have been discussed in chapter 3. The African and Asian continents are rich in neem raw materials, therefore, what is the status of neem research in these areas and what is the outlook for the new millennium is the content of chapters 4 and 5. Various risk implications of neem and various regulatory processes involved for the development of neem based products are the topics which have been included as chapters 6 and 7 and make a very important component of biopesticide development process.

The specific topics ignored in earlier volumes have been included in this book. Natural enemies are very important in tritrophic interaction in an ecosystem, therefore, it is important to know the impact of neem on these beneficial organisms. Accordingly chapter 8 discusses neem versus entomopathogens and natural enemies of crop pests and various strategies adopted to overcome the hazards, if any. Chapter 9 also discusses the use of neem for plant pathogenic fungal control.

The biotechnological approach has been only developed in the last 15 years for neem. Micropropagation can be carried out successfully on different usable protocols. This is very important because of the recalcitrant nature of the neem seeds and also if one wants to maintain a healthy plantation. Also, if it can be proven that elite producing trees exist this will be a very powerful tool for multiplying "true-to-type" plantlets. This aspect has been comprehensively dealt with in chapter 10.

Mode-of-action is still not very clear due to mani-fold activities of neem allelochemicals and azadirachtin in particular. Therefore, chapter 11 describes the mode of action of azadirachtin against insects, at the whole animal level as well as the cellular level where the basic lesion(s) occur. It is important to compare efficacies of azadirachtin action across species, to highlight, when using azadirachtin as a neem insecticide, the important safety margins between insects and vertebrates. The last chapter of this book gives an overall synthesis of neem biotechnology as it stands today with future outlook in the new millennium.

We are thankful to all contributors for the meticulous job they have done in preparing their respective chapters within the stipulated period. We are also grateful to the reviewers of various manuscripts for taking their time to give useful suggestions for the improvement of the chapters. It is hoped that the book would prove to be of immense use to all those who are involved in neem research, insect plant interactions, toxicology, biological control, chemical ecology, behaviour and integrated pest management (IPM). It would also stimulate further research in the vital areas of mode-of-action, micropropagation, and IPM.

OPENDER KOUL
SEEMA WAHAB

PROFILE EDITORS

Opender Koul, Fellow of the National Academy of Agricultural Sciences and the Indian Academy of Entomology, is an insect toxicologist/physiologist/chemical ecologist and currently the Director of the Insect Biopesticide Research Centre, Jalandhar, India. After obtaining his Ph.D. in 1975 he joined the Regional Research Laboratory (CSIR), Jammu and then became Senior Group Leader of Entomology at Malti-Chem Research Centre, Vadodara, India (1980-1988). He has been a visiting scientist at the University of Kanazawa, Japan (1985-1986), University of British Columbia, Canada (1988-1991), Institute of Plant Protection, Poznan, Poland (2001). His extensive research experience concerns insect-plant interactions, spanning toxicological, physiological and agricultural aspects. Honoured with an Indian National Science Academy medal (INSA) and the Kothari Scientific Research Institute award, he has authored over 140 research papers and articles, and is the author/editor of the books *Insecticides of Natural Origin*, *Phytochemical Biopesticides*, *Microbial Biopesticides, Predators and Parasitoids and Integrated Pest Management: Potential, Constraints and Challenges*. He has also been an informal consultant to BOSTID, NRC of USA and at ICIPE, Nairobi.

Seema Wahab, Adviser, Department of Biotechnology, Ministry of Science and Technology, is a biotechnologist involved in promoting research and development and coordinating programmes in the area of biological control. Her involvement and contribution in the development of biopesticide technologies in the country, it's validation through large scale demonstrations suited to different agroclimatic zones and their large scale adoption by the farmer's is a pioneering step towards the promotion of IPM concept for sustainable agriculture. Dr Wahab after obtaining her PhD from Lucknow University has worked in different institutes including Forschungs Institute, Germany, Institue Pasteur, Paris and National Institute of Health, Maryland, USA. Honoured with Distinguished Achievement award (1998) and Dr. M. R. Siddiqui Gold Medal (2002), Dr. Wahab has published more than 80 research papers and about 30 theme papers in biotechnology. She has also participated as a member of several delegations representing India in Russia, Poland, Australia, China, and Sri Lanka.

Contributors

Ramesh Arora
Department of Entomology
Punjab Agricultural University
Ludhiana- 141 004
E.mail: rarora@redifmail.com

G. S. Dhaliwal
Department of Entomology
Punjab Agricultural University
Ludhiana- 141 004
E.mail: gsd251@redifmail.com

N. Ganapathy
Tamil Nadu Agricultural University
Department of Agricultural Entomology
Agricultural College and Research Institute
Madurai 625 104, India

Germina Giagnacovo
ENEA, C.R. Casaccia
Rome, Italy

Geetha Gopalakrishnan
Centre for Natural Products
SPIC Science Foundation, New 64
Mount Road, Chennai 600 032, India
E.Mail: geethagopal@hotmail.com

Murray B. Isman
Faculty of Agricultural Sciences
University of British Columbia
Vancouver, BC, Canada V6T 1Z4
E.Mail: murray.isman@ubc.ca

Hubertus Kleeberg
Trifolio-M GmbH, Sonnenstrasse 22
D-35633 Lahnau, Germany
E.Mail: info@trifolio.de

Opender Koul
Insect Biopesticide Research Centre
30 Parkash Nagar, Jalandhar- 144 003, India
E.mail: koul@jla.vsnl.net.in

Selladurai Masilamani
Centre for Natural Products
SPIC Science Foundation, New 64
Mount Road, Chennai 600 032, India

A. J. Mordue (Luntz)
Department of Zoology, University of Aberdeen
Tillydrone Avenue, Aberdeen
AB24 2TZ, UK
E.Mail: a.j.mordue@abdn.ac.uk

E. David Morgan
Chemical Ecology Group
Lennard-Jones Laboratory
Keele University, Staffordshire
England ST5 5BG
E.Mail: e.d.morgan@chem.keele.ac.uk

S. Raguraman
Tamil Nadu Agricultural University
Department of Agricultural Entomology
Agricultural College and Research Institute
Madurai 625 104, India
E.Mail: raman63@eth.net

Ramesh C. Saxena
Chairman, Neem Foundation
J-1041 Palam Vihar
Gurgaon 122 017, Haryana, India
E. Mail: susaxena@satyam.net.in

John D. Stark
Washington State University
Puyallup Research and Extension Center
Puyallup, Washington 98371, USA
E.Mail: starkj@wsu.edu

Govindaraghavan Suresh
Centre for Natural Products
SPIC Science Foundation, New 64
Mount Road, Chennai 600 032, India
E.Mail: sureshgovind55@hotmail.com

S. Andrew Van Der Esch
ENEA, C.R. Casaccia
Rome, Italy
E.Mail: vanderesch@casaccia.enea.it

T. Venkatesan
Project Directorate of Biological Control
P.B. No. 2491, H.A. Farm Post
Hebbal, Bangalore-560 024, India

Seema Wahab
Department of Biotechnology
Block 2, 6th Floor
C G O Complex, Lodhi Road
New Delhi- 110 003, India
E.Mail: *seema@dbt.nic.in*

Chapter 1

NEEM: A GLOBAL PERSPECTIVE

OPENDER KOUL
Insect Biopesticide Research Centre
30 Parkash Nagar, Jalandhar- 144 003, India

1. INTRODUCTION

Neem, *Azadirachta indica* A. Juss, a versatile tree of family Meliaceae, has its origin in the forests of Karnataka (South India) or the dried island forests of Myanmar (Gamble, 1902). The subject of origin is still controversial. Roxburgh (1874) gave its origin in India when Myanmar (Burma) was a part of India. Brandis (1921) and Jacobs (1961) describe its origin in dry regions of Upper Myanmar (Irrawady valley, upper region of Prome). Its natural distribution range extends upto Shivalik hills in India (Duthie, 1903; Kanji Lal, 1928). Some authors also suggest the place of origin of neem tree in parts of South India, such as Karnataka (Troup, 1921; Vartak and Ghate, 1990). However, neem is a large-sized evergreen tree growing up to a height of 20m and a girth of 2.5m. Today neem trees are found in nearly 80 countries worldwide and global estimate is about 91 million trees. South Asian and sub-Saharan regions constitute the main areas of distribution. In the past century the tree was introduced in east Africa, the Caribbean islands, Fiji, Mauritius and other areas, and it now grows in the islands of the South Pacific, the West Indies, Haiti, Surinam, the Dominican Republic, Cuba, Nicaragua and in some areas of Mexico. It was introduced into selected regions of California, Southern Florida, Oklahoma, Arizona (USA) and Queensland (Australia) (Puri, 1999). At the US Department of Agriculture Experimental Station in Mayaguez, Puerto-Rico, neem trees are now 20 years old. These trees were grown from greenhouse transplants when they were close to 1m in height. In Arizona, the goal of developing a neem tree with frost resistance to $-8^{O}C$ using seeds from Northern India was undertaken in the late 1980s (Jacobson, 1987). In South America, neem trees grow in Venezuela, Colombia, Ecuador and Bolivia. Recently, more than 300,000 trees were planted in dry areas of Savannah in north-east Brazil as part of huge afforestation programmes; there is a plan to have ten million neem trees by the year 2003.

O. Koul and S. Wahab (eds.), Neem: Today and in the New Millennium, 1–19.

Today it is estimated that there are several million neem trees along the east coast of Africa from Eritrea and Somalia via Kenya and Tanzania northwards to Mozambique. Neem is widely distributed in the dry, southwestern regions of Madagascar and throughout the island of Mauritius (GTZ report, 2000). In Queensland in the northeast of Australia no less than half a million neem trees have been planted in recent years. Besides reforestation and erosion protection, the Australians have set high hopes on the use of neem as a non-synthetic pesticide. In the People's Republic of China the first neem trees were planted in the 1980s on the peninsula of Hainan and the southern part of Guangdong. Today nearly 100,000 trees exist in various parts of China (Forester, 1998). In western Asia neem occupies the southern low-lying areas of Iran, the Euphrates-Tigris Valley in Iraq, the Arabian Peninsula and Mecca in Saudi Arabia (Schmutterer, 1995) (more details discussed in Chapter 5)

During last three decades neem has been established as a multi-purpose tree that could be used for agriculture, forestry, medicine and household purposes (Koul *et al.*, 1990; Koul, 1996a). Neem could adapt itself to a dry, harsh and hostile climate and degraded soil, particularly in arid zones of the world. It could also help in soil reclamation (Sastry and Kanathekar, 1990). The tree could provide shade to cattle and man, and the leaves could be used as fodder for ruminants. The wood can be used as fuel, timber for household furniture, and for agricultural implements. The seeds can provide oil for use in household lamps, as a lubricant for agricultural machinery, as a pest control material, for disease control and for the manufacture of soaps (Axtell and Fairman, 1992). The seed cake, after washing, can be used in small amounts in poultry and cattle feed, as a source of organic manure, for conservation of nitrogenous fertilizers and for the elimination of nematodes (Koul, 1996a).

The major study of neem materials has been its recognition as a source of valuable plant allelochemicals, specifically for the insecticidal, insect repellent, antifeedant and growth regulatory properties of neem kernel extracts, which have attracted worldwide attention. The key active ingredient is azadirachtin, a tetranortriterpenoid that exhibits classical insect growth regulatory (IGR) effects on the immature stages of insects for which the molecular mechanisms of action (Mordue and Blackwell, 1993; Koul, 1996b) are still being evaluated (see chapter 11).

2. NEEM IN REFORESTATION AND AGROFORESTRY

Neem is a very valuable forestry species in India and Africa and is also becoming popular in tropical America, the middle-east countries and Australia. Being a hardy, multipurpose tree, it is ideal for reforestation programs and for rehabilitating degraded, semiarid and arid lands. Neem is useful as windbreaks and in areas of low rainfall and high windspeed. In the Majjia Valley in Niger, over 500 km of windbreaks comprised of double rows of neem trees have been planted to protect millet crops which resulted in a 20 per cent increase in grain yield (Benge, 1988).

Neem windbreaks on a smaller scale have also been grown along sisal plantations in coastal Kenya. Large scale planting of neem has been initiated in the Kwimba Afforestation Scheme in Tanzania.

In Somalia and Mauritiania, neem has been used for halting the spread of the Sahara desert. Also, neem is a preferred tree along avenues, in markets, and near homesteads because of the shade it provides. However, it is best planted in mixed stands and has all the good characters for various social forestry programs. Neem is an excellent tree for silvipastoral systems involving production of forage grasses and legumes. However, according to some reports (Radwanski and Wickens, 1981), the tree cannot be grown among agricultural crops due to its aggressive habit. Others say that neem can be planted in combination with fruit cultures and crops such as sesame, cotton, hemp, peanuts, beans, sorghum, cassava, etc., particularly when the trees are still young. The neem tree can be pruned to reduce shading and to provide fodder and mulch. Recent advances in tissue culture and biotechnology should make it possible to select neem phenotypes with desirable height and stature for use in intercropping and various agroforestry systems (see chapter 10).

A recent report for the Rural Industries/Land & Water Australia and the Forest and Wood Products Research and Development Corporations (Chudleigh, 2001) gives an elaborative picture in the Australian context that would apply to many developed countries. The biggest challenge is to develop a production system that allows profitable production and distribution of formulated products. A particular focus is to assess the prospects for neem trees in the low rainfall areas of Australia where agroforestry that targets commercial production as well as providing a sustainability function is required. Further analyses incorporating extraction, manufacturing and marketing operations could be useful to assess the likely economic viability of a prospective growing operation under present conditions. A form of extensive production (without irrigation and fertilizer) but utilizing mechanical harvesting would be intuitively attractive provided growth rates and seed yields would be high enough to justify the overall investment. The economics of harvesting could work against low yielding extensive plantings. Plantations as far north in Australia as possible, and with irrigation available, would be more likely to produce seed in a competitive manner with other parts of the world. Neem trees may have to be produced under intensive conditions in Australia in order to obtain the seed and azadirachtin yields, and low mechanical harvesting costs, necessary to make an extraction and marketing venture financially attractive. While the neem tree is reasonably well adapted to different environments and may be useful in drawing down water tables, little information is available on how it might produce in terms of growth rates. According to the outputs from the PLANTGRO model, the ideal average temperature for the neem tree is 33°C. It will grow down to 10-14°C and will not tolerate much above 53°C. Overall, if the neem tree is to be commercialized in Australia, it is more likely to be grown in the more favourable conditions in northern Australia rather than in more temperate conditions. If Rural Industries Research and Development Corporations were to support an R&D program on neem, areas for R&D would need to include the economics and potential economics of production, selection, clonal propagation, planting densities and cultural practices such as irrigation needs and fertilizer requirements, harvesting methods, extraction processes and testing and data assembly for registration. Specific priorities would need to be developed in conjunction with private interests pursuing neem development.

3. BIOMASS PRODUCTION AND UTILIZATION

Full-grown neem trees yield between 10 to 100 tonnes of dried biomass/ha, depending on rainfall, site characteristics, spacing, ecotype or genotype. Leaves comprise about 50 per cent of the biomass while fruits and wood constitute one-quarter each. Improved management of neem stands can yield harvests of about 12.5 cubic meter (40 tonnes) of high quality solid wood/ha. Neem wood is hard and relatively heavy, and is used to make religious icons in some parts of India. The wood seasons well, except for end splitting. Being durable and termite resistant, neem wood is used in making fence posts, poles for house construction and furniture. There is a growing market in some European countries for light-colored neem wood for making household furniture (Koul *et al.*, 1990). Pole wood is especially important in developing countries; the tree's ability to resprout after cutting and to regrow its canopy after pollarding makes it well suited to pole production (National Research Council, 1992). Neem grows fast and is a good source of firewood and fuels; the charcoal has high calorific value.

4. NEEM PROCESSING

India is perhaps the only country with the facilities to procure raw materials in large quantities for processing and seeds arrive at the market between July and October. Using neem for various purposes it is essential to have a quality and economically acceptable raw material. Therefore, to have quality material utmost care is required for processing at every step. The seeds are the most important part of the tree. The fruit should be collected ripe from the trees when it starts becoming yellow in colour. Before drying, pulp should be removed from the seeds by washing and scrubbing. The seeds should be shade-dried for several days and checked every day for any mould. Only good quality dry seeds can be stored for longer duration. The moisture content of seeds should remain around 8-10 per cent and storage temperature around 20°C. This will keep the seeds in good condition for more than a year (Puri, 1999).

For cosmetics and repellents, neem oil should have a high concentration of active ingredients and a high degree of purity. It is necessary that the oil be analyzed. According to various regulations, the permitted concentration of Aflatoxins is 4 μg/g, fungi and bacteria 500 units/g and no contamination of *Enterobacter, Staphylococcus and Pseudomonas* (Tewari, 1992; Uniyal and Uniyal, 1996; Puri, 1999).

When leaves and bark are used as raw materials the best harvesting period for high quality leaves is just after the main seed harvest. This is also the best time to make a formation cut on trees, so that leaves, twigs and bark are available automatically (GTZ report, 2000).

The market for neem-based products in Australia is restricted due to the difficulty of registering azadirachtin in Australia, questions of efficacy, and the likely price required to cover the costs of production. There is potential for the market to grow due to the preference for natural insecticides, both in Australia and elsewhere, but the relative costs of production of neem seed in different locations could work either for or against production and/or processing in Australia. Quality neem seed in commercial quantities could be produced from Australian plantations for processing in Australia in the future if the production economics were favourable and an extraction operation were to be established. At present neither of these conditions appear likely. The lack of Australian registration of products currently works against any processing operation in Australia, either from seed produced in Australia or elsewhere. If Australian neem seed producers were to supply export seed markets they would face competition from existing producers, as well as from other sizable plantings of neem trees taking place or planned in other areas including southern China, Florida and Haiti.

5. POTENTIAL USES OF NEEM

There is a general concept that in those countries where neem has been grown for ages, it has also been used for a long time. According to a report from GTZ, Germany (2000), neem has been used for more than 50 years in various Asian and African countries, whereas in industrialized countries neem has been utilized for the last 20 to 25 years only. Commercial neem products have only gained greater significance on the Indian subcontinent where there are commercially marketed products for virtually all types of usage (Koul, 1996b). In all other countries commercial neem products account for only a modest share of the market, and globally most emphasis is on neem-based pesticides.

The neem tree has many potential uses and all parts in one way or the other could be used for various purposes (Fig. 1). In fact, the oldest known use of neem is as a medicinal plant having a long tradition in Indian Ayurveda and Unani medicine dating back more than 3000 years. Today, neem is well known as a traditional medicinal plant for remedies of intestinal complaints, malaria attacks, skin diseases, bacterial infections, inflammations, diabetic conditions (Koul, 1996b) and analgesic action (Khanna *et al.*, 1995). The latest addition to this list is its use as a contraceptive, affecting sperm and inhibiting egg implantation (Riar, 1993; Talwar *et al.*, 1993). The contraceptive preparations are already used successfully in India (Talwar *et al.*, 1993).

Dental hygiene using the thin twigs of neem trees is a traditional use in India and established to help prevent paradontosis and other gum related diseases (Koul *et al.*, 1990). Stomatitis is also cured by an extract from bark of the neem tree. Nimodent, a product of Hamdard Co., Pakistan and neem tooth paste and powder made by Calcutta Chemicals, India is effective dentifrice products (Koul, 1996b). Keimdrat Gmbh, Ausberg, Germany manufactures Dr. Grandel's neem toothpaste containing an extract of neem bark. Soap production from neem-seed oil is widespread in India and Africa, containing about 50 - 80 per cent neem oil. Neem leaf extracts are added to soap to give natural greenish colour and to enhance beneficial effects on the skin. In India leaves and purified neem oil are added to face creams and in Germany leaf extracts are found in hair tonics and shampoos.

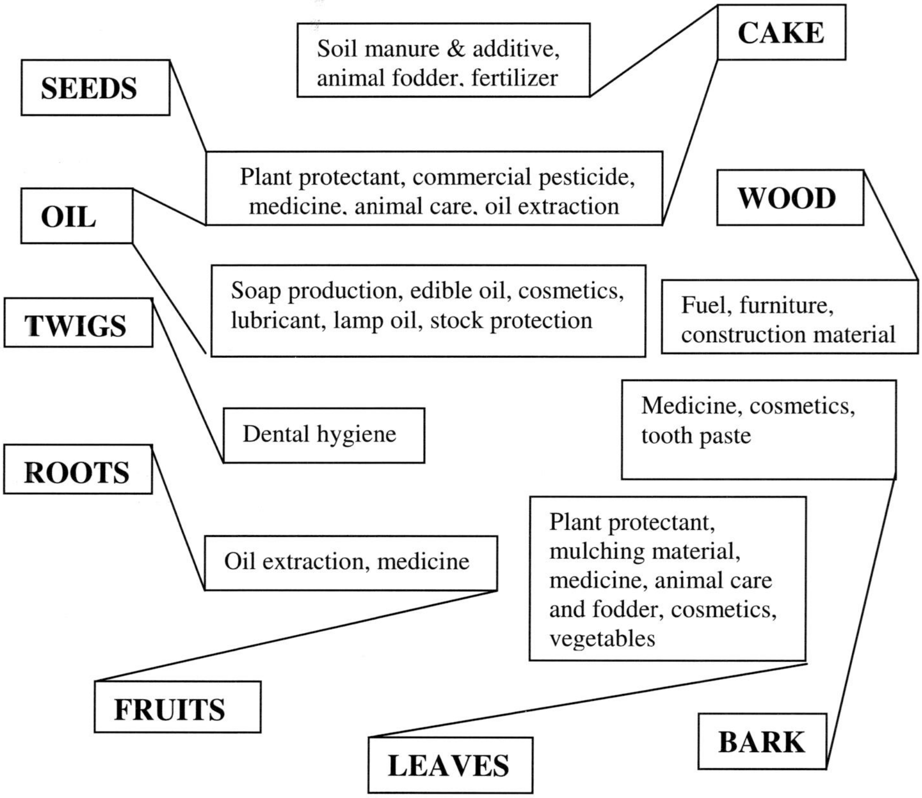

Figure 1. Parts of neem tree and their potential usage

Due to the minimal ecological demands on soil and water, neem tree plantation has helped in preventing erosion. These broad crowned trees with dense foliage planted in cities, villages, gardens and other locations provide suitable shade in tropical regions of the world and the main fuel wood in some African countries. Due to its termite resistant properties, the timber is used in construction and furniture making. The quality index based on the quality performance is 114 as compared to 100 for teak.

Neem pressed cake produced from seeds after removing oil has been shown to act as an organic fertilizer that prevents nitrogen leaching and acts as a source of nitrogen for plants. Leaves can also be used in the soil as mulching material. Besides regulating weed growth and controlling nematodes, neem cake helps to improve the availability of nutrients (Randhawa and Parmar, 1996). The press cake can also be used in animal fodder; it contains 35 per cent digestible protein and is used as a fodder additive for ruminants in India. Goats and camels in West Africa eat fresh leaves and shoots. Simple aqueous extracts of dried seeds are used in animal hygiene, and commercial products are obtainable in India, Thailand, Australia and Germany (Koul *et al.*, 1990; Randhawa and Parmar, 1996).

6. NEEM AS A BIOPESTICIDE

Various results obtained globally have shown that neem and its allelochemicals have a variety of effects on pests. More than 140 active principals have been identified to date that occur in different parts of the tree. The most important components identified have been the tetranortriterpenoids, the azadirachtins. These occur at concentrations of 0.1 to 0.9 per cent in the seed core and it has been established that 30 to 60 g azadirachtin per hectare suffice to combat and repel the key pests of various crops (see chapter 2). It seems that approximately 20 to 30 kg of neem seeds are required per hectare if 2 g of azadirachtin per kg of seed is obtained. This will incur a cost in the range of US $ 1 to 60, although in most countries the range may narrow down to $ 5 to 20. Neem has been shown to control key pests in varied ways (Fig. 2). It has a high level of efficacy, low risk of pest resistance due to different mode-of-action, specific effects on pests, safety for non-target organisms, biodegradable nature and is easily obtained from a renewable source.

It is only in the past decade that the pest control potential of neem, which does not kill pests like neurotoxins but affects their behaviour and physiology, has been recognized. Though subtle, neem's effects such as repellency, feeding and oviposition deterrence, growth inhibition, mating disruption, chemo-sterilization, etc. (Schmutterer, 1995, 2002) are now considered far more desirable than a quick knock-down in integrated pest management programs as they reduce the risk of exposing pests natural enemies to poisoned food or starvation.

In spite of high selectivity, neem derivatives affect ca. 400 to 500 species of insects belonging to Blattodea, Caelifera, Coleoptera, Dermaptera, Diptera, Ensifera, Hetroptera, Homoptera, Hymenoptera, Isoptera, Lepidoptera, Phasmida, Phthiraptera, Siphonoptera, Thysanoptera, on species of ostracod and several species of mites. Neem preparations also act as nematicides against endoparasitic species of *Meloidogyne* and *Globodera*, ectoparasite species of *Hoplolaimus* and *Tylenchorhynchus* and semiendoparasitic species of *Rotylenchus* and *Pratylenchus* nematodes (Musabyimana and Saxena, 1999). Similarly as a fungicide neem products are effective against a number of fungal pathogens (see chapter 9). Water snails as vectors of diseases such as *Melinia scabra* (schistosomiasis) and phytophagous land-snails in greenhouses and horticulture are killed by neem preparations (West and Mordue, 1992). The neem products also control many acarines of *Tetranychus* genus, bacterial plant pathogens and animal and plant viruses (Mansour *et al.*, 1987; Hunter and Ullman, 1992; Schmutterer, 1995).

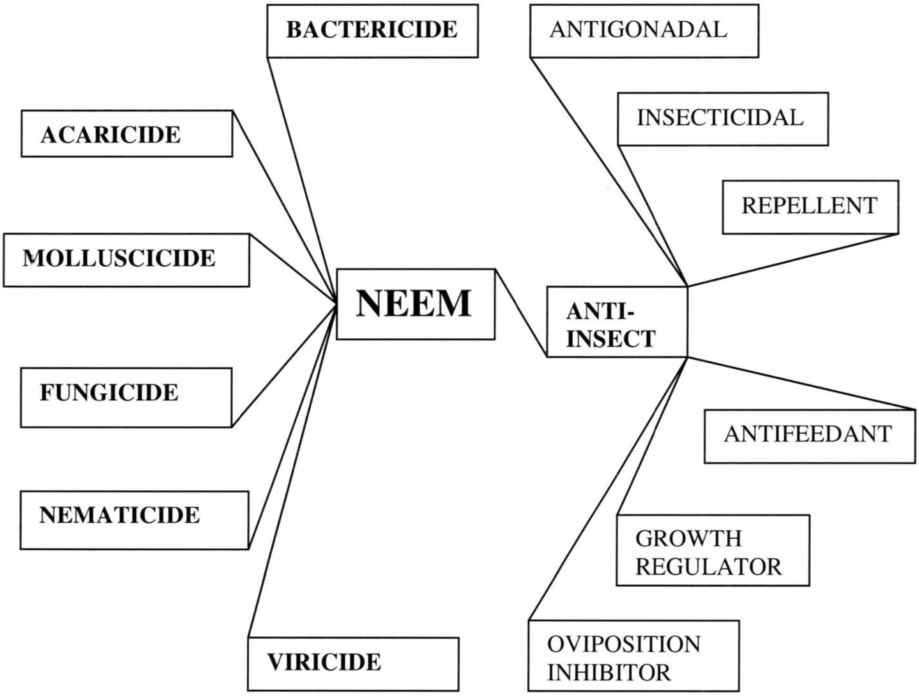

Figure 2. Neem as pesticide

6.1 Commercial Neem-Based Biopesticides

Vikwood Industries was the first company to register a neem formulation in 1985 in USA and subsequently the technology was transferred to W. R. Grace and the product was marketed as Margosan 'O', containing a 0.3 g AI per litre of azadirachtin as an EC formulation. Further modifications resulted in a formulation of 2.5 g AI per litre EC in early 1990s, followed by 45 g AI formulation in 1995 that is marketed as Neemix and NeemAzal (Immaraju, 1998). In 1996 W.R. Grace sold the biopesticide unit to Thermo Trilogy, a subsidiary of ThermoEcotek Inc. In 1992 another commercial unit, AgriDyne Technologies provided a registered product based on 30 g AI per litre EC formulation that was sold as Azatin, Turplex and Align. In 1996 the same company produced a 45 g AI formulation based on hydrogenated azadirachtin derivative. However, following subsequent mergers and closures of the companies, the azadirachtin registration in USA belongs to ThermoTrilogy. AMVAC has received registration for neem products in USA in 1999 and is offering neem-based pesticides of its partner Fortune Biotech, India. Although no neem-based pesticide is registered in Australia, there are some experimental products purchased from developing countries. Neem cake is sold in Australia as a fertilizer. The University of Queensland is trying to develop a neem cake-based nematicide for tomato growers and is supported by private sector as well. Numerous products are available globally with a varied registration status (Table 1).

As mentioned above, various commercial formulations are based on azadirachtin content as the active ingredient. Azadirachtin content in the kernel peaks at fruit maturity. In the process of manufacture of a biopesticide from such seeds the hulls and kernels are separated in decorticating units. The kernels are then passed on to mechanical expellers where they are soft-pressed to provide expeller oil and oil rich cake (Puri, 1999). This is followed by solvent extraction to produce azadirachtin. At present there are several patents with procedures and compositions that claim to confer such preparations with unique stability of azadirachtin in EC formulations. The technical powder containing 10 to 26 per cent azadirachtin is easily prepared, and higher concentrations could be manufactured but at disproportionately higher costs. These powders are then formulated in different ways keeping in view the shelf life of the formulations. During last decade serious effort has been made to develop stable formulations; however, it has been demonstrated that higher azadirachtin concentration in the final formulation results in greater instability. Formulations having less than 10 g / liter azadirachtin content have flatter degradation curves than formulations containing 20 g AI / liter or more (Immaraju, 1998).

In practice, three types of neem-based biopesticides are employed:

- Simple domestic biopesticide
- Unfinished products
- Ready to use neem-based biopesticides

Simple domestic preparations are used as part of farmer-oriented techniques that involve all aspects from planting of trees to applying the aqueous extracts in the field. Farmers harvest neem fruits, depulp them, dry seeds, grind them, and prepare aqueous extracts at the rate of 50 g of ground seed per litre of water; 10 to 20 kg of seed is required for one treatment per hectare of field crops. This simple technology is beyond doubt an important method in many tropical and semi-arid areas of developing countries.

Unfinished products involve basic standardization and are marketed commercially in small units. This method involves crushed vacuum-packed neem seeds in quality plastic bags stored at low temperature (10-25°C) in dark and dry

Table 1. Commercial neem-based biopesticides in various parts of the world

Country	Number of products	Nature of product	Status of registration
Australia	2	Formulated biopesticide	About to be registered
Austria	3	Formulated biopesticide and oil, seeds, pressed cake	Registration available for eco-cropping
Benin	1	Formulated oil	Unclear
Brasil		Formulated biopesticide and oil	Imported from Germany, registration pending
Canada	1	Formulated extract	Temporary registration for forest pests
China	1	Alcoholic extract	Provisional registration
Columbia	3	Alcoholic extract and oil	Unclear
Costa Rica	2	Formulated alcohol extract Formulated oil	Registered
Cuba	3	Seed, pressed cake, formulated oil	Registered
Denmark	2		Under registration
Dominican Republic	3	Seeds, press cake, formulated oil	Registered
Dutch Antilles	2	Formulated alcohol extract, formulated oil	Registered
Ecuador	4	Formulated oil and alcohol extract	Locally produced, also imported from USA
Egypt	5	Formulated insecticides and fungicides	Registered, some imported from Israel, Germany and Sweden
Fiji	1	Seed kernel powder	Registered

Germany	5	Formulated pesticides seeds, press cake, and formulated oil	Registered Offered as raw material
Ghana	3	Alcoholic extract and formulated oil	Provisional registration Imported from India, USA and Germany
India	~ 100	All sorts of products, formulated biopesticides, oils, fertilizer-based formulations.	Limited number of products registered
Indonesia	1	Extracts from India seeds, oil and cake available	Registration pending No registration required
Israel	3	Formulated pesticides Fungicide	Registered and from USA, Locally produced
Italy	2	Formulated pesticides	Registered product from India and about to be registered product from Germany
Kenya	8	Formulated enriched oil & standardized press cake Formulated insecticides Other 4 products	Registered Provisional registration Imported from India and USA
Mauritius	1	Formulated pesticide	Registered and imported from India
Myanmar	3	Formulated alcoholic extract and press cake	Registered
Nepal	1	Formulated pesticide	Registered and imported from India
Nicaragua	4	Seed, press cake, formulated oil	Registered
Niger	2	Enriched alcoholic extract, formulated oil	About to be registered

Pakistan	2	Formulated pesticides	Unclear
Saudi Arabia	4	Formulated pesticides	Registered, Imported from USA
Senegal	2	Formulated oil Alcoholic extract	Provisional registration
Spain	2	Formulated pesticides Seeds	Registered Unclear
Sri Lanka	2	Formulated pesticide Formulated powder Locally produced press cake, oil and seed	Registered, from India Unclear Local trials
Sweden	2	Formulated pesticides	Registered
Switzerland	3	Formulated pesticides Seeds, formulated oil	Registered for fruit crops Offered as raw material
Thailand	3	Locally produced formulated alcoholic extract and oil Several products from India	Registered Registration pending
Togo	2	Seeds, press cake, oil	Registration not required
Uganda	2	Formulated enriched oil Standardized press cake	Registered Imported from Kenya
USA	5	Formulated pesticides	Registered
Venezuela	3	Alcoholic extracts	Unclear, imported from USA

Modified from Status report on global neem usage (2000), GTZ, Germany

Places, These products are used to control soil borne pests and as fertilizers. Normally, 30 g / cm^2 is added to the soil a few weeks before sowing.

Ready to use neem-based biopesticides are oil-and-water emulsions based on the principal active ingredient azadirachtin, vegetable oils, detergents and stabilizers. Normally the mixture of oil and emulsifier is used diluted to between 0.5 and 1 per cent in water. Formulated neem biopesticides can also be made using advanced technology. Such manufacturing is carried out in semi-industrial extraction plants (Foerster, 1999; Foerster *et al.*, 2000) that involve alcoholic extraction, purification and concentration and finally formulation and stabilization. Such products have to pass through pesticide regulations and require comprehensive tests to ensure the effectiveness against pests and safety for humans, environment and non-target organisms (see chapter 7).

Table 2. Commercial neem applications used against key crops and pests

Category	Pests	Crops
Vegetables	Aphids, caterpillars, fruit flies, leaf miners, thrips, white flies, nematodes	Aubergine, cabbage, cucumber, garlic, lettuce, okra, onion, tomato
Fruits	Aphids, caterpillars, fruit flies, leaf miners, pallid scale insects, spider mites, thrips, nematodes	Apple, avocado, banana, grape, lemon, mango, melon, papaya, strawberry
Grains and basic foodstuffs	Beetles, bugs, caterpillars, gall midges, grubs, leaf and plant hoppers, locusts, soil borne pests, stalk borers, termites, nematodes, fungi	Beans, grain, maize, millet, potatoes, rice
Stock protection	Bruchids, corn borers, moths, weevils	Beans, maize, rice, various grains
Others	Aphids, bollworms, caterpillars, locusts, stalk borers, white flies, nematodes	Cotton, ornamental plants, sugar cane, tobacco

In view of the ecofriendly nature of neem products, neem is now being accepted by subsistence farmers, fruit and vegetable producers, grain producers, animal producers and cash-crop producers; however, about 1/5 of all responses are from subsistence farmers and 4/5 by the producers. Various key crops and pests for neem application are shown in Table 2. Apparently the use of neem-based biopesticides seems to be more acceptable for vegetables and fruit crops.

Neem biopesticide formulations are well suited for "Integrated Pest Management" (IPM) Programs because of the following salient features:

- Neem-based biopesticides are natural products, absolutely non-toxic to higher animals, 100 per cent biodegradable and environmentally friendly.
- It is suited for mixing with other synthetic pesticides and in fact enhances their action.
- As less synthetic pesticides need to be used, thereby reducing the environmental load.
- Synthetic pesticides being single chemical compounds may cause development of resistant pests. Neem formulations contain several compounds; hence development of resistance is restricted to a greater extent.
- Neem does not destroy natural enemies of pests thereby allowing a check on the pest population.
- Neem has systemic action; seedlings can absorb and accumulate the neem compounds to make the whole plant resistant to pests.
- Neem has a broad spectrum of action, active on more than 400 species of pests.

7. FUTURE PROSPECTS

Today it is evident that neem has great potential, particularly as a natural pesticide. Some countries do see the potential of neem as a biopesticide, but in some regions it is still underexploited. Even in India only 30-35 percent of the neem seeds are harvested. The efforts of GTZ to work out practice-related concepts to exploit the potential of neem trees has been outstanding and is contributing extensively in plant protection and storage in order to bring consistency in neem usage. The earlier concept of self-development of products by the farmer is changing its course. Presently farmers prefer ready-to-use biopesticides, which should be locally available at a reasonable cost. However, ready-to-use neem products are used in limited quantities due to high price of the products manufactured in industrialized countries. This is why neem biopesticides are mainly used in niche markets, such as organic farming, private gardens, and in those cases where pests are difficult to control by conventional pesticides. The future of neem may be in developing countries where cheap labour is available and production costs will be extremely low, such as Kenya, Thailand or India where tremendous competition exists. It is, therefore, imperative to generate region-wise economic data on processing, marketing strategies and marketing potential. According to Praneetvatkul (1999) there are key questions to be considered for the future of neem :

- Is neem processing profitable for small-scale entrepreneurs in developing countries?
- What are the key factors that determine the profitability of neem processing?
- To what extent can neem products replace/substitute the conventional insecticides?
- What are the bottlenecks for neem products to gain a greater market share?
- What factors determine the price of neem products?
- What is the potential market share for neem products?

This type of economic assessment will help in making neem processing a profitable business, producing a cheaper biopesticide by 25 to 40 per cent, and a better marketable product. In addition to the agricultural potential, the results of trials in health and vector control are promising. The potential of neem in health projects and medicine has not been exploited systematically and would also be a promising field for technical cooperation. There is also ample scope for the systematic use of neem products in the fields of livestock, veterinary services and fisheries where indigenous knowledge is reported from India and Sri Lanka.

8. OUTLOOK

In most countries it has become obligatory to follow FAO code-of-conduct for proper trading of pesticides. Obviously, what we have to look for is effective and selective pesticides with low mammalian toxicity, and low persistence. Neem-based biopesticides have these characteristics and are suitable for organic farming and integrated pest management systems. It has become necessary to follow these regulations in view of the hazardous pesticides that have alarmingly damaged the environment; many such pesticides have been banned.

Azadirachtin-based biopesticides have the most convincing potential market in future, as many more products are being prepared and registered. Azadirachtin-based products are gaining market due to:

- cancellation of older registrations and the lack of support for re-registration due to loss of patent protection,
- less chance of resistance and cross resistance,
- regular cuts in synthetic insecticides world-wide,
- faster approval of neem-based biopesticides in comparison with synthetic insecticides (US-EPA has exempted neem from residue tolerance requirements on food crops as long as the dosage does not exceed 50 g AI

per hectare. The suggested rate of azadirachtin generally range from 12.5 g to 40 g AI per hectare),

- technological advancement and improved performance, and
- continued public awareness on pesticide safety issues and shifting trend to use biorational pesticides.

Many neem industries are starting up in developing countries because neem offers a low toxic alternative to standard broad-spectrum pesticides. There is the need to standardize products, develop new concepts and develop sustainable products. Non Government Organizations (NGOs) can play a vital role in this regard by providing specific consultancies locally and exploit neem in the social interest.

ACKNOWLEDGEMENTS

I am grateful to Michael J. Smirle, Agriculture and Agri-Food Canada for his helpful comments on earlier draft of this chapter.

REFERENCES

Axtell, B.L. and Fairman, R.M. (1992), Minor oil crops, Part I Edible oil, Part II Non-edible oil, Part III Essential oil. *FAO Agriculture Science Bulletin,* No. 94.

Benge, M.D. (1988), Cultivation and propagation of the neem tree, in *Focus on Phytochemical Pesticides*, Vol. 1, *The Neem Tree*, ed. M. Jacobson, CRC Press, Boca Raton, FL, pp. 1-18.

Brandis, D. (1921), *Indian Trees*, Constable & Co., London (reprinted 1971).

Chudleigh, P. (2001), Growing neem trees in Australia: a feasibility study. A report for the Rural Industries/Land & Water Australia and Forest and Wood Products Research and Development Corporations Joint Venture Agroforestry Program. RIRDC Publication No 01/061

Deutsche Gesellschaft fur Technische Zusammenarbeit (GTZ) GmbH (2000), *Status report on global neem usage*, GTZ, Eschborn, Germany.

Duthie, J.F. (1903), *Flora of Upper Gangetic Plains and Adjacent Shivalik and Sub-Himalayan Track*, Vol. 1, Government Press, Calcutta.

Foerster, P. (1998), *Mission report on a fact-finding study on neem in people's Republic of China*, GTZ/UNIDO, 12 pp.

Foerster, P. (1999), Development process of a small-scale neem processing plant:

Experiences of a GTZ project in Kenya, in *Natural Crop Protection in the Tropics – Let in Information Come to Life*, ed. G. Stoll, Verlag Markgraf, Weikersheim, 400 pp.

Foerster, P., Leypolz, W., Quentin, H. and Praneetvatakul, S. (2000), Economics of neem processing by small scale entrepreneurs in Thailand, Kenya and the Dominican Republic, in *Practice Oriented Results on Use and Protection of Neem Ingredients and Pheromones IX*, eds. H. Kleeberg and P.W. Zebitz, Trifolio, Germany, pp. 47-54.

Gamble, J.S. (1902), *A Manual of Indian Timbers* (Reprint 1972) Bishen Singh Mohinderpal Singh, Dehradun, India.

Hunter, W.B. and Ullman, D.E. (1992), Effect of the neem products, RD-Repellin, on settling behaviour and transmission of zucchini yellow mosaic virus by the pea aphid, *Acyrthosiphon pisum* (Harris) (Homoptera: Aphididae), *Annals of Applied Biology*, **120**, 9-15.

Immaraju, J.A. (1998), The commercial use of azadirachtin and its integration into viable pest control programmes, *Pesticide Science,* **54**, 285-289.

Jacobs, M. (1961), The genetic identity of *Melia excelsa. Garden Bulletin (Singapore),* **18**, 71-75.

Jacobson, M. (1987), Neem research and cultivation in the Western Hemisphere, in *Natural Pesticides from the Neem tree (Azadirachta indica A. Juss) and Other Tropical Plants*, eds. H. Schmutterer and K.R.S. Ascher, Proc. 3rd Int. Conf., Nairobi, GTZ, Eschborn, Germany, pp. 33-44.

Kanji Lal, U.N. (1928), *Forest Flora of Chakrata. Dehradun and Saharanpur Forest Division*, U.P. Manager of Publications, Delhi.

Khanna, N., Goswami, M., Sen, P. and Ray, A. (1995), Antinociceptive action of *Azadirachta indica* (neem) in mice: Possible mechanisms involved, *Indian Journal of Experimental Biology*, **33**, 646-850.

Koul, O. (1996a), Advances in neem research and development – present and future scenario, in *Supplement to Cultivation and Utilization of Medicinal Plants*, eds. S.S. Handa and M.K. Koul, National Institute of Science Communication, CSIR, New Delhi, pp. 583-611.

Koul, O. (1996b), Mode of azadirachtin action, in *Neem*, eds. N.S. Randhawa and B.S. Parmar, New Age International Publishers Ltd., New Delhi, pp. 160-170.

Koul, O., Isman. M.B. and Ketkar, C.M. (1990), Properties and uses of neem, *Azadirachta indica, Canadian Journal of Botany*, **68**, 1-11.

Mansour, F., Ascher, K.R.S. and Omari, N. (1987), Effect of neem (*Azadirachta indica*) seed kernel extracts from different solvents on the predacious mite *Phytoseiulus persimilis* and the phytophagous mite *Tetranychus cinnabarinus*. *Phytoparasitica*, **15**, 125-130.

Mordue (Luntz), A.J. and Blackwell, A. (1993), Azadirachtin: an update, *Journal of Insect Physiology*, **39**, 903-924.

Musabyimana, T. and Saxena, R.C. (1999), efficacy of neem seed derivatives against nematodes affecting banana, *Phytoparasitica*, **27**, 43-49.

National Research Council (1992), *Neem: A tree for solving global problems*, National Academy Press, Washington, DC.

Praneetvatkul, S. (1999), *Economics of small scale industrial production of neem-based pesticides in Thailand,* GTZ report, Germany.

Puri, H.S. (1999), *Neem, the Divine Tree Azadirachta indica*, Harwood Academic Publishers, Amsterdam.

Radwanski, S.A. and Wickens, G.E. (1981), Vegetative fallows and potential value of the neem tree (*Azadirachta indica*) in the tropics, *Economic Botany*, **35**, 398-414.

Randhawa, N.S. and Parmar, B.S. (1996), *Neem*, New Age International Publishers Ltd., New Delhi.

Riar, S.S. (1993), Antifertility and other medical applications, in *Neem Research and Development*, eds. N.S. Randhawa and B.S. Parmar, Society of Pesticide Sciences India, New Delhi, pp. 220-226.

Roxburg, W. (1874), *Description of Indian Plants*, Today and Tomorrow, New Delhi.

Sastry, T.C. and Kanathekar, K.V. (1990), *Plants for Reclamation of Wastelands*, Publication and Information Directorate, CSIR, New Delhi.

Schmutterer, H. (1995), *The Neem Tree and Other Meliaceous Plants, Source of Unique Natural Products for Integrated Pest Management, Medicine, Industry and Other Purposes*, VCH Verlagsgesellschaft, Weinheim.

Schmutterer, H. (2002), *The Neem Tree, 2nd Edition*, The Neem Foundation, Mumbai, India.

Talwar, G.P., Upadhyay, S.M., Garg, S., Kaushic, C., Kaur, R. and Dhawan, S.

(1993), Induction of cell mediated immunity in genital tract, in *Neem Research and Development*, eds. N.S. Randhawa and B.S. Parmar, Society of Pesticide Sciences India, New Delhi, pp. 227-234.

Tewari, D.N. (1992), *Monograph on Neem Azadirachta indica A. Juss*, International Book Distributors, Dehradun, India.

Troup, R.S. (1921), *The Silviculture of Indian Forest Trees*, Vol 1, Claredon Press, Oxford, UK.

Uniyal, K. and Uniyal, D.P. (1996), Mycoflora of neem seed, *Indian Forester*, **122**, 423.

Vartak, V.D. and Ghate, V. (1990), Ethanobotany of neem, *Biology India*, **1**, 55-59.

West, A.J. and Mordue (Luntz), A.J. (1992), The influence of azadirachtin on the feeding behaviour of cereal aphids and slugs, *Entomologia Experimentalis et Applicata*, **62**, 75-79.

Chapter 2

THE PLACE OF NEEM AMONG MODERN NATURAL PESTICIDES

E. DAVID MORGAN
Chemical Ecology Group, Lennard-Jones Laboratory, Keele University, Staffordshire, England ST5 5BG

1. INTRODUCTION

We have become accustomed to think of pesticides as synthetic products of the chemical industry. However, on reflection we realise that natural pesticides have a longer history. We tend to accept the proposition that all advances in science are for the better, but then, have synthetic pesticides been wholly a success or has something been lost along the way? We appear to be in a time of re-assessment of pesticides, and we need to re-assess neem and its place in the control of pests.

Plants and some insects have co-existed on the earth for about three and a half million years, which has permitted them both plenty of time to evolve offensive and defensive strategies. Plants have developed many strategies to protect themselves from being attacked by predators, most of which are not the subject of this discussion. The plant strategy of producing compounds that are toxic to insects is the one of interest here (Jacobson and Crosby, 1971; Warthen and Morgan, 1985; Arnason *et al.*, 1989; Morgan and Wilson, 1999). Thousands of such toxic compounds are known, many of them, for one reason or another, are not suitable to use as pesticides, but there still remains a large number that are suitable and have been used or have potential for use in agriculture against insect pests. Many have been used in the past and some are in use today. It will be instructive to look more closely at a few examples later so that useful comparisons can be made with neem products.

Natural compounds from plants were used since ancient times, more or less effectively to give protection from insect pests. In the 19th century they became scientifically established, and widely applied in the first half of the 20th century. I well remember as a child helping my father to spray our currant bushes with a

O. Koul and S. Wahab (eds.), Neem: Today and in the New Millennium, 21–32.

product called "Blackleaf 40", a 40 per cent aqueous solution of nicotine sulphate. The discovery in 1940 that p, p'-dichlorophenyltrichloroethane (DDT), a compound that had been synthesized many years earlier was a very effective moth-proofing agent was the beginning of the era of synthetic pesticides. The phenomenal success of DDT during wartime to control a typhus plague in Naples and the dramatic reduction of malaria in the southern USA drew attention to the possibilities for more synthetic pesticides. The cheapness and effectiveness of the synthetic insecticides quickly threw natural compounds into the shade, but very soon other shades began to appear. In 1962 the appearance of the book "Silent spring" (Carson, 1962) showed that all was not well with the new pesticides, that residues were building up in the ecosystem, with detrimental effects on wildlife and beneficial insects. The large scale death of fish in the lower Mississippi River the next year and the tracing of its cause to synthetic pesticides was only one piece of evidence to confirm the Carson case. Soon a whole new area of science grew up around pesticide residue analysis and toxicity tests. More slowly did another problem appear, the growing increase of resistant strains of insects, requiring the use of larger and larger doses to maintain the control of some pests.

The greatest legacy for us today of this period from the 1940s to the 1990s was the growth in government regulation and tighter licensing laws for new pesticides. The result of that period of indiscriminate use of persistent synthetic pesticides hinders us today in attempts to replace the persistent and resistance-prone synthetics with more benign and ecologically acceptable natural substances.

2. NATURAL PESTICIDES

The dried flower heads of several species of *Chrysanthemum* known as pyrethrum have been used since classical times in Asia, its original home, as an insecticide and introduced into Europe in 1828 (Cassida, 1973). It is a good representative of natural pesticides with a long pedigree. It has been used through the period of dominance of synthetic pesticides for special purposes where its fast knockdown effect was desired. It is in use today, in spite of its noted lack of persistence. It is rapidly hydrolysed by acid and alkali and is sensitive to light and oxygen. It consists of a mixture of six related compounds (not isomers), which differ considerably in their effect on insects, and no attempt is made to separate them in use. It is interesting to note that all the possible isomers and enantiomers of the six compounds have been synthesized and it has been found that the naturally occurring ones are the most insecticidally active. Pyrethrin is chiefly found in three species of Compositae, *Chrysanthemum cinerariaefolium, C. coccineum* and *C. marshalii.* Pyrethrin has been the model for some highly successful synthetic pesticides, but resistance to the synthetic analogues has developed very quickly in some pests, while resistance to the natural mixture has not been serious. Pyrethrin readily penetrates the insect cuticle but is relatively non-toxic by ingestion. Synergism is an important property with pyrethrum. Substances which themselves are not toxic can increase the toxicity of the expensive pyrethrum by inhibiting its detoxification. Pyrethrin compounds act as nerve poisons, binding to cell membranes so that sodium channels are held open, and nerves fire continuously, as happens with DDT. It is relatively harmless to mammals because they can hydrolyse the compounds very quickly. The similarities and contrasts between pyrethrin and neem limonoids will become apparent in this discussion.

Tobacco also has a long history of use for insect control. Watery extracts of tobacco (*Nicotiana tabacum*) leaves are recorded as being sprayed on crops to destroy pests in 1690, and it was regularly used throughout the eighteenth century as a spray or dust (Schmeltz, 1971). The active principle is a quite complex mixture of alkaloids, at least 45 compounds, with nicotine, anabasine, anabaseine, nornicotine and nicotyrine usually the principal ones. Its vapour penetrates insect cuticle and human skin, so it is as toxic to mammals as it is to insects and is most effective against small, soft-bodied insects. Because of its volatility it leaves no residues on food. It too is a nerve poison. It is expensive to produce, unpleasant to handle, very toxic and does not have a wide spectrum of activity nevertheless it was only slowly displaced by synthetic insecticides. There are a number of alkaloids in addition to the nicotine group, which have been used in the past or are still in use as natural insecticides. These include anabasine, hellebore, sabadilla, quassia and ryania.
Rotenone in its crude form as powdered derris root has been used as an insecticide for 150 years. It has been in continuous use for special purposes throughout the high period of synthetic insecticides. There are about 40 compounds with related structure which also have insecticidal properties, and it or its relatives have been isolated from at least 68 species of plants chiefly *Derris, Lonchocarpus, Millettia* and *Tephrosia* (all Leguminosae) and one species of *Verbascum* (Scrophulariaceae). Rotenone itself is the most active. It is most effective against leaf-eating insects, and is only moderately toxic to mammals. It is decomposed by light and oxygen and is sensitive to alkali. Rotenoids interfere with electron transport in the oxidation of NADPH to $NADP^+$ causing a sharp, fall in respiration and oxygen uptake. It has been synthesized but synthesis is not commercially practical and there are no obvious synthetic pesticides based on its structure.

There are many other plant compounds or mixtures that have been used or been considered for use as insecticides (Crosby, 1971, Dev and Koul, 1997). Although most users regard the products described above as non-toxic because of their plant origin, nicotine is highly toxic, and the oral LD_{50} for pyrethrin and rotenone would place them in the moderately toxic range (Doull, 1976).

3. PEST STRATEGY TODAY

Now we are emerging from the thrall of the synthetic pesticides. We learned first about their advantages and only later about their disadvantages. Today we hear all about integrated pest control, sustainable agriculture, biological control or the "push-pull" or stimulodeterrent diversionary strategy (Miller and Cowles, 1990; Khan *et al.*, 1997). Old texts like that of Sweetman (1936) on biological control have been resurrected and studied for ideas. As an indication of modern trends, The Biopesticides Manual (Copping, 1998) lists some 800 products, including natural products, pheromones and insect predators, etc. that are used or have potential for use as pest control agents. Biological control alone is too asthenic. Integrated pest management is the route ahead.

What then are we to use as the pesticidal part of this strategy? Many new plant products have come to light since the decline of interest in them in the 1940s. Principal among them is an extract of the seeds of the neem tree (*Azadirachta indica* A. Jussieu, Geraniales: Meliaceae), but the regulations put in place in the 1960s and 1970s, designed for synthetic pesticides, have made the introduction of new natural products much more difficult. Today natural products are said to claim about 1 per cent of the world insecticide market (Isman, 1997).

4. NEEM AMONG NATURAL PESTICIDES

The history of neem, like that of many other botanicals, began with a long history of use as an insecticide through folklore. The modern study on the effect of crude neem seed extracts on crop pests began chiefly in the hands of Schmutterer in the Sudan in the 1960s (Schmutterer, 1995) and later Saxena (Saxena *et al.*, 1981) and the US Agricultural Research Service at Beltsville, Maryland (Warthen, 1979). The scientific study of neem required the identification and isolation of its active substances. That began with the isolation of azadirachtin from neem seeds, using a locust-feeding bioassay (Butterworth and Morgan, 1968), and the demonstration that there was another effective compound present in smaller amounts (Butterworth and Morgan, 1971). It was soon demonstrated that azadirachtin was effective systemically (Gill and Lewis, 1971) and where insects consumed azadirachtin it had a toxic effect, interrupting growth and development (Ruscoe, 1972). In subsequent work, as predicted by Butterworth and Morgan (1971) other compounds with antifeedant effects were isolated in smaller amounts from the seeds (for reviews, see Koul, 1992; Morgan and Wilson, 1999, Kraus, 2002). Today about 10 minor compounds in neem seeds have properties similar to azadirachtin.

Today, through the work of many experimentalists, we have the results of laboratory, greenhouse and field studies that amply demonstrate the activity of neem compounds or extracts against a wide variety (at least 400 species) of plant-feeding insects. Why is it then that neem products have been so slow to come into general use and why are they not better known and more widely used around the world?

5. THE TERM AZADIREX

We do not have a good term to describe the mixture of insecticidally active compounds from neem seeds. The term “neem extract” is used very loosely. It may refer to products from fruit, seeds or leaves, and has been used of insecticidal, nematicidal, bactericidal and herbal medicine products, and may or may not contain azadirachtin. “Azadirachtin” on the other hand refers only to one compound, and the active mixture from the seeds contains more than this. The important compounds for insecticidal action of the neem seed extract are a group of highly oxidized limonoids. If there are others, they have not been discovered, and at least 160 triterpene compounds have been isolated from one or more parts of the neem tree, including leaves, bark, twigs and roots. About one third of these have been examined for biological properties. The possibilities of finding new compounds with insecticidal effect are now very slim.

To try to avoid imprecise descriptions, the term azadirex (an extract containing azadirachtin) is suggested here. This coined word refers to an insecticidally active extract (however obtained) of the seed kernels of the neem tree that contains azadirachtin as its principal active component, but also contains other related biologically active compounds of the limonoid group of triterpenes from the seeds. This azadirex product may or may not contain other inactive limonoids such as nimbin and the marginally active compound salannin. The word azadirex is used where possible throughout this chapter.

6. COMPARATIVE VALUE

If we make a comparison of the properties and advantages of azadirex with those of conventional natural pesticides, such as those outlined earlier, azadirex compares very favourably with them all, when considering properties in turn.

- *Spectrum of activity.* Azadirex has a broad spectrum of activity against leaf-eating insects and stored grain and seed pests. It is not very useful against insects feeding deep inside plants, such as those inside fruit, or against spiders or the valuable parasitic and carnivorous insects. Few other products are so advantageous to beneficial insects.

- *Systemic.* Azadirex works systemically in plants. It is taken up from the soil and translocated to the leaves and growing tips, which are the parts most susceptible to damage. Hardly any other products have this advantage.

- *Resistance.* Nature, in its wisdom, almost always makes a mixture of active products to prevent induced resistance. Pyrethrum, for example, consists of a mixture of six active compounds. Resistance to it is limited whereas resistance or tolerance to some of the synthetic pyrethroids has developed quickly in some tropical environments. At least eleven compounds are known to be highly active in azadirex, therefore development of resistance is highly improbable, as shown by the experiments of Feng and Isman (1995).

- *Mammalian toxicity.* In all the tests so far published (which are unfortunately on poorly defined neem seed extracts, we cannot here refer to azadirex), neem has displayed remarkably low mammalian toxicity, far lower than that of other natural insecticides.

- *Stability.* Azadirex has unjustly acquired a reputation for low stability on storage and use. That arose from ignorance of its properties. We have shown that azadirachtin, salannin and nimbin, as representative compounds of azadirex are stable indefinitely in neutral organic solvents, and stable for many days in weakly acid water (Jarvis *et al.*, 1998). The stability after spraying on crops is not really known, since proper tests have not been made public. Azadirachtin is relatively stable to UV light and oxygen in laboratory conditions and no field trials including UV stabilizers appear to have been published (see Sunderam, 1996).

- *Residues.* Residues of azadirex on food plants present no problem. Azadirachtin is rapidly destroyed by boiling in pH neutral water (Jarvis, unpublished). Related to this is its stability in streams, it has a half-life of about 15 days in a temperate climate (Sunderam, 1996), and there is no known toxicity to fish, in contrast to rotenone, which is highly toxic.

- *Availability of supply.* There is a ready source of the material, for the neem tree grows widely throughout the drier tropics, especially in Asia and Africa. The tree is now found in at least 78 countries with an estimated total of 60 to 90 million trees, with 30 to 45 million of them in Africa. Large new plantations have been started or are planned in several countries, including Saudi Arabia, China and Brazil. Other advantages of the tree, such as its ability to grow on poor and eroded soil, its modest need for rainfall, its resistance to drought, and the usefulness of its timber need not be elaborated here, but they have a positive influence on economics of production. The fact that the seeds are a renewable part of the plant is also an advantage, compared to rotenone, where the roots have to be harvested, or tobacco, where also the plant is killed and new plants have to be sown. Against this is the labour-intensivity of present methods of harvesting seed, practical only in low-wage countries.

- *Extraction.* There is no doubt that preparing an active extract of neem seeds is more difficult and costly than producing pyrethrum, nicotine extract, or powdered derris (for rotenone). This has a negative influence on cost of production, and is considered later.

6.1 Comparison with Synthetic Pesticides

The same exercise can be made of comparing azadirex with synthetic pesticides in general.

- *Spectrum of activity.* The synthetic pesticides generally have a wider spectrum of activity, but that is good and bad. Beneficial insects are destroyed along with the pests, and no pesticide is much use against insects once the pest is inside fruit, plant stems or under tree bark.

- *Systemic.* Very few synthetic pesticides act systemically, a serious disadvantage for them compared to azadirex.

- *Resistance.* Synthetic pesticides, because they are single compounds, are very open to development of tolerance or resistance in the pest. Azadirex and most natural pesticides have the advantage here.

- *Mammalian toxicity.* Azadirex has a great advantage in terms of mammalian toxicity.

- *Stability* and *Residues.* Their greater stability is both an advantage and disadvantage to synthetic pesticides. Their persistence on food crops is the reason why we have such strict rules about licensing and permitted levels of residues, and has led to the image in the minds of the general public of the dangers of pesticide use.

- *Availability of supply.* Synthetic pesticides are made ultimately from simple petroleum hydrocarbons. Their supply is therefore comparatively unlimited and the cost is determined by the price of petroleum to only a small extent. The manufacturer creates large added value by their production. Their advantage is considerable here.

- *Extraction.* No direct comparison can be made between natural product extraction and large-scale synthesis.

6.2 Economics

Too little consideration of the economics of azadirex production has been made in published considerations of its use. The producer of a synthetic pesticide has complete control of its production for a known period of years, through patent protection. It can therefore calculate its potential profits and make allowance for the cost of assessing the market, field trials, spectrum of activity, stability, formulation, toxicity, licensing and many other factors. The cost of toxicity tests required by the licensing authority can be enormous, and this cost is for a single compound with ancillary substances used in the formulation, such as wetting agents, solvent, UV stabilizers. Isman (1997) has estimated the cost of studies in support of registration of a new pesticide is at least US $250,000 and can exceed US $2 million.

It has been suggested that the principal barriers to marketing a new natural pesticide are threefold (Isman, 1997).

1. Scarcity of the natural resource, which is not a problem for azadirex;
2. Standardization and quality control of the product. There is as yet no standard method for the determination of azadirachtin (although HPLC is much used) and certainly no standards for the lesser compounds in azadirex (but see Jarvis and Morgan, 2000), and no agreed standards of quality control. An important consideration for azadirex is the control of aflatoxin in the product.
3. Registration, this is a major problem. The source, structure and pesticidal properties of azadirachtin and azadirex are in the public domain and therefore unpatentable. Only limited patent protection is available for processing or formulation. Therefore the possibilities of recovering costs of toxicity tests for registration are much less certain.

While regulations continue to require the complete analysis of a pesticide and toxicity data on all its components (and this still applies in many countries), no new natural pesticides can be licensed, because a complete analysis can never be given and the cost of toxicity tests, if it could be analysed, would be astronomical. Only with the easing of regulations for natural products, as has been done in a few countries, notably the USA, but also Canada (see Isman, 1997), Germany, Spain and Sweden, can a azadirex product be licensed.

The economics of production of azadirex is shrouded in confidential information. The price of clean, dried seeds varies from US $0.05 to $1.60 per kilogram, possibly pesticide quality would cost a minimum of US $0.30-0.40 per kilogram (Förster and Moser, 2000). Each hectare of crops would require 20-60 g azadirachtin per treatment and assuming 2 g azadirachtin per kilo of seed that means 10-30 kg seed per treatment at a cost of US $1.00-60.00, although it is estimated it would be in the narrower range of US $5.00–20.00 (Förster and Moser, 2000). The efficiency of recovery is also important. It was very low in early products such as Margosan-O. If 50 per cent of the azadirachtin is lost during isolation, the cost of the product doubles. If a strain of neem trees could be obtained by selective breeding that gave a mean of 4 g of azadirachtin per kilo, the price could be halved. The price that can be obtained for by-products helps to lower the cost. There is not yet a market for these except the oil. Much is said of the use of neem oil for soap-making. It is the view of this author that the oil is of little value. Soap made from neem oil alone is too water-soluble. Neem oil with some residual azadirex has possibly great potential in specialized plant protection, animal hygiene or impregnated textile applications.

The technology of production in India has certainly been achieved, with ten plants in operation there. There is a high probability of plants in Australia and China in the near future. A high tech process using microwaves and computer control is available in Australia. Can such processes give a product competitive on price with synthetics? It is possible that an azadirex product can only be priced competitively if all the fractions (oil, azadirex and residual seed cake) from processing can find markets (O'Shea, personal communication). Otherwise its market is limited to higher priced products that appeal to consumers in richer countries concerned about pollution. The economic analysis conducted by GTZ concluded that neem pesticides must be 25-40 per cent cheaper than that offered at present to gain widespread use. It is nevertheless incorrect to compare cost of azadirex with synthetic pesticides without taking into consideration the environmental cost and the added natural biological control that azadirex provides.

6.3 Acceptance

The German organization for technical assistance to Third World countries, GTZ, has done much to encourage the use of crude neem seed extract among poor farmers in the tropics since the 1970s, showing them how to store and use neem seeds with the minimum of technical equipment. GTZ have concluded after a detailed study that there has been a poor acceptance of this "low-tech" use of neem. After some twenty years of their efforts less than half of the farmers who had access to the seeds and knowledge of how to use them in fact did apply the extract on their crops. More preferred to use bought synthetic pesticides (Förster and Moser, 2000). The same conclusion was reached by Childs *et al.* (2001) from work in Ghana and India. GTZ have subsequently shifted emphasis to the promotion of formulated neem-based pesticides with greater market potential.

GTZ circulated a questionnaire on global neem usage. Their conclusions for the lack of acceptance of neem are worth quoting at length, but it must be pointed out that they had in mind chiefly poor farmers in the third world. Their points are:

- knowledge of neem is not widespread.
- Neem trees not available in areas that could use the product.
- Harvesting and processing are laborious and require careful attention to raw material quality.
- Neem harvesting and processing come at times with other heavy demands on labour.
- Neem products are more expensive and synthetics are easier to buy and use.
- Doubts about neem's efficacy, as there is no "knock-down" effect.
- Homemade neem pesticides do not have a high social status.
- There are technical problems in producing a homemade product.
    - Storage of neem seeds in humid tropical conditions is difficult.

Many of these problems (except availability, harvesting and processing) can be overcome with the help of their simple illustrated booklet "Neem, a natural insecticide" (GTZ, no date, Pesticide Service Project, Eschborn, pp 34). Although the interest from poor farmers has declined, GTZ has found there is a boom in enquiries about neem from many different sources (Forester and Moser, 2000).

The "organic foods" movement of recent years could be a catalyst for the acceptance of azadirex use. It is worth noting that growers of "organic" crops are permitted to use pyrethrin and quassia to control pests, as well as *Bacillus thuringiensis* toxins and "azadirachtin extracted from *Azadirachta indica*", but for azadirachtin "only on the mother/parent plants for seed and vegetative reproduction material and on ornamental crops" (Soil Association standards for organic food and farming, 1999, Revision 12, Section 3, pp 29-31). Derris (rotenone) is restricted (i.e., approval must be obtained from the certification department before use) and nicotine is prohibited. Laws in individual countries overrule these standards, for example where azadirex is not registered. A programme of education of the organic foodstuff producers is necessary to have the rules on azadirachtin use changed so it refers to azadirex and its direct use on foods permitted as for pyrethrin.

7. CONCLUSIONS

We have to consider that earlier natural pesticides were introduced at a time of little or no regulation. It is highly unlikely that either tobacco or rotenone would be licensed for use as pesticides if they had been discovered only today. In many countries, including Great Britain, both are still licensed for use. Neem is the first new natural pesticide (excluding special products like *Bacillus thuringiensis* toxin) to come into use since the strict government controls of the 1970s and 1980s. Any return to the use of natural pesticides from the present low proportion is bound to favour neem, because it has great advantages over earlier natural pesticides.

There was little recognition among early enthusiasts of the large capital outlay consequentially necessary to carry out the required testing, stability studies, formulation, and advertising to launch a new pesticide. We have passed this phase and the work and cost of introducing a new product are being recognized. Tests never designed for multicomponent natural pesticides are still a serious hindrance to licensing in some countries. This will only change when governments see they are missing out on a valuable market.

A higher level of understanding is required for azadirex over synthetics when being used by poor farmers.

For the near future, the use of azadirex will be mostly on vegetables and fruit, and on ornamentals and gardens in developed countries. Organically grown foods offer a special route for azadirex to enter greater use. The long-term benefits of azadirex will only be brought into the economic calculations when neem products are much more widely recognized.

There is still a need for standardization of neem products. We still do not have an agreed standard analytical method to measure azadirachtin content.

According to GTZ, its surveys show how vital and promising natural pesticides are for the future (Förster and Moser, 2000). Neem is ready and waiting.

ACKNOWLEDGEMENTS

I thank my colleagues Graeme Jones and Tessa Grasswitz for their comments and former students S. Johnson and A. P. Jarvis for all their contributions to our knowledge of azadirex.

REFERENCES

Arnason, J.T., Philogène, B.J.R. and Morand, P. (1989), *Insecticides of Plant Origin*, American Chemical Society, ACS Symposium series 387, Washington DC.

Butterworth, J.H. and Morgan, E.D. (1968), Isolation of a substance that suppresses feeding in locusts, *Journal of Chemical Society Chemical Communications*, 23.

Butterworth, J.H. and Morgan, E.D. (1971), Investigation of the locust feeding inhibition of the seeds of the Neem tree (*Azadirachta indica*), *Journal of Insect Physiology*, **17**, 969-977.

Carson, R.L. (1962), *Silent spring*, Houghton Mifflin, Boston, USA.

Cassida, J.E. (1973), *Pyrethrum, the Natural Insecticide,* Academic Press, New York

Childs, F.J., Chamberlain, J.R., Antwi, E.A., Daniel, J. and Harris, P.J.C. (2001), *Improvement of neem and its potential benefits to poor farmers,* HDRA Publishing, Coventry, England.

Copping, L.G. (1998), *The Biopesticides Manual*, British Crop Protection Council, Bracknell, England.

Crosby, D.G. (1971) Minor insecticides of plant origin, in *Naturally Occurring Insecticides*, eds. M. Jacobson and D.G. Crosby, Marcel Dekker, New York, pp. 177-239.

Dev, S. and Koul, O. (1997), *Insecticides of Natural Origin*, Harwood Academic Publishers, Amsterdam, The Netherlands.

Doull, J. (1976) Insecticide toxicity, in *Insecticide Biochemistry and Physiology,* ed. Wilkinson, C.F., Plenum Press, New York and London, pp. 649-667.

Feng, R. and Isman, M.B. (1995), Selection for resistance to azadirachtin in the green peach aphid, *Myzus persicae*, *Experientia,* **51**, 831-833.

Förster, P. and Moser, G. (2000), *Status Report on Global Neem Usage*, GTZ, Eschborn, Germany.

Gill, J.S. and Lewis, T. (1971), Systemic action of an insect feeding deterrent, *Nature,* **232**, 402-403.

Isman, M.B. (1997), Neem and other botanical insecticides: barriers to commercialization, *Phytoparasitica,* **25**, 339-344.

Jarvis, A.P. and Morgan, E.D. (2000), Analysis of smalll samples of limonoids of neem (*Azadirachta indica*) using solid phase extraction from tissue culture, *Phytochemistry Analytical,* **11**, 184-189.

Jarvis, A.P., Johnson, S. and Morgan, E.D. (1998), Stability of the natural insecticide azadirachtin in aqueous and organic solvents, *Pesticide Science*, **53**, 217-222.

Khan, Z.R., Ampong-Nyarko, K., Chiliswa, P., Hassanali, A., Kimani, S., Lwande, W., Overholt, W.A., Pickett, J.A., Smart, L.E., Wadhams, L.J. and Woodcock, C.M. (1997), Intercropping increases parasitism of pests, *Nature*, **388**, 631-632.

Koul, O. (1992), Neem allelochemicals and insect control, in *Allelopathy, Basic and Applied*, eds. S.J.H. Rizvi and V. Rizvi, Chapman & Hall, London, pp. 389-412.

Kraus, W. (2002), Azadirachtin and other triterpenoids, in *The Neem Tree*, ed. H. Schmutterer, Neem Foundation, Mumbai, India, 39-111.

Miller, J.R. and Cowles, R.S. (1990), Stimulo-deterrent diversion: a concept and its possible application to onion maggot control. *Journal of Chemical Ecology,* **16**, 3197-3212.

Morgan, E.D. and Wilson, I.D. (1999), Insect hormones and insect chemical ecology, in *Comprehensive Natural Product Chemistry*, Vol. 8, ed. K. Mori, pp. 263-375.

Ruscoe, C.N.E. (1972), Growth disruption effects of an insect antifeedant. *Nature New Biology,* **236**, 159-160.

Saxena, R.C., Liquido, N.J. and Justo, H.D. (1981), Neem seed oil, a potential antifeedant for the control of the rice brown planthopper, *Nilaparvata lugens*, in *Natural Pesticides from the Neem Tree (Azadirachta indica A. Juss.),* eds. H. Schmutterer, K.R.S. Ascher and H. Rembold, GTZ, Eschborn, Germany, pp. 171-188.

Schmeltz, I. (1971), Nicotine and other tobacco alkaloids, in *Naturally Occurring Insecticides*, eds. M. Jacobson and D.G. Crosby, Marcel Dekker, New York, pp. 99-136.

Schmutterer, H. (1995), *The Neem Tree, Source of Unique Natural Products for Integrated Pest Management, Medicine, Industry and other Purposes*, VCH Publishers, Weinheim, Germany.

Sunderam, K.M.S. (1996), Azadirachtin biopesticide: a review of studies conducted on its analytical chemistry, environmental behaviour and biological effects. *Journal of Environmental Science and Health, Part B,* **31**, 913-948.

Sweetman, H.L. (1936), *The Biological Control of Insects,* Comstock Publishing, Ithaca, NY.

Warthen, D.J. (1979), *Azadirachta indica: A Source of Insect Feeding Inhibitors and Growth Regulators,* US Department of Agriculture, Agricultural Review Manuals, ARM-NE4, Beltsville, Maryland.

Warthen, J.D. and Morgan, E.D. (1985), Insect feeding deterrents, in *Handbook of Natural Pesticides*, Vol. 4, *InsectAttractants andRepellents*, eds. E.D. Morgan and N.B. Mandava, CRC Press, Boca Raton, FL, pp.23-149.

Chapter 3

FACTORS LIMITING COMMERCIAL SUCCESS OF NEEM INSECTICIDES IN NORTH AMERICA AND WESTERN EUROPE

MURRAY B. ISMAN
Faculty of Agricultural Sciences, University of British Columbia
Vancouver, BC, Canada V6T 1Z4

1. INTRODUCTION

In 1959, Heinrich Schmutterer, stationed in the Sudan, observed that the only green plants that avoided the depredations of a plague of desert locusts at the time were trees of the introduced species *Azadirachta indica*. He decided to investigate this phenomenon, and indeed it became his life's work. However, it is unlikely that he could have ever imagined at the outset that a botanical insecticide derived from the Indian neem tree (*A. indica*) would generate more interest, investigation and commercial development than any other botanical product in the 20th century.

This interest spawned at least seven international conferences specifically focussed on neem, beginning in Germany in 1980, with numerous other workshops and symposia at other scientific gatherings. In turn, research activity across the globe, but largely concentrated in India, Germany, the United States, the United Kingdom and Canada gave rise to an abundant body of scientific literature dealing with neem insecticides, culminating in the seminal volume *The Neem Tree*, edited by Schmutterer himself and first published in 1995.

The attributes of neem-basd insecticides are widely known. They include both behavioural (antifeedant) and physiological (growth disruptant) effects on insects, efficacy against a broad spectrum of insect pests, systemic action in some plants, minimal disruption of natural enemies, minimal impact on other non-target organisms, (especially pollinators), rapid breakdown in the environment, and

O. Koul and S. Wahab (eds.), Neem: Today and in the New Millennium, 33–41.

essentially no toxicity to mammals, conferring a wide margin of safety to the user and to consumers. Not surprisingly then, neem has been considered a modern paradigm for botanical insecticides (Isman, 1997).

The global pesticide market has been recently estimated at more than US$ 30 billion, although insecticides account for slightly less than one-third of that total (US$ 9 billion) (National Research Council, 2000). However, North America and Western Europe together represent 55 per cent of the entire market. For example, the U.S. insecticide market was valued at US$ 3.6 billion in 1997. Clearly, North America and Western Europe represent lucrative markets for the pesticide manufacturer, and the heightened awareness and standards for safety in these regions would be expected to provide an additional competitive advantage for neem-based products. In this paper I discuss why neem insecticides have largely failed (to date) to meet expectations for their success in these markets.

2. HISTORY OF NEEM IN THE UNITED STATES AND EUROPE

As a commercial entity, neem based product was introduced to the United States by Robert Larsen, who under the name Vikwood Botanicals, received approval from the Environmental Protection Agency (EPA) for the use of Margosan-O^R for non-food uses in 1985. Developed through collaboration with the U.S. Department of Agriculture, Margosan-O^R was an ethanolic extract of neem seeds containing 0.3 per cent azadirachtin as the stated active ingredient and about 10 per cent neem seed oil. Three years later W.R. Grace & Co. purchased the rights from Larson, improved the formulation and introduced Margosan-O^R into the greenhouse industry in 1990 (Walter, 1999).

In 1992, Agridyne Technologies received EPA registration for the use of Azatin^{TM} on greenhouse plants and the companion product Turfplex^{TM} for turf. These products were oil-free and contained 3 per cent azadirachtin as the active ingredient. One year later the EPA approved the same formulation, under the trade name Align^{TM}, for use on food crops, with an exemption from residue tolerance when used at <20 g azadirachtin/acre. In the same year, Grace received approved for Neemix^{TM} (containing 0.25% azadirachtin) for use on food crops, and in 1994 introduced Neemix 4.5^{TM} (containing 4.5% azadirachtin). Azatina^R (same as Azatin^{TM}) was registered in Mexico in 2000.

In 1996, ThermoTrilogy (a subsidiary of ThermoEcotek) acquired the neem biopesticide technology from W.R. Grace, and less than a year later acquired the neem technology of Agridyne, at the time owned by Biosys who had subsequently gone into bankruptcy. In the most recent transaction involving these technologies, Certis U.S.A., a subsidiary of Mitsui Trading of Japan, acquired ThermoTrilogy in early 2001. Certis is marketing the following neem products in the U.S.: Azatin^R Insect Growth Regulator, Neemix^R Botanical Insecticide and Neemix 4.5 Insect Growth Regulator, and two fungicide/miticide/insecticide products based on extracts of neem seed oil, Triact^R and Trilogy^R.

In 2000, in a joint venture with Fortune Biotech Ltd. (India), Amvac Chemical Co. received registration from the EPA and introduced three products containing 3 per cent azadirachtin: Ecozin^R for use on food crops, Ornazin^R for ornamentals, and Amazin[R] for mushrooms. In India comparable products are marketed under the trade name Fortune-Aza. In the same year, The Gowan Co. introduced Aza-Direct[TM], an insecticide containing 1.2 per cent azadirachtin, and produced by E.I.D. Parry (India) Ltd. And these products were joined by AgroNeem[TM] (0.15% azadirachtin), produced by Ajay Biotech in India and marketed in the U.S.A. by AgroLogistic Systems. At this writing then, there are four companies selling neem insecticides in the United States, namely Certis, Amvac, Gowan and AgroLogistics, the first mentioned marketing two lines of neem products previously developed independently.

In October 2000, Health Canada granted temporary registration to Neemix 4.5, exclusively for the control of sawflies in forestry in Canada. This represents the first registration of a neem insecticide in Canada, following eight years of review, initiated by W.R. Grace for registration of Margosan-O. The product will be used on 5000-6000 ha of forest in Canada in 2001, and the treated area is expected to rise sharply in the future.

The situation in Western Europe mirrors that in Canada. In spite of three decades of research effort in Germany, spearhead by Heinrich Schmutterer, Wolfgang Kraus, Heinz Rembold and others, and generously funded by the Deutsche Gesellschaft fur Technische Zusammenarbeit (GTZ) (Schmutterer and Ascher, 1987), neem insecticides were not registered in that country until 2000 when approval was granted to Trifolio-M GmbH for NeemAzal-T/S (containing 1% azadirachtin) for use on both food and nonfood crops. The toxicological and environmental data required in support of German registration was voluminous. This product is also being marketed in Germany under the Celaflor[R] tradename. Align[R] (3% azadirachtin) represented the first neem product approved for use on food crops in Europe, having been granted registration in Spain in 1999.

In Sweden, Gabrol Produkter received registration in 1997 for Bionim, an alcoholic extract of neem kernels containing 0.3 per cent azadirachtin, for insect control on ornamental plants. Additional registrations were granted in early 2001 for neem-based mosquito repellents, fly repellents for application to horses, and a shampoo for cats and dogs. This and other companies are seeking registration for neem-based insecticides in France. Surprisingly, no neem products have been approved for use in the United Kingdom, nor has significant commercial development of neem products proceeded there.

3. COST OF NEEM INSECTICIDES

One might expect that with their high standards of living and high monetary returns for produce, growers in the U.S. and western European countries would best be able to afford expensive pesticides. On the other hand, high input costs and foreign competition are eroding the profits in many commodities. At the same time, these growers place the greatest demands for performance on their crop protectants. In producing a consistent, high quality neem insecticide (i.e. 1% or more of azadirachtin by weight), technical shortcuts cannot be taken, and the end result is that neem insecticides are expensive to manufacture and therefore expensive for the end user. When neem seeds as a commodity in India began to be diverted to insecticide production from soap and industrial oil production, the cost of seeds doubled (from approximately US$ 100 to US$ 200/tonne).

To realize a reasonable profit, a manufacturer's cost of approximately US$ 0.40/gm formulated azadirachtin is required. Factoring in costs of equipment and other manufacturing assets, costs of data acquisition in support of registration, and distribution/marketing costs, it is hardly surprising that neem insecticides sell for US$ 125/liter, which, based on a formulation containing 3 per cent azadirachtin, equates to a price of US$ 2.60/gm azadirachtin to the end user. As such, neem is 2.5-3 times more expensive than synthetic pyrethroids. At US$ 125/liter, a grower will be paying US$ 25/treatment/acre, compared with the US$ 8-10/treatment/acre normally spent on field crops and cotton. And as the costs of *Bacillus thuringiensis*-based pesticides (Bt) have dropped in price, to compete with these products for control of lepidopteran pests in large markets, would require producing formulated azadirachtin at US$ 0.25-0.30, i.e. 25-38 per cent less than the current price (for economics of neem products, see chapter 2 also). On the other hand, neem can be used in combination with Bt-pesticides, for example in tree fruits.

This helps explain why neem insecticides in the U.S. are currently positioned with an emphasis on high value row crops (e.g. fresh market tomatoes) and greenhouse crops. In the latter case, growers will accept high treatment costs on extremely high value ornamentals. Azatin[R] is currently the top-selling IGR in greenhouses in the U.S. Another approach is to recommend neem in rotation with other (synthetic) insecticides.

At approximately US$ 35/liter, the price of neem insecticide containing 1 per cent azadirachtin in Germany is comparable to that of the products marketed in the United States. How future supplies of neem seeds from competing regions (Australia, Mexico, South America) will influence the ultimate price to the user is a subject of considerable speculation.

4. EFFICACY AND GROWER ACCEPTANCE

Apart from safety to the user and the environment, major attributes of neem-based insecticides include their broad spectrum-of-action against pests, their feeding deterrent action, and their systemic action in plants. In practice though, there are limitations to each of these attributes that influence when and how neem insecticides can be used successfully.

Many reviews of neem for insect control provide long lists of pests susceptible to neem, some of which are based exclusively on laboratory tests, with the total number of susceptible pest species reputed to be in excess of 400. Neem is unquestionably effective, under specific conditions, against certain pests, particularly lepidopteran and coleopteran larvae. On the other hand, neem has failed to show efficacy against other pests, e.g. some species of tephritid flies (apple maggot, cherry fruit fly). And owing to its limited persistence on plants, multiple applications are necessary to achieve acceptable control against some important pests with a wide flight period (e.g. codling moth on apple, bollworm on cotton), which may not be economically feasible. For example, when applied weekly as a stand alone insecticide (at rates of 25-50 g aza/ha), Align^R provides commercially competitive insect control, but acceptable control is not always achieved when applications are made two weeks apart, or if weekly applications are made at rates less than 25 g aza/ha (Wood *et al.*, 1995).

Aphids were shown to be highly variable in their susceptibility to neem: some aphid species are quite susceptible while others are relatively tolerant. In this case the host plant appears to play a significant role, as the aphids must acquire azadirachtin through feeding on plant tissues (Lowery and Isman, 1994; Koul, 2003). Though there is a popular notion that azadirachtin moves systemically in all plants, systemic action has in fact only been demonstrated in a few agronomic plants (tomato, potato, rice, maize) and these may prove to be the exception rather than the rule.

While azadirachtin (and neem seed extracts) are potent antifeedants to many species of insects there is again a wide range or susceptibilities, even among related insects. Studies with noctuid larvae indicate that behavioural responses to azadirachtin are much more variable between species than physiological responses (Isman, 1993). Furthermore, the antifeedant response in insects can be modified by experience – repeated exposures or continuous exposure leads to demonstrated habituation in both tobacco cutworms and in adult Japanese beetles (Bomford and Isman, 1996). And even though azadirachtin is the most potent antifeedant for desert locusts discovered to date, North American grasshoppers can eat neem-treated plants with impunity, although they subsequently suffer from the physiological effects of azadirachtin (Champagne *et al.*, 1989; Reynolds *et al.*, 2001).

In terms of grower acceptance, neem insecticides suffer from the same weakness as other insect growth regulators – they lack contact action and work slowly. Insects on treated plants may indeed cease feeding shortly after treatment, but they can remain alive for days, a situation disheartening to growers used to using synthetic pyrethroids or other contact toxins that kill pests in a matter of hours. Neem is also relatively ineffective against late stages or instars of pests and many types of adult insects. Under moderate to high insect pressure, neem may not act rapidly enough to prevent economic damage. Where neem does display excellent efficacy is against pests capable of explosive population growth, such as aphids and whiteflies. In these pests the survival of a particular generation is far less noticeable than the overall population trend.

5. REGULATORY ISSUES

Outside of the United States, registration has proven a formidable barrier to the successful commercialization of neem-based pesticides. In its review of the first neem insecticide in North America, Margosan-O^R, the Environmental Protection Agency chose to recognize azadirachtin as the sole active ingredient, deeming the remaining chemical constituents of the neem kernel extract as 'inert' ingredients. This approach vastly simplified the decision-making process for approval of this, and subsequent, neem insecticides in the U.S.A. This philosophical view of neem extract has been held even though some members of the agency acknowledge that there are other azadirachtin analogs (e.g. 3-tigloylazadirachtol, also known as 'azadirachtin B') in lesser amounts that likely contribute to the overall biological activity of a neem product.

Although the bioactivity of the dozen or so azadirachtin analogs isolated to date is well documented, the contribution of the remaining limonoid constituents of neem kernels to overall efficacy of neem insecticides remains controversial (Isman, 2002). Unfortunately, this controversy has become a point of confusion and uncertainty for regulatory agencies charged with evaluating neem as a pesticide. Suggesting that numerous triterpenoids in neem are effectively 'active ingredients' has created a Pandora's box for government toxicologists. For example, if a large number of compounds in neem kernels are considered active ingredients, then regulatory agencies may be justified in requiring identification of all major constituents in a neem preparation, and perhaps in requiring some simple biological tests for these compounds in isolation. In Canada, the Pest Management Regulatory Agency agreed to issue an experimental use permit for aerial application of a neem insecticide for management of a forest pest based on the identification (via HPLC) of the seven major terpenoids in the technical grade concentrate, which together made up approx. 75 per cent of the concentrate by weight.

A report indicating that a minor constituent of neem kernels may be generally cytotoxic does little to allay the fears of regulatory authorities. While it makes sense to evaluate the safety of neem pesticides based on toxicological evaluation of the technical grade concentrate as a whole, regulatory agencies in some jurisdictions remain concerned with the chemical variability of neem products and the lack of chemical standardization.

Regulatory approval has greatly limited the introduction of neem insecticides in Europe, particularly on food plants, owing to residue and environmental data requirements. The situation in Japan is very similar. These costly requirements preclude small-scale entrepreneurial companies from the marketplace, although there have been some successes with products for ornamental plants and for veterinary use.

6. RECENT TRENDS AND PROSPECTUS

Neem insecticides in the United States are registered for use on a wide range of food crops including tree fruits and nuts, bush fruits and berries, and vegetables, including potatoes. However, neem is not equally efficacious on all crops or against all pests on any individual crops, and as a result, neem insecticides must be introduced and markets developed with personalized attention. Grower education is essential, as is the development of practical strategies for incorporation of neem insecticides into existing IPM programs. Having been approved for use in organic agriculture by the Organic Materials Review Institute, neem should be particularly competitive in organic production, but is gaining favour with conventional growers as well. Trilogy^{R}, consisting of clarified neem oil containing little or no azadirachtin, is gaining in popularity as a fungicide for use on grapes, stone fruits and nuts, and as an acaricide in citrus and cotton.

The State of California maintains perhaps the most detailed database on pesticide use of any jurisdiction in the world, and given that California accounts for about 30 per cent of all insecticide use in the United States, the data therein merit attention (Calif. EPA, 2001). In 1998, neem insecticides (listed under 'azadirachtin') had reported uses on over 60 food crops, with the total amount of active ingredient applied amounting to over 370 kg. However, two crops, lettuce (67 kg) and tomato (83 kg) together accounted for 40 per cent of total use, and represented the predominant uses. From 1997 to 1999, the total volumes applied to lettuce and tomato increased by 80 and 77 per cent, respectively. Data for 1998 indicate the use of azadirachtin on tomato, at 80 kg, accounted for only 0.25 per cent of insecticide use on that crop, paling in comparison to *Bacillus thuringiensis* products (780 kg), carbaryl (1464 kg), dimethoate (2980 kg), methamidophos (6870 kg), methomyl (13114 kg), and permethrin (865 kg). Even more recently introduced insecticides (imidicloprid, 705 kg and spinosad, 373 kg) were more heavily used than neem. In spite of a wealth of research activity, neem has not enjoyed the rapid growth of spinosad, whose reported use in California increased from 4500 kg in 1997 to 20,100 kg in 1999, an average increase of 220 per cent per year.

Azadirachtin is up against strong competition in North America and Western Europe. Spinosad, though also relatively expensive, has a toxicological and environmental profile comparable to neem (carrying the 'Caution' signal word; minimum allowable re-entry interval following treatment – 4 hours), but with excellent efficacy against lepidopteran pests. Indoxacarb is another recently introduced 'reduced-risk' insecticide with a good toxicological and environmental profile and excellent efficacy against lepidopterans. Pymetrozine is a selective aphicide with systemic action that also meets the criteria for 'reduced-risk'.

Along with the spectacular growth of internet commerce has emerged a new class of competitors to the producers of registered, high quality neem insecticides. These are the small-scale privateers, many in India and Australia, offering extremely inexpensive neem insecticides with exaggerated claims. Most of these products are simple formulations of crude neem oil, likely with little, if any azadirachtin. The trade in these products is entirely unregulated and in some regions growers will buy these because they are cheap. The danger exists that their poor efficacy, let alone the potential for spray injury, will tarnish the image of neem that the legitimate manufacturers have worked hard to establish.

So what is the future for neem insecticides in North America and Western Europe? Most pest management and pesticide industry experts expect neem to show continued sales growth, in the range of 5-20 per cent per annum, as new markets continue to be developed. Neem is likely to experience broad acceptance by organic producers, and therefore growth in neem insecticide sales may partially parallel the growth of organic food production in the United States and Europe. The gradual elimination of organophosphate, carbamate and synthetic pyrethroid insecticides (together holding 73% of the world insecticide market) under the Food Quality Protection Act in the United States should also lend encouragement to those manufacturing and promoting neem insecticides. But overall, for reasons outlined in this chapter, neem is unlikely to assume a position of dominance as a crop protectant in these highly developed countries, and instead will be best accepted as one among a number of safe pesticides with efficacy in niche markets. Neem might be expected to become a dominant insecticide only in situations where grower use of crop protectants is determined by government or industry mandate rather than by individual choice (e.g. future use on cotton in China?).

ACKNOWLEDGMENTS

The following individuals generously provided information relevant to this review: Michael Dimock (Certis, USA), Ramon Georgis (EcoSMART Technologies, USA, previously with ThermoTrilogy Corp.), Elisabeth Hammarberg (Gabrol Produktor, Sweden), John Immaraju (Amvac Chemical Co., USA), Hubertus Kleeberg (Trifolio-M GmbH, Germany), Dave Martin (EcoSMART Technologies, USA, previously with Agridyne Technologies), Jim McNutt (Gowan Corp., USA). Drs. John Stark (Washington State University) and Michael Dimock (Certis, USA) kindly reviewed the manuscript.

REFERENCES

Bomford, M.K. and Isman, M.B. (1996), Desensitization of fifth instar *Spodoptera litura* (Lepidoptera: Noctuidae) to azadirachtin and neem, *Entomologia Experimentalis et Applicata*, **81**, 307-313.

California Environmental Protection Agency, Department of Pesticide Regulation. Website address: www.cdpr.ca.gov

Champagne, D.E., Isman, M.B. and Towers, G.H.N. (1989), Insecticidal activity of phytochemicals and extracts of the Meliaceae, in *Insecticides of Plant Origin*, eds. J.T. Arnason, B.J.R. Philogene and P. Morand, American Chemical Society, Washington DC, pp. 95-109.

Isman, M.B. (1993), Growth inhibitory and antifeedant effects of azadirachtin on six noctuids of regional economic importance, *Pesticide Science*, **38**, 57-63.

Isman, M.B. (1997), Neem insecticides, *Pesticide Outlook*, **8**, 32-38.

Isman, M.B. (2002), Lepidoptera: butterflies and moths, in *The Neem Tree, 2nd Edition*, ed. H. Schmutterer, The Neem Foundation, Mumbai, India, pp. 378-401.

Koul, O. (2003), Variable efficacy of neem-based formulations and azadirachtin to aphids and their natural enemies, in *Biopesticides and Pest Management*, eds. O. Koul, G.S. Dhaliwal, S.S. Marwaha and J.K. Arora, Campus Books International, New Delhi, pp. 64-74.

Lowery, D.T. and Isman, M.B. (1994), Insect growth regulating effects of neem extract and azadirachtin on aphids, *Entomologia Experimentalis et Applicata*, **72**, 77-84.

National Research Council (2000), The *Future Role of Pesticides in US Agriculture.* National Academy Press, Washington.

Reynolds, E.R., Graceffa, L. and Majumdar, S.K. (2001), Effect of neem-based bioactive compounds on the development of three laboratory strains of *Drosophila melanogaster*, *International Journal of Ecology and Environmental Sciences*, **27**, 147-154.

Schmutterer, H. (1995), *The Neem Tree Azadirachta indica A. Juss and Other Meliaceous Plants: Sources of Unique Natural Products for Integrated Pest Management, Medicine, Industry and Other Purposes.* VCH, Weinheim, Germany.

Schmutterer, H. and Ascher, K.R.S. (1987), *Natural Pesticides From the Neem Tree and Other Tropical Plants,* GTZ, Eschborn, Germany.

Walter, J.F. (1999), Commercial experience with neem products, in *Methods in Biochemistry, vol. 5: Biopesticides: Use and Delivery*, eds. F.R. Hall and J.J. Menn, Humana Press, Totowa, pp. 155-170.

Wood, T., Ruggero, W. and Nelson, R. (1995), Performance profile for the neem-based insecticide 'Align™', in, *The Neem Tree Azadirachta indica A. Juss and Other Meliaceous Plants: Sources of Unique Natural Products for Integrated Pest Management, Medicine, Industry and Other Purposes*, ed. H. Schmutterer, VCH, Weinheim, pp. 445-453.

Chapter 4

NEEM FOR ECOLOGICAL PEST AND VECTOR MANAGEMENT IN AFRICA: OUTLOOK FOR THE NEW MILLENNIUM

RAMESH C. SAXENA
Chairman, Neem Foundation, J-1041 Palam Vihar
Gurgaon 122 017, Haryana, India

1. INTRODUCTION

African countries are facing an uphill task in producing sufficient food for their rapidly expanding populations (about 3% per year in the late 1980s) and at the same time gain economic independence. The situation is particularly grim in Sub-Saharan Africa (Goliber, 1989). Extremes of climate, a high proportion of eroded or degraded land, and a high percentage of people suffering from severe mal-nourishment characterize the region. Other ailments and diseases, such as malaria, yellow fever, intestinal worms, etc., make human lives extremely stressful and less productive. The harsh geophysical conditions have repeatedly led to famines, particularly during the last 25 years in Sudan, Ethiopia, Somalia, and northern Kenya. The tragic consequences for want of food and readily available and inexpensive medication have been deaths of multitudes of humans and livestock. These tragedies need to be averted.

To increase yields of production systems and improve the health status of the populations in Africa are a daunting challenge. About 75 per cent of the regions' population lives in villages; agriculture is the largest economic activity. The adoption of sustainable agro-ecological practices that would help the great mass of resource-poor farmers, often pushed to marginal lands, is therefore urgently needed. Such practices should contribute to achieving self-sufficiency in food, reduce the reliance on purchased agrochemical inputs, and rebuild the productive capabilities of farmer landholdings and households.

O. Koul and S. Wahab (eds.), Neem: Today and in the New Millennium, 43–64.

Although "green revolution technologies" have more than doubled the yield potential of rice, wheat, and maize, especially in Asia, these high-input production systems cannot be sustained over generations in African countries. To a great extent, future food security and economic independence of African countries would depend on improving the productivity of biophysical resources through the application of sustainable production methods, by improving tolerance of crops to adverse environmental conditions, and by reducing crop and post-harvest losses caused by pests and diseases. Appropriate technologies would have key roles to play in ensuring food security, in improving public and animal health, and in rehabilitating and conserving the environment. Instead of striving for more "green revolutions" with emphasis on miracle seeds, hard-hitting, synthetic and engineered pesticides, and increased use of fertilizers, the future must look to natural ways and processes for augmenting agricultural productivity. In fact, all development efforts and activities, including pest and vector management, should be within well-defined ecological rules rather than within narrow economic gains. Sustainable agricultural systems must be efficient and ecologically sound for long-term food sufficiency, equitable in providing social justice, and ethical in respecting both future generations and other species. For developing countries, especially in Africa, the use of the neem tree may provide a key component in ensuring sustainable pest and vector management.

2. NEEM: POTENTIAL FOR ACTION

Neem, a member of the Meliaceae family, is a botanical cousin of mahogany. According to a report of an ad hoc panel of the Board on Science and Technology for International Development, "this plant may usher in a new era in pest control, provide millions with inexpensive medicines, cut down the rate of human population growth and even reduce erosion, deforestation, and the excessive temperature of an overheated globe" (National Research Council, 1992). Neem's other descriptions, such as "nature's bitter boon," "nature's gift to humankind," "the tree for many an occasion," the tree that purifies," "the wonder tree," "the tree of the 21st century," and "a tree for solving global problems," are a recognition of its versatility.

During the last century, neem was introduced in arid zones of Africa. Today, it is grown in many Asian countries, in tropical regions of the New World, in several Caribbean and in some Mediterranean countries (see chapter 1). The tree most likely made its way to Africa as a result of British colonialism at the start of this century and today it occurs in considerable numbers from Eritrea (an estimated 500,000 trees; Beraki and Foerster, 1998) and Somalia via Kenya and Tanzania northwards to Mozambique. Many neem trees do occur in some central regions of East Africa, e.g. in Uganda (25,000), Kenya, Tanzania and Malawi (Schmutterer, 1995). Neem is frequently encountered in sub-Saharan Africa as many million trees are located through the entire area from Ethiopia, Sudan, Senegal and Mauritania due to favourable hot climatic conditions with a precipitation level of 500-1200 mm/a; Nigerian record is of about 10 million trees followed by Snegal and Ghana approximately 6 million trees and Mali, Burkina Faso and Niger about 2.5 million trees (Strozok, 1992; Schmutterer, 1995).

Neem is an evergreen, tall, fast-growing tree, which can reach a height of 25m and 2.5 m in girth in African region (Fig. 1). It has an attractive crown of deep-green foliage (which can spread 10 m across) and masses of honey-scented flowers. The tree thrives even on nutrient-poor dry soil. It tolerates high to very high temperatures, low rainfall, long spells of drought, and salinity. It is propagated by seed; 9 to 12 month-old seedlings transplant well. Birds and fruit bats also disperse the seed. Fruiting begins, in 3 to 5 years. In coastal Kenya, fruiting occurs in March and April; some off-types also fruit in November or December. The fruit is about 2 cm long and, when ripe, has a yellow fleshy pericarp, a white hard shell, and a brown, oil-rich seed kernel. Fruit yields range from 30 to 100 kg per tree, depending on rainfall, insolation, soil type, and neem ecotype or genotype. Fifty kg of fresh fruit yields 30 kg of seed, which gives about 6 kg of oil and 24 kg of seed cake. Seed viability ranges from 6 to 8 weeks, but thoroughly cleaned and properly dried and cooled seeds remain viable up to 6 months.

Neem is bitter in taste. The bitterness is due to the presence of an array of complex compounds called "triterpenes" or more specifically, "limonoids." More than 100 bioactive compounds have been isolated from various parts of the neem tree; still more are being isolated. This formidable array of highly bioactive compounds makes neem a unique plant with potential applications in agriculture, animal care, public health, and for regulating even human fertility. The limonoids in neem belong to nine basic structure groups: *azadirone* (from oil), *amoorastaitin* (from fresh leaves), *vepinin* (from seed oil), *vilasinin* (from green leaves), *gedunin* (from seed oil and bark), *nimbin* (from leaves and seed), *nimbolin* (from kernel), and *salannin* (from fresh leaves and seed), and the *aza* group (from neem seed) (Kraus, 2002). *Azadirachtin* and its analogs have fascinated researchers for the past 30 years because of phagorepellency, growth inhibition, and chemosterilzing effects on insect pests (Saxena, 1989; Schmutterer, 1990, 1995). The azadirachtin content in neem could vary considerably due to edaphic, climatic, or genotypic differences.

Figure 1. Neem tree - Nature's gift to man-kind. The versatile tree is widespread in Asia and Africa, but lack of awareness of its potential in Africa led someone to fell a full-grown tree *(foreground)*, which could have been a rich source of natural pest control materials and other useful products *(photo by R.C. Saxena)*

3. NEEM FOR ECO-FRIENDLY PEST AND VECTOR MANAGEMENT

3.1 Crop Pests

Pest control as practiced today in most developing countries relies mainly on the use of imported pesticides. This dependence has to be reduced. Although pesticides are generally profitable on direct crop returns bases, their use often leads to the contamination of terrestrial and aquatic environments, damage to beneficial insects and wild biota, accidental poisoning of humans and livestock, and the twin problems of pest resistance and resurgence. Almost 500 arthropods pest species have become resistant to one or more insecticides (Georghiou and Lagunes-Tejada, 1991). Resistance of the cotton bollworm in India and Pakistan, of the Colorado potato beetle in the USA to all available insecticides, and of the diamondback moth to all classes of insecticides, including *Bacillus thuringiensis*, in Hawaii, Malaysia, the Philippines, Taiwan, and Thailand, illustrate the complexity of the problem. Shifts in pest status - from minor to major, and resurgence of pests, such as white flies, caused by direct or indirect destruction of pests' natural enemies are other unwelcome developments associated with pesticide use. A World Health Organization and United Nations Environment Program report estimated that there are 1 million human pesticide poisonings each year in the world, with about 20,000 deaths, mostly in developing countries (Levine, 1986). The problem is rendered even more difficult because few, if any, new compounds are coming to replace old insecticides. The cost of developing and registering new pesticides is staggering almost US$ 60 million, and pesticide manufacturers are unwilling to risk investments on products whose market life could be shortened by development of pest resistance.

For ecologically sound, equitable, and ethical pest and vector management, there is a need for control agents that are pest-specific, nontoxic to humans and other biota, biodegradable, less prone to pest resistance and resurgence, and relatively less expensive. Among various options, neem has been identified a source of environmentally "soft" natural pesticides.

Neem has had a long history of use primarily against household and storage pests and to some extent against crop pests in the Indian sub-continent. As early as 1930, neem cake was applied to rice- and sugarcane fields against stem borers and white ants. Early observations that swarming locusts did not attack neem leaves have been confirmed in laboratory studies and attributed to neem's antifeedant activity against locusts.

The pest control potential of neem in developing countries, however, remained largely untapped due to the advent of broad-spectrum synthetic insecticides. Also, publicity given to slogans such as "the only good bug is a dead bug" and identifying traditional uses of neem as backward, gradually weaned people away from using neem. It is only in the past decade that the pest control potential of neem has been appreciated. Though subtle, neem's effects such as repellence, feeding and oviposition deterrence, growth inhibition, mating disruption, chemo-sterilization, etc. are now considered far more desirable than a quick knock-down in integrated pest management programs as they reduce the risk of exposing pests' natural enemies to poisoned food or starvation. In spite of high selectivity, neem derivatives affect ca. 400 to 500 species of insect pests belonging to different orders (Schmutterer and Singh, 1995), one species of ostracod, several species of mites, and nematodes, and even noxious snails and fungi, including aflatoxin-producing *Aspergillus* spp. Results of field trials in some major food crops will illustrate the value of neem-based pest and vector management.

3.1.1 Rice

Saxena (1989) has reviewed the efficacy of neem derivatives against major pests of rice and virus diseases transmitted by them and a corresponding increase in rice grain yield. Although most of this work was done in South and Southeast Asia, African countries could also benefit from this information. In field trials conducted in the Philippines, application of a 2:10 neem cake-urea mixture at 120kg/ha reduced the incidence of ragged stunt, grassy stunt, and tungro viruses and significantly increased the rice yield more in both dry and wet seasons. Also, weekly ultra-low volume spray application of 50 per cent neem oil-custard-apple oil mixture in 4:1 proportion (vol/vol) at 8*l*/ha from seedling to the maximum tillering stage decreased the tungro incidence and increased the yield. The low input cost of the treatment contributed to a high net gain compared with the insecticide treatment. In India, neem treatments controlled populations of the green leafhopper, the yellow stem borer, the rice gall midge, and grasshoppers.

As neem materials reduce or disturb insect feeding activity on treated plants, they also have scope in management of rice virus vectors. Survival of *Nilaparvata lugens*, the vector of grassy stunt- and ragged stunt viruses, decreased progressively at 3 days after exposure on rice seedlings sprayed with neem oil at increasing concentrations (Saxena and Khan, 1985). Compared with successful transmission in untreated plants, the planthopper failed to transmit the viruses to plants sprayed with 50 per cent neem oil. Likewise, when seedlings were grown in soil applied with neem cake at $\geq$150kg/ha, then only 2 per cent seedlings were infected with rice tungro spherical virus (RTSV), 13 per cent with rice tungro bacilliform virus (RTBV), while 30 per cent had RTBV+RTSV infection (Saxena *et al.*, 1987). In contrast, 58 per cent of the untreated seedlings had RTBV+RTSV infection; 32 per cent had RTBV only. Protection with neem cake at 250kg/ha was on par with application of carbofuran 3G at 0.75kg (a.i.)/ha. In another study, although application of carbofuran at 1- or 2kg (a.i.)/ha cause 98 to 100 per cent mortality of *Nephotettix virescens* adults, yet tungro infection was 28 per cent at 1kg (a.i.)/ha (Abdul Kareem *et al.,* 1989). In contrast, plants treated with a mixture of neem kernel-carbofuran (1 kg (a.i)/ha) (1:1 proportion) had only 2 per cent tungro infection.

3.1.2 Maize, sorghum and millet

In trials conducted at the Field Station of the International Centre of Insect Physiology and Ecology (ICIPE) and in farmers' fields at Mbita in Kenya, foliar

Table 1. Tassel breakage by *Chilo partellus* larvae and grain yield in plots planted to stemborer- susceptible 'Katumani' maize cultivar and applied with neem seed powder (NSP) or Furadan. ICIPE Field and farmer's field, Mbita, short-rains cropping season 1992 (Saxena, unpubl.)[1]

	ICIPE Field Station		Farmer's Field	
Treatment	Tassel breakage (%)	Yield (kg/ha)	Tassel breakage (%)	Yield (kg/ha)
NSP (basal)	17.0b	4530b	12.0b	3570b
NSP(foliar)	2.0a	6430a	4.0a	5480a
NSP(basal+foliar)	2.0a	5870a	2.0a	5630a
Furadan 5G	0.3a	6400a	0.3a	6130a
Untreated (control)	21.0b	3370b	17.0c	3850b

[1] Within a column, means followed by a common letter are not significantly different at the 5% level by the LSD test; averages of 4 replications.

Table 2. Infestation and plant damage by *Chilo partellus* and *Eldana saccharina* larvae and grain yield in plots planted to moderately resistant ICZ5 maize cultivar and applied with neem cake (NC) or an insecticide. ICIPE Field Station, Mbita, long-rains cropping season 1994 (Saxena, unpubl.)[1]

Treatment	Damag[2] 9WE	Plant height (cm) 14WE	Tunnel length (cm) 7WE	Tunnel length (cm) 14WE	*Chilo* (no.)[3]	*Eldana* (no.)[4]	% tassel breakage[5]	Yield (kg/ha)
Fresh NC	2.4a	180ab	0.6a	15.6a	1.5a	7.8a	8.6a	7458a
1-yr-old NC	2.4a	186a	0.7a	13.0a	1.5a	6.5a	8.1a	7760a
2-yr-old NC	2.4a	185a	0.6a	18.5a	0.3a	9.4a	9.4a	7469a
Dipterex	2.1a	177ab	0.5a	20.3ab	1.5a	17.0a	17.0a	7271a
Untreated (control)	4.6b	167b	7.8b	31.6b	14.3b	25.5b	35.5b	5088b

[1] Within a column, means followed by a common letter are not significantly different at the 5% level by LSD test; averages of 4 replications.
[2] Foliar damage scored visually on 1-9 scale (1= no damage, 9 = completely damaged). [3,4,5] *Chilo* larvae were recorded at 9 WE; *Eldana* larvae at 11 WE, and tassel breakage at 11 WE.

application of powdered neem seed at 3 g/plant or powdered neem cake at 1g/plant once at 4 weeks after crop emergence (WE) or twice at 4- and 6 WE to maize, which had been infested with the spotted stem borer, significantly reduced the foliar damage, stem tunneling, tassel breakage, and populations of borer larvae. Grain yield in neem-treated maize plots was as high as that obtained with insecticides and significantly higher than that in untreated control plots (Table 1). Storage of neem cake up to 2 years in the dark did not reduce its pest effectiveness (Fig. 2, Table 2). Similar reduction in pest damage, including their body size as measured by the width of larval head capsules, and increase in yield were obtained when neem cake was applied to the sorghum crop (Table 3). In trials conducted in Mali, the use of local neem extract resulted in a significant increase in yield of early and main season millet as a result of the control of millet head pests, blister beetle, and head miner.

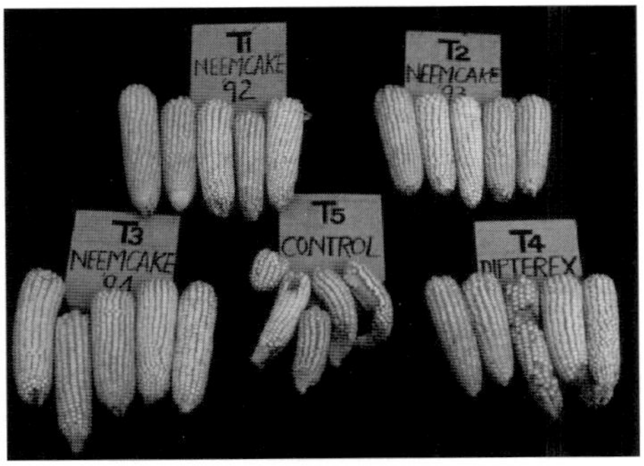

Figure 2. Cobs harvested from neem-treated (T1-T3), insecticide-treated (T4), and untreated maize fields (T5). Compared with cobs in untreated (T5) or insecticide-treated plants (T4), the grain quality was much superior in cobs harvested from pants treated with fresh (T3), 1-year-old (T2), or 2-year-old neem cake (T1) (photo by RC Saxena)

3.1.3 Banana

The banana weevil, *Cosmopolites sordidus*, and parasitic nematodes are major pests of banana and plantain. They often occur together in the same plant and may destroy the corm and the root system, resulting in loss of fruit yield. Most of the highland bananas in Eastern Africa are highly susceptible to the weevil and nematode infestations. Soil applications of neem seed powder or neem cake at 100g/ plant at planting and, subsequently, at 3-month intervals, reduced the populations of the root-lesion nematode, *Pratylenchus goodeyi* and the root-knot nematodes, *Meloidogyne* spp., on par with Furadan 5G applied at 40g/ plant at planting and then at 6-month-intervals to banana plants grown in 100 *l* containers with controlled levels of banana nematode infestations (Musabyimana and Saxena, 1999). Eight months after planting, banana plants treated with neem cake, neem seed powder, kernel powder, or with oil had 7 to 95 times less parasitic nematodes than the untreated control. However, only neem cake or neem seed powder applied to unpared banana plants kept the nematode population below the economic threshold (Table 4). At 8 months after incorporation into the soil, neem cake or neem seed powder application was still effective against banana nematodes, while the nematicidal activity of Furadan seemed to decline. Weevil larvae fed little on or avoided altogether neem-treated corms, while extensive feeding and damage occurred on untreated corms (Musabyimana *et al.,* 2001). Few larvae survived when confined for 14 d on neem-treated banana pseudostems. Females deposited up to 75 per cent fewer eggs on neem-treated corms. Also, egg hatching was reduced on neem-treated corms. The higher the rate of application of neem materials, the more severe the effect.

Neem treatments effectively controlled the banana nematodes and the banana weevil in field trials conducted in Kenya with a susceptible banana cultivar, 'Nakyetengu' (Musabyimana *et al.*, 2000). Regardless of soil fertility levels, incorporation around the plant base of powdered neem seed or cake at 60 – 100g/mat at 4-month interval, gave better control of the banana weevil and of parasitic nematodes than that achieved with soil application of Furadan 5G (carbofuran) at 60g/mat at 6-month intervals. Compared with untreated control, fruit yield in most neem treatments was significantly higher, particularly during the second cycle of crop production (Fig. 3). Neem application conferred a net economic gain, whereas Furadan application proved uneconomical (Table 5).

Figure 3. Difference in size of 'Nakyetengu' banana bunches harvested from an untreated plot *(left)* and from neem-treated plot *(right)* in a farmer's field in Oyugis, western Kenya, 1998 *(Photo by T. Musabyimana)*

3.1.4 Grain legumes and vegetables

Because of high profitability, especially of vegetables, farmers tend to overuse chemicals, which results in hazard to the environment and health of producer and consumer, as well as serious resistance problems. However, neem can provide satisfactory control of insect pests affecting grain legumes and vegetables. In trials conducted at ICIPE's Experiment Station and in a farmer's field in Kenya, applications of 2 or 3 per cent neem seed extract at *200l*/ha at 38, 47, and 51 days after emergence (DE) of cowpea crop or ULV (ultra-low volume) spray application of 5, 10, or 20 per cent neem seed extract at 10*l*/ha significantly reduced the number of larvae of the flower thrips (Saxena and Kidiavai, 1997). Cowpea grain yield was significantly higher in plots sprayed with 20 per cent neem seed extract than in untreated control plots and was at par with that obtained with cypermethrin (Table 6). Because of the low cost of neem seed extract treatment the net gain was often more when cowpea was sprayed with NSE than with the insecticide. Also, grain quality was superior in neem-treated plots than in cypermethrin-treated plots (Fig. 4).

In common beans, high volume spray applications of 2 per cent neem kernel extract at 11-day intervals effectively controlled the chrysomelid beetle, *Ootheca benningseni* (Karel, 1989). Neem derivatives also proved effective against pod borers and bollworms on Bengal gram, against the leaf roller and flea beetles on okra, and against pod borers and the pod fly on pigeon pea (Saxena, 1989). Weekly spray applications of 2.5 or 5 per cent aqueous neem seed kernel extract 100 per cent protected cabbage against the diamondback moth in Africa and was superior to Dipel at 900 g/ha; even a 1.25 per cent kernel extract was quite effective (Dreyer, 1987). Similar results have been obtained in Asia. Other lepidopterous pests of cabbage and aphids are also controlled with neem. In trials conducted in Togo, weekly high volume spray applications of a 4 per cent methanolic neem kernel extract (Adhikary, 1985) or even 2.5 to 5 per cent aqueous neem kernel extract (Dreyer and Hellpap, 1991) almost completely protected the cabbage. In field trials conducted in Kenya, weekly ULV spray applications of 20 per cent neem seed extract provided excellent control of the diamondback moth on kale crop and improved crop yield (Saxena, unpubl.). The population of spiders, which are important predators of the pest larvae, was as high as in neem-treated plots as in untreated control, while it was much lower in cypermethrin-treated plots.

In Africa, the root-knot nematodes, *Meloidogyne* spp., and the fruit borer, *Helicoperva armigera*, are the most damaging pests of tomato. As nematodes are 'unseen enemies,' their role in limiting tomato production is generally overlooked. Rössner and Zebitz (1987) reported nematicidal effects of neem materials in tomato. In field trials conducted at ICIPE's Experiment Station in Kenya, weekly spray applications of 5 per cent aqueous neem seed extract controlled the fruit borer damage and increased the marketable fruit yield. Application of powdered neem seed or cake at 3g/hill at planting significantly reduced the number of galls per plant on par with Furadan in farmers' fields at Mbita and increased production of high

Table 3. Infestation and plant damage by *Chilo partellus* larvae and grain yield in plots planted to stemborer-susceptible 'Serena' sorghum cultivar and applied with neem cake (NC) once at 4 weeks after crop emergence (WE) or twice at 4- and 6 WE, or with Dipterex. Mbita, short-rains cropping season, 1994 (Saxena, unpubl.)[1]

	Experiment Station						Farmer's Field					
Treatment	Damage[2] 9WE	Plant height (cm) 9WE	Tunnel length (cm) 15WE	Larvae (no.) 9WE	Head width (mm)	Yield (kg/ha)	Damage[2] 9WE	Plant height (cm) 9WE	Tunnel length (cm) 15WE	Larvae (no.) 9WE	Head width (mm)	Yield (kg/ha)
NC once	2.9ab	110a	27.5ab	7.4ab	0.66a	6182ab	2.9a	114a	25.4b	18.8ab	0.64a	5242a
NC twice	2.9ab	106ab	21.3a	2.2a	0.67a	7312a	3.0a	112a	18.9a	16.8a	0.63a	5052a
Dipterex	2.3ab	117a	20.3a	7.0ab	0.79ab	6523ab	2.5a	111a	22.9ab	19.4ab	0.96c	5043a
Untreated (control)	3.5b	94b	30.3b	10.2b	0.84b	6056b	4.1b	99b	27.9b	25.8b	0.90b	3908b

[1] Within a column, means followed by the same letter are not significantly different at the 5% level by the LSD test; averages of 4 replications.

[2] Foliar damage scored visually on 1 to 9 scale (1 = no damage; 9 = completely damaged).

Table 4. Effect of soil application of neem seed powder (NSP), neem cake (NC), neem kernel powder (NKP), or treatment with neem oil (NO) on population of banana nematodes at 2 and 8 months after treatment of pared or unpared suckers planted in drums. Mbita Point Field Station, Kenya (Musabyimana and Saxena 1999) [1]

	Nematode population at (No./100g of roots) ± SEM	
Treatment	2 months	8 months
Pared	1200 ± 489a	22200 ± 3747a
Unpared +NSP	300 ± 300a	3600 ± 490a
Pared + Furadan	0 ± 0a	16800 ± 2135a
Pared + NC	0 ± 0a	12000 ± 2135a
Pared+ NSP	0 ± 0a	22500 ± 2265a
Pared + NKP	0 ± 0a	81600 ± 23510b
Unpared + NC	300 ± 300a	1200 ± 0a
Unpared + NO	0 ± 0a	5700 ± 1025a
Unpared + NKP	125 ± 125a	27600 ± 3730a
Unpared (untreated)	25050 ± 4057b	114000 ± 4673b
CV%	95.7	50.4
Difference	**	**

[1] Means in columns followed by the same letter do not differ significantly (*P* <0.05; Tukey's test); avg. of 4 replicates; ** = *P* < 0.01 (Tukey's test).

Okra pests, such as the leaf-eating caterpillar, *Sylepta derogata*, were quite susceptible to spray applications of even at 0.25 per cent aqueous neem kernel extract (Dreyer, 1987) or more (Cobbinah and Olei-Owusu, 1988). Also, the cotton aphid, *Aphis gossypii*, was well controlled on okra by four weekly sprays of 0.5 per cent aqueous neem seed extract or 2 per cent neem oil; the effects being on par with butacarboxim insecticide (Dreyer and Hellpap, 1991).

In Niger, foliar applications of aqueous neem seed extract 0.25, 0.5 or 1 per cent to amaranth fields strongly repelled *Spodoptera exigua*, while a soil drench of 0.5 per cent neem seed extract repelled *Spodoptera littoralis* (Ostermann, 1992). Spray applications at 0.5 per cent or 1 per cent neem seed extract reduced the foliar damage by *S. exigua*, while pre- and post-sowing soil drenches with 0.5 per cent neem seed extract at 1000*l*/ha stopped the immigration of *S. littoralis* larvae into treated fields and almost doubled the leaf yield over that in untreated plots.

Table 5. Effects of neem seed powder (NSP), neem cake (NC), Furadan or a mixture of NC and Furadan applied at different rates to banana plants grown in a field with a moderate level of pest infestation and fertile soil on pest infestation, plant damage, fruit yield, and net gain during the 2nd crop cycle. Kabondo, Kenya, 1996 to 1999 (Musabyimana *et al.*, 2000)

Treatment (g/mat)[1]	Necrosis index[2]	Nematodes (no./100g roots)	Weevils (no./plot)	PCI[3] outer	Fruit yield (t/ha)	Yield value ($/ha)[4]	Treatment cost ($/ha)[5]	Net gain ($/ha)[6]
NSP 60	0.3ab	3533ab	3.0b	0.0a	19.9ab	1990	252	1738
NSP 80	0.0a	2667a	0.2a	0.7a	24.4a	2440	318	2122
NSP 100	0.0a	1800a	0.8ab	0.0a	19.9ab	1990	384	1606
NC 60	0.0a	2233a	1.8ab	0.2a	24.7a	2470	351	2119
NC 80	0.5ab	4733ab	2.1ab	0.0a	27.0a	2700	450	2250
NC 100	0.0ab	3530ab	1.0ab	2.3ab	18.7ab	1870	549	1321
Furadan 60	0.8ab	7273a	0.5a	3.0ab	15.1b	1510	801	709
Furadan30+ NC30	3.0ab	2800a	0.5a	1.2a	24.0a	2400	777	1623
Control (untreated)	1.1b	29533c	0.3a	5.6b	15.3b	1530	0	1530

[1] Within columns, means followed by a common letter do not differ significantly ($P<0.05$; Student-Neuman-Keuls test); averages of six replicates.
[2] During the crop cycle, NSP, NC, or NC+Furadan was applied thrice; Furadan alone was applied twice.
[3] Price of banana fruit = $0.1/ kg (source: Market Information Branch, Ministry of Agriculture, Kenya).
[4] Treatment cost: NSP = $1/ kg; NC = $1.5/ kg; Furadan = $5.8/ kg; labour = $18/ ha.
[5] Net gain = yield value – treatment cost.

Table 6. Comparative yield and value of cowpea grain after deducting the cost of neem seed extract (NSE) or cypermethrin applied thrice to cowpea crop. Mbita, long-rains cropping season, 1994[1]

	Experiment Station				Farmer's Field			
Treatment[2]	Yield (kg/ha)	Yield value ($/ha)[3]	Treatment cost ($/ha)[4]	Value of yield minus treatment cost ($/ha)	Yield (kg/ha)	Yield value ($/ha)[3]	Treatment cost ($/ha)[4]	Value of yield minus treatment cost ($/ha)
NSE 5%	1160ab	580	1.5	578.5	1450d	725	1.5	723.5
NSE 10%	1280ab	640	3.0	637.0	1630c	815	3.0	812.0
NSE 20%	1480a	740	6.0	734.0	1760b	880	6.0	874.0
Cypermethrin	1480a	740	108.0	632.0	2130a	1065	108.0	957.0
Untreated	1050b	525	0.0	525.0	1290e	645	0.0	645.0

[1] Means followed by a common letter do not differ significantly (*P* <0.05; yan-Einot-Gabriel-Welsch Multiple Range Test).
[2] Using ULV applicators, NSE or cypermethrin was applied 3 times at 10 l/ha at 31, 39, and 49 DE.
[3] US$ = KSh 54; cost of cowpea grain = US$ 0.50/kg.
[4] Treatment cost includes only cost of NSE or cypermethrin; neem seed @ US$ 0.5/kg, cypermethrin @ US$ 36/l.

quality tomatoes (Fig. 5) (Saxena, unpubl.). Likewise, in Niger, weekly spray applications of 5 per cent neem seed extract reduced the tomato fruit borer damage and increased the marketable fruit yield (Ostermann, 1992).

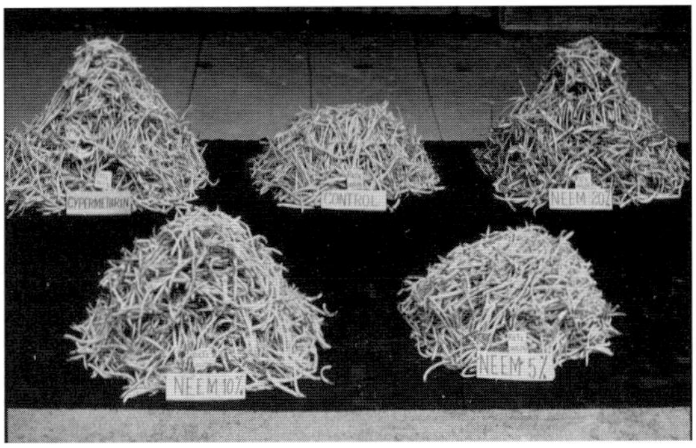

Figure 4. *(clockwise)* Cowpea pods harvested from plots sprayed with cypermethrin, 'Teepol' liquid detergent (control), or with 20%, 5% or 10% neem seed extract (NSE). Pods harvested from plots sprayed with 20% NSE have a reddish pigmentation with uniformly ripened grains, which had an appealing sheen; pods from cypermethrin-treated plots had grains, which did not ripen evenly *(Photo by RC Saxena).*

Figure 5. Galled roots and healthy roots from untreated and neem-treated tomato plants *(left)*; difference in quality of fruits harvested from neem-treated plots *(middle)* and untreated plots *(right) (photos by R. C. Saxena)*

In Sudan, remarkable results were obtained with neem products in the control of the sweet potato whitefly, *Bemisia tabaci* and the leafhopper, *Jacobiasca lybica* on potato (Siddig, 1987,1991). Two high volume applications of 2.5 per cent aqueous neem kernel extract sprayed at fortnightly intervals reduced the pest populations to <50 per cent of the control and increased the yield. The potato tuber moth *Phthorimaea operculella*, was unaffected in the field but spray applications of

0.05 and 0.1 per cent neem oil strongly deterred oviposition and prevented damage in the stored products (Siddig, 1988).

3.1.5 Agroforestry and tobacco

Insects and nematodes also affect trees and crops in agroforestry. In collaborative trials conducted by International Centre for Research in Agroforestry (ICRAF) in Shinyanga in Tanzania in 1995-1996, application of powdered neem cake at 2g/plant to a hybrid maize, 'Cargill,' at 4 and 5 weeks after sowing, registered a 30 per cent yield increase over the untreated control (ICIPE, 1998). Application of neem cake at 135kg/ha also reduced the termite damage and significantly increased the grain yield of hybrid maize over the Furadan-treated or untreated crop.

In trials conducted in Tabora, Tanzania, although application of neem seed powder or cake at 15g/m^2 was not as effective as ethylene dibromide at 62ml/m^2 in reducing the root galling index in tobacco plants, the tobacco yield increased significantly with neem treatments (ICIPE, 1998).

3.2 Stored Products Pests

Post-harvest losses are notoriously high in developing countries, especially in Africa. Worldwide annual losses in store reach up to 10 per cent of all stored grain, i.e. 13 million tons of grain lost due to insects or 100 million tons to failure to store properly. Saxena (1995) has reviewed the potential of neem against pests of stored products: grain legumes, maize, sorghum, wheat, rice and paddy, and potato tubers. At farm level storage and warehouses, the application of neem derivatives to bags and stored grains has provided protection against insect pests. Powdered neem seed kernel mixed with paddy (1 to 2%) significantly reduced infestation in warehouses. Neem leaves mixed with paddy (2%), bags treated with 2 per cent neem extract, or 20- to 30-cm dried neem leaf barrier between the bags and storage floor significantly reduced insect infestation and damage to grain during a 3-month storage period; the effectiveness being comparable to methacrifos dust. Likewise, neem seed extract at 7.2g/90 kg capacity jute bag (100 x 60 cm) controlled 80 per cent of the population of major insects and checked the damage to wheat up to 6 months. The treatment was effective up to 13 months and provided more than 70 per cent protection as compared with untreated control. The neem seed extract treatment was as effective as that of 0.0005 per cent primiphos methyl mixed with the grain. Using this technology in Sind, Pakistan, high benefit-cost ratios were obtained by small-, medium, and large-scale farmers.

The effectiveness of neem oil alone or in combination with fumigation was evaluated against five major species of stored grain pests infesting rice and paddy grains in a warehouse trials conducted in the Philippines. Rice grain treated with 0.05 to 0.1 per cent neem oil or treated with neem oil after fumigation with 'Phostoxin,' and stored for 8 months had significantly less red flour beetle adults than in untreated control. Both kinds of neem treatments were as effective as the bag treatment with 'Actellic' at 25 $\mu g/cm^2$ or grain treatment with Actellic at 0.0005 per cent, and suppressed the pest population by 60 per cent. The population build-up also was reduced when either fumigated or non-fumigated rice was stored in bags treated with neem oil at ≥1 mg/cm^2. The lesser grain borer, the rice weevil, the saw-toothed grain beetle, and the rice moth were similarly affected by neem treatments alone or in combination with prior grain fumigation. Fumigation and Phostoxin were effective only for about 2 months against the lesser grain borer, and for up to 6 months against other pest species, while neem oil treatments were effective up to 8 months. Compared with the pest damage to untreated or fumigated rice, neem oil treatment significantly reduced the damage to rice grain. At 8 months after storage, weevil attacked grains in neem treatments were 50 per cent of those in the fumigated rice and 25 per cent of those in the untreated rice. Neem treatments also reduced the pest populations and the damage in paddy.

In studies conducted in Kenya, the growth and development of 1st instars of the maize weevil was completely arrested in maize grain treated with neem oil at 0.02 per cent, while the weight loss of treated cobs was less than 1 per cent as compared with a 50 per cent reduction in weight of untreated cobs stored for 6 months. (Kega and Saxena, 1996).

While neem treatments cannot replace completely chemical pesticides used in stored product preservation, the amounts of pesticides needed could be reduced, thereby decreasing the pesticide load in food grains. With proper timing and innovative methods of application, their use could be integrated in stored products management.

3.3 Blood-Sucking Pests

The effects of neem on hematophagous insects affecting humans and livestock have been reviewed (Ascher and Meisner, 1989). Application of a paste made from neem leaves and turmeric in 4: 1 proportion to the skin cured 97 per cent of the patients suffering from scabies caused by the itch mite in 3-15 d. Likewise, repeated application of neem oil cured scabies, which is common among children in rural Africa. Monthly sprays of ethanolic extracts of neem or weekly bathing in azadirachtin-rich aqueous 1:20 'Green Gold' controlled the bush tick and the cattle tick in Australia, but were less effective against the brown dog tick (Rice, 1993). In Jamaica, neem kernel extract controlled ticks on cattle and dogs. In Kenya, engorgement duration by larvae and nymphs of *Amblyomma variegatum* and larvae of *Rhipicephalus appendiculatus* were significantly prolonged due to slowed feeding on rabbit host sprayed with neem oil (Table 7) (Kaaya *et al.*, 2003). Neem treatment also led to a reduction in engorgement weight of larvae, nymphs, and adults of *A. variegatum, R. appendiculatus* and *Boophilus decoloratus* feeding on neem-treated rabbits and fewer larvae and nymphs molted to the next developmental stage. Egg masses produced by neem-treated ticks weighed significantly less while hatchability of their eggs was adversely affected. Regardless of tick species, attachment by larvae also was significantly reduced on neem oil-treated rabbit. In trials conducted in pastures in Kenya, application of neem oil on cattle repelled all stages of *R. appendiculatus*, *B. decoloratus,* and *A. variegatum* (Kaaya *et al.*, 2003).

Neem products also repel and affect the development of mosquitoes. Two per cent neem oil mixed in coconut oil, when applied to exposed body parts of human volunteers, provided complete protection for 12 h from bites of all anophelines (Sharma *et al.,* 1993). Kerosene lamps containing 0.01-1 per cent neem oil, lighted in rooms containing human volunteers, reduced mosquito biting activity as well as catches of mosquitoes resting on walls in the rooms; protection was greater against *Anopheles* than against *Culex.* Effectiveness of mats with neem oil against mosquitoes has also been demonstrated; the vaporizing oil repelled mosquitoes for 5-7 h at almost negligible cost. The sandfly also was totally repelled by neem oil, mixed with coconut or mustard oil, throughout the night under field conditions. Application of neem cake at 500 kg/ha, either alone or mixed with urea, in paddy fields was very effective and reduced the number of pupae of *Culex tritaeniorhynchus*, the vector of Japanese encephalitis, and also resulted in higher grain yield.

4. PEST RESISTANCE TO NEEM MATERIALS

A few herbivorous insects, including some sucking insects, some beetles, and some moths do survive on neem but, largely, the tree is free from serious pest problems. Some insects can adapt to limonoids, but in laboratory tests two genetically different strains of the diamondback moth treated with a neem seed extract showed no sign of resistance in feeding and fecundity tests up to 35 generation (Völlinger, 1987). In contrast, deltamethrin-treated lines developed resistance factor of 20 in one line and 35 in the other. There was no cross-resistance between deltamethrin and neem seed extract in the deltamethrin-resistant lines. The diversity of neem compounds and their combined effects on insect pests seem to confer a built-in resistance prevention mechanism in neem. However, wisdom demands that users should refrain from exclusive and extended application of single bioactive materials, such as azadirachtin.

5. OUTLOOK FOR THE NEW MILLENNIUM

For nearly the past two decades, neem has come under close scientific scrutiny as a source of novel, natural insecticides and more than 2000 scientific papers have been published to date on neem. Several international conferences have been held in the past two decades both in developing and industrialized countries.

The interest in neem in the developed world is attributable to the fact that neem-based pest control products with diverse modes of action not only are effective against pests, but also inherently safer, less persistent in the environment, and less prone to the problem of pest resistance than the synthetics. Today, technical grade neem active ingredients, principally azadirachtins, fetch the highest price, about US$ 375/kg as compared with US$ 75/kg for pyrethrum (Isman, 1995).

In that context, tropical countries of Asia and Africa could become major exporters of the raw material or even value-added finished products. In fact, introduction of neem has been very fast in some of the African countries. It grows well in Egypt, Sudan, Ethiopia, Somalia, Kenya, Uganda, Tanzania,

Table 7. Growth, development, and fecundity of *A. variegatum*, *R. appendiculatus*, and *B. decoloratus* on rabbit hosts treated with neem oil.

Treatment	Engorgement duration (d) Larva	Engorgement duration (d) Nymph	Engorgement weight (mg) Larva	Engorgement weight (mg) Nymph	Engorgement weight (mg) Adult	Molting (%) Larva	Molting (%) Nymph	Weight (mg) per egg mass	Hatchability (%)
					A. variegatum				
Neem oil	10a	16a	43±19a	44±1b	211±8a	66±1.3a	94±3.3a	940±1a	31±1.2a
Peanut oil (control)	6b	9b	43±4a	35±1a	306±6b	85±0.7b	94±3.3a	1320±20b	68±1.8b
					R. appendiculatus				
Neem oil	5a	5a	8±0.6a	9±0.1a	295±10a	50±0.8a	94±0.9a	100±10a	52±5.8a
Peanut oil (control)	3b	5a	8±0.1a	9±0.3a	388±10b	87±1.2b	94±0.9a	182±10b	100±2.3b
					B. decoloratus				
Neem oil	--[2]	--[2]	--[2]	--[2]	--[2]	2±0.1a	--[2]	39±0.2a	43±0.6a
Peanut oil (control)	--[2]	--[2]	--[2]	--[2]	--[2]	86±0.6b	--[2]	57±0.1b	88±1.2b

[1] For a particular species, means followed by the same letter within a column are not significantly different ($P < 0.05$; REGW Multiple Range Test). [2] Not tested.

(*Source:* Kaaya *et al.*, 2003)

Mozambique and Chad in the northern and eastern part of Africa and in Mauritania, Snegal, Mali, Ghana, Ivory Coast, Togo, Nigeria and Cameron in the rest of Africa. However, from African perspective it seems that knowledge about neem is not widespread. Another aspect is the costs involved. For instance, low prices are paid in remote poor areas in Africa, e.g. Nigeria, Benin, Niger and Medagascar. At these sites quality is also a problem, the trading through middlemen and transport costs increase the prices for those who want to manufacture neem seed-based pesticides and need high quality seed. Principally NGO's could play an important role in spreading the knowledge about neem and its uses. This strategy is slowly but definitely working, as in most west African countries considerable number of neem trees occur in Benin and NGOs and women groups were trained to produce neem oil and neem soap and similarly pesticide development strategies could be made possible locally. In fact, some NGOs are actively promoting the use of simple neem preparations such as neem water extracts in Ghana, Guatemala, Mozambique and Uganda. Usage of neem as natural pesticide in Kenya by Kenyan Neem Foundation is a promising development and it is also creating awareness on neem in schools and community groups in Mombassa area. This should help in promoting neem-based technologies in the African region in the new millennium.

REFERENCES

Abdul Kareem, A., Saxena, R.C. and Boncodin, M.E.M. (1989), Neem-carbofuran mixtures against *Nephotettix virescens* (Distant) (Hom., Cicadellidae) and its transmission of tungro associated viruses, *Journal of Applied Entomomlogy,* **108**, 68-71.

Adhikary, S. (1985), Results of field trials to control the diamond-back moth, *Plutella xylostella* L. in Togo by application of crude methanolic extracts of leaves and kernels of the neem tree, *Azadirachta indica* A. Juss. in Togo, *Zeitschrift fur Angewandte. Entomologie,* **100**, 27-33.

Ascher, K.R.S. and Meisner, J. (1989), The effects of neem on insects affecting man and animal, in *Focus on Phytochemical Pesticides. Vol. 1, The Neem Tree*, ed. M. Jacobson, CRC Press, Boca Raton, FL, pp. 113-131.

Beraki, G. and Foerster, P (1992), *Pre-factfinding study of neem in Eritrea, Ethiopia and Sudan*, GTZ report, Germany.

Cobbinah, J.R. and Olei-Owusu, K. (1988), Effects of neem seed extracts on insect pests of eggplant, okra and cowpea, *Insect Science and Applications,* **9**, 601-607.

Dreyer, M. (1987), Field and laboratory trials with simple neem products as protectants against pests of vegetables and field crops in Togo, in *Natural Pesticides from the Neem Tree (Azadirachta indica A. Juss.) and other Tropical Plants*, eds. H. Schmutterer and K.R.S. Ascher, Proc. 3rd Int. Neem Conf., Nairobi, GTZ, Eschborn, pp. 431-447.

Dreyer, M. and Hellpap, C. (1991), Neem - a promising natural insecticide for small-scale vegetable production in tropical and subtropical countries, *Zeitschrift fur Pflkrankh PflSchutz,* **98**, 428-437.

Georghiou, G.P. and Lagunes-Tejada, A. (1991), *The occurrence of resistance to pesticides in arthropods: an index of cases reported through 1989*, FAO, Rome.

Goliber, T.J. (1989), Africa's expanding populations: old problems, new solutions. *Population Bulletin,* Vol. 4, No. 3, Population Reference Bureau, Washington, DC.

ICIPE (International Centre of Insect Physiology and Ecology) (1998), Natural pesticides from neem, *Annual Report*, 1995-1997, ICIPE, Nairobi, pp. 93-110.

Isman. M.B. (1995), Leads and prospects for the development of new botanical insecticides, *Review Pesticide Toxicology,* **3**, 1-20.

Kaaya, G.P., Saxena, R.C. and Solomon, G. (2003), Potential of neem producst for control of three species of ticks affecting livestock in Africa, *Journal of Economic Entomology,* **96**, (in press).

Karel, A.K. (1989), Response of *Ootheca bennigseni* (Coleoptera: Chrysomelidae) to extracts from neem, *Journal of Economic Entomology,* **82**, 1799-1803.

Kega, K.M. and Saxena, R.C. (1996), Neem derivatives for management of *Sitophilus zeamais* Motschiulsky (Coleoptera: Curculionidae). *Paper presented at the International Neem Conference, Australia, 1996* (abstract).

Kraus, W. (2002), Biologically active ingredients: azadirachtins and other triterpenoids, in *The Neem Tree*, ed. H. Schmutterer, Neem Foundation, Mumbai, India, pp. 39-111.

Levine, R.S. (1986), Assessment of mortality and morbidity due to unintentional pesticide poisoning, Unpubl. Doc, WHO/VBC, pp. 186-929.

Musabyimana, T. and Saxena, R.C. (1999), Efficacy of neem seed derivatives against nematodes affecting banana, *Phytoparasitica,* **27**, 43-49.

Musabyimana, T., Saxena R.C., Kairu, E.W., Ogol, C.K.P.O. and Khan, Z.R. (2000), Powdered neem seed and cake for management of the banana weevil, *Cosmopolites sordidus*, and parasitic nematodes, *Phytoparasitica,* **28**, 321-330.

Musabyimana, T., Saxena, R.C., Kairu, E.W., Ogol, C.K.P.O. and Khan, Z.R. (2001), Effects of neem seed derivatives on behavioral and physiological responses of *Cosmopolites sordidus* (Coleoptera: Cucrulionidae), *Journal of Economic Entomology,* **94**, 449-454.

National Research Council. (1992), *Neem - A Tree for Solving Global Problems.* Report of an ad hoc panel of the Board on Science and Technology for International Development, National Academy Press, Washington DC.

Ostermann, H. (1992), Zur Wikung und Anwendung einfacher Niemproduckte in Kleinbauerlichen Tomaten-, Vignabohnen- und Amaranthkulturen in Niger, Ph.D. thesis, University of Giessen, Germany.

Rice, M. (1993), Development of neem research and industry in Australia, in *Souvenir, World Neem Conference, Bangalore, 1993*, pp. 8-24

Rössner, J. and Zebitz, C.P.W. (1987), Effect of neem products on nematodes and growth of tomato (*Lycopersicon esculentum*) plants, in *Natural Pesticides from the Neem Tree (Azadirachta indica A. Juss) and Other Tropical Plants*, eds. H. Schmutterer and K. R. S. Ascher, Proc. 3rd Int. Neem Conf., Nairobi, GTZ, Eschborn, pp. 611-621.

Saxena, R.C. (1989), Insecticides from neem, in *Insecticides of Plant Origin*, eds. J. T. Arnason, B.J.R. Philogene and P. Morand, *ACS Symposium Series 387*, American Chemical Society, Washington, DC, pp. 110-135.

Saxena, R.C. (1995), Pests of stored products, in *The Neem Tree, Azadirachta indica A. Juss., and Other Meliaceous Plants: Source of Unique Natural Products for Integrated Pest Management, Medicine, Industry and Other Purposes*, ed. H. Schmutterer, VCH, Weinheim, pp. 418-432.

Saxena, R.C. and Khan, Z.R. (1985), Effect of neem oil on survival of *Nilaparvata lugens* (Homoptera: Delphacidae) and on grassy stunt and ragged stunt virus transmission, *Journal of Economic Entomology,* **78**, 647-651.

Saxena, R.C., Khan, Z.R. and Bajet, N.B. (1987), Reduction of tungro virus transmission by *Nephotettix virescens* (Homoptera: Cicadellidae) in neem-cake treated seedlings, *Journal of Economic Entomology,* **80**, 1079-1082.

Saxena, R.C. and Kidiavai, E.L. (1997), Neem seed extract applications as low-cost inputs for management of the flower thrips in cowpea crop, *Phytoparasitica,* **25**, 99-110.

Schmutterer, H. (1990), Properties and potential of natural pesticides from the neem tree, *Azadirachta indica. Annual. Review of Entomology,* **35**, 271-297.

Schmutterer, H. (1995), *The Neem Tree, Azadirachta indica A. Juss., and Other Meliaceous Plants: Sources of Unique Natural Products for Integrated Pest Management, Medicine, Industry and Other Purposes,* VCH, Weinheim.

Schmutterer, H. and Singh, R.P. (1995), List of insect pests susceptible to neem products, in *The Neem Tree, Azadirachta indica A. Juss., and Other Meliaceous Plants: Source of Unique Natural Products for Integrated Pest Management, Medicine, Industry and Other Purposes*, ed. H. Schmutterer, VCH, Weinheim, pp. 326-365.

Sharma, V.P., Ansari, M.A. and Razdan, R.K. (1993), Mosquito repellent action of neem (*Azadirachta indica*) oil. *Journal of American Mosquito Control Association,* **9**, 359-360.

Siddig, S.A. (1987), A proposed pest management program including neem treatments for combating potato pests in Sudan, in *Natural Pesticides from the Neem Tree (Azadirachta indica A. Juss) and Other Tropical Plants*, eds. H. Schmutterer and K. R. S. Ascher, Proc. 3rd Int. Neem Conf., Nairobi, GTZ, Eschborn, pp. 449-459.

Siddig, S.A. (1988), Cultural means of controlling potato tuber moth (*Phthorimaea operculella* Zell.) and improvement of potato yield and quality in Sudan. *Acta Horticulture,* **218**, 281-287.

Siddig, S.A. (1991), Evaluation of neem seed and leaf water extracts and poweders for the control of insect pests in the Sudan. *Tech. Bulletin # 6. Agric. Res. Corp.*, Shambat Research Station, 39p.

Strozok, P.P. (1992) *Cost/benefit analysis of neem tree product technologies: South Asia, Haiti and the Sahelian zone, West Africa*, USAID report.

Völlinger, M. (1987), The possible development of resistance against neem seed kernel extract and deltamethrin in *Plutella xylostella*, in *Natural Pesticides from the Neem Tree (Azadirachta indica A. Juss) and Other Tropical Plants*, eds. H. Schmutterer and K.R S. Ascher, Proc. 3rd Int. Neem Conf., Nairobi, GTZ, Eschborn, pp. 543-554.

Chapter 5

NEEM RESEARCH IN ASIAN CONTINENT: PRESENT STATUS AND FUTURE OUTLOOK

G.S. DHALIWAL[1], RAMESH ARORA[1] AND OPENDER KOUL[2]
[1]*Department of Entomology, Punjab Agricultural University, Ludhiana-141 004, India;* [2]*Insect Biopesticide Research Centre, Jalandhar-144 003, India*

1. INTRODUCTION

As we enter the new millennium, we face the gigantic task of increase in human population and degradation of environment. The global population is projected to rise to over 8 billion in 2025 and much of this increase is expected to occur in developing countries. Over half of the world population growth will occur in Asia and one-third will be in Africa. Many countries in these regions are already facing problems of starvation, malnutrition, land degradation, water shortages and loss of biodiversity. Thus, providing adequate food and environmental security to the expanding populations in developing countries are the twin challenges facing mankind today. Nature has provided mankind a rich repository of plants, which are a source of versatile compounds. It is estimated that there are about 2,50,000 to 5,00,000 different plant species in the world today. Only 10 per cent of these have been examined chemically indicating a vast resource, which still remains untapped. One of the most useful of these trees is the neem, *Azadirachta indica* A. Juss., which has been credited as a tree for solving global problems in view of its potential of improving pest control, bettering health, assisting reforestation and perhaps checking over-population (NRC, 1992).

2. ORIGIN AND DISTRIBUTION

In chapter 1 (this volume) the origin and distribution of neem tree has been discussed, however, looking from the Asian perspective it is said that Muslim travelers and settlers to Hind (India) from West Asia bestowed the admiring title of

O. Koul and S. Wahab (eds.), Neem: Today and in the New Millennium, 65–96.

Azad dirakhat-I-Hind (noble/free tree of India) on this plant and neem's scientific name, *Azadirachta indica*, is derived from this title (Ahmed, 1995). The great variability in the shape of its leaves and other morphological features in Myanmar (Oo, 1989) also support the hypothesis that the neem tree could have originated in this region. At present, the tree is widely distributed by introduction, mainly in the arid tropical and subtropical countries of Asia, Africa, the Americas, Australia and the South Pacific Islands. The tree may continue to spread to new areas in 21st century wherever it can exist. There are many common names for the neem tree in different Asian countries (Table 1).

Table 1. Common names of neem tree in Asian countries

Country	Common name(s)
India	*Limba, Limbo, Neem, Nim, Nimb, Nimba, Verbu, Vepa, Veppam,* etc. (more than 100)
Indonesia	*Imba, Intaran, Mimbo, Mindi*
Iran	*Azad-drakht-I-hindi* (Free tree of India, Persian), *Nib*
Malaysia (West)	*Mambu*
Myanmar (Burma)	*Tamarkha*
Pakistan	*Nimmi*
Singapore	*Nimbagaha*
Sri Lanka	*Kohomba*
Thailand	*Dao, Kwinin, Sadao India*
Yemen	*Meraimarah*

Source: Schmutterer (1995b)

A. indica has been planted in many parts of Asia, i.e. Bangladesh, Combodia, India, Indonesia, Iran, Malaysia, Myanmar, Nepal, Pakistan, Sri Lanka, Thailand and Vietnam. It has recently been introduced into Saudi Arabia, the northern parts of Yemen and China (Hainan Island) (NRC, 1992). In India, the estimates of their number vary from 14 million trees (Ketkar, 1976) to 16-18 (Hegde, 1996), 15-25

(Rembold, 1996) and 15-20 (Walia *et al.*, 2002) million trees. However, the highest number of trees seems to occur in Uttar Pradesh (53.4%), followed by Tamil Nadu (17.0%), Karnataka (5.3%), Madhya Pradesh (4.9%), Maharashtra (4.7%), Andhra Pradesh (4.4%) and Gujarat (4.2%) (Table 2). Neem is a popular tree in drier parts of Pakistan. In Sri Lanka, neem tree is widespread but mostly in the drier regions of northern and southern provenances of the island, whereas in Nepal neem grows wild as well as in homesteads in Terai and inner Tarai regions up to an altitude of 900m. However, the tree has established well in the drier parts of the western region, compared to the wet eastern region. In Bangladesh,

Table 2. Distribution of neem trees in different states of India

State	No. of trees ('000)	Total seed potential ('000t)	Total oil potential ('000t)	Actual collection of seeds (%)
Andhra Pradesh	653.9	12.2	2.5	27
Gujarat	636.2	21.0	4.2	1
Karnataka	790.6	20.1	4.0	20
Madhya Pradesh	735.6	18.2	3.6	2
Maharashtra	710.1	28.2	5.6	1
Orissa	48.7	1.2	0.2	--
Punjab	391.3	12.0	2.4	--
Rajasthan	183.8	3.9	0.8	--
Tamil Nadu	2544.1	57.1	11.4	29
Uttar Pradesh	7972.6	265.9	53.2	--
West Bengal	273.0	2.5	0.5	27
Total	14939.9	442.3	88.5	24

Source: Ketkar (1976), Rajasekaran (1991)

neem is a household tree, widely distributed in the northwest part of the country, particularly in the central parts of the country, i.e. Rajashahi, Sirajganj, Bogra and Dinajpur districts. In Myanmar, neem is very common in the central parts of the country, i.e. in Mandalay, Magwe and Sagaing divisions (Oo, 1987). In China, *A. indica* was planted for the first time a few years ago on the subtropical island of Hainan (Chiu and Zhang, 1989). Neem was also planted several years ago in small plantations in central Vietnam.

In Indonesia, neem is found mainly in the low-lying northern and eastern parts of Java and in the drier islands to the east (Bali, Sumbawa) from seacoast up to an altitude of 400m. The tree was introduced in 1940s by migrating Indians to Malaysia (Hegde, 1996) and it exists mainly in Penang Island, Langkawi Island and in the northern provinces of Perlis and Kedah. There are some very old trees in Melaca. The recent survey also indicates the greatest abundance of neem trees in northern Peninsular Malaysia, which also produced higher yield of fruits compared with that of southern and east coast of Peninsular Malaysia (Kadir *et al.*, 1998). In the Philippines, neem was introduced during the last decade and planted in most of the bigger islands, using seeds from India, Africa and Indonesia. However, its growth on Luzon Island and on the Visayas is hampered due to typhoons and a fungal disease, but it thrives on Mindanao, where typhoons normally do not occur (Schmutterer, 1995b). Thailand has many *A. indica* trees as well as its own *A. indica siamensis*.

Neem was introduced into Saudi Arabia more than 50 years ago and has acclimatized remarkably well to the hot and arid conditions. About 50,000 neem trees were planted on the plains of Arafat near Mecca to provide relief from the blazing summer sun to the pilgrims (Ahmed *et al.*, 1989). This is probably the world's largest neem plantation on about 10 km^2 area (Dube, 1996). Other regions of this area that grow neem trees are coastal areas up to Chat el Arab in Iran and various avenues in Qatar where they are planted under irrigation. Numerous neem trees also exist in the southern parts of Yemen, particularly in the Abyan valley east of Aden.

3. DIVERSITY IN *AZADIRACHTA* SPP.

Azadirachta has two species, i.e. *A. indica* A. Juss. (syn. *Melia indica* Brandis, *Melia azadirachta* L.) and *A. excelsa* (Jack) Jacob (Syn. *A. integrifolia* Merr., *Melia excelsa* Jack). *A. indica siamensis* Val., indigenous to Thailand where it is widely used as a source of insecticide, is also found in nearby countries and probably in the adjoining Myanmar (Oo, 1987). The distinguishing features of *A. indica* and *A. indica siamensis* are given in Table 3.

The *A. indica* tree in India grows widely under different ecological regimes and fruiting takes place during different months of the year with consequent variation not only in morphological features but also in extract yields and chemical constituents. Trees growing in northern India are bigger in size than those growing in southern India (Singh *et al.*, 1998). There are also vast differences in extract yields and bioefficacy of neem ecotypes, i.e. those growing in dry areas have higher bioactivity than those growing near the sea (Singh, 1987) and major bioactivity occurs due to the presence of a limonoid, azadirachtin.

Table 3. Comparison of features of Indian and Thai neem

Feature	Indian neem	Thai neem
External		
Shape of tree	Crown dense, branches numerous	Crown relatively open, branching moderate
Shape of leaflets	Smaller and thinner	Wider, longer and thicker
Margin and tip of leaflet	Distinctly and regularly serrate, tip pointed. Basal part of anterior part of leaflet strongly curved to mid vein.	Irregularly crenate to entire, tip relatively blunt. Basal part of anterior part of leaflet slightly curved to mid vein.
Inflorescens and flowers	Panicles loose, open and long; flowers usually axillary, smaller.	Panicles dense; flowers stout and bigger, often non-axillary.
Flowering period	March, or any other time during the year.	November/December and rarely March.
Fruit/seed	Narrower; no dark green layer under brown testa of seed kernel.	Wider; dark green layer under brown, parchment paper like testa of seed kernel.
Internal		
Roots	Pores of vessels of root wood and bole with gum.	Pores of vessels of root wood and bole without gum.
Leaves	Tannin content higher, number of stomata $4.15/mm^2$	Tannin content lower, number of stomata $2.45/mm^2$

Source: Sombatsiri *et al.* (1995)

Seed is the most important part of *Azadirachta* spp. As most biologically active materials are concentrated in this part of the tree and hence seed parameters and yield are very important. The differences in the seed yield and size of the seeds in different species of *Azadirachta* are depicted in Table 4 (Singh *et al.*, 1998).

There are quantitative variations in the major active ingredient, azadirachtin content in different species of *Azadirachta* and also within *A. indica* from different locations. Singh *et al.* (1998) reported 0.35 per cent azadirachtin A in *A. indica* and *A. indica siamensis*, and 0.38-0.56 per cent in *A. excelsa*. There have been vast differences in azadirachtin content of seeds obtained from different countries (Ermel

Table 4. Variations in seed size and yield of seeds of *Azadirachta* spp.

Parameter	*A. indica*	*A. indica siamensis*	*A. excelsa*
Length (cm)	1.3–1.7	1.5–2.2	1.7–3.2
Width (cm)	0.8–1.1	1.0–1.6	1.2–2.5
Weight of 1000 seeds (g)	230	320	1140
Weight of kernel (g)	120	180	584
Yield (kg)	15	>15	50

Source: Singh *et al.* (1998)

et al., 1984, 1987; Ermel, 1995). However, the highest amount of azadirachtin was detected in samples from south and southeast Asia, i.e. India, Myanmar and Thailand (Table 5). Sombatsiri *et al.* (1995) reported considerable variation in azadirachtin A content in samples from 7 provinces of Thai neem. The azadirachtin content varied from 1.8-5.3 μg/g. Ketkar and Ketkar (1993) reported that azadirachtin content in neem from different parts of India varied from 1 to 8g/kg. Similarly, Koundal *et al.* (2003) found wide variation in azadirachtin content in neem seeds from 5 states of India, varying from 1.32 to 5.69 mg/g (Table 6).

Rengasamy *et al.* (1996) reported that azadirachtin content in neem seed kernels from eight agroecological regions of India varied from 0.14 to 1.66 per cent. The ecotypes growing in regions with moderate climate, red laterite and shallow medium black soils and altitudes less than 500m above mean sea level were rich in azadirachtin content, whereas ecotypes growing in high altitude alluvial soils with extreme hot and cold climates had very low azadirachtin content. The azadirachtin content of neem fruits collected from a agroclimatic zones in Rajasthan varied from 0.194 to 0.670 per cent (Gupta and Prabhu, 1997).

Similarly, azadirachtin content in neem seeds collected from 12 different locations in Tamil Nadu, India varied from 3.47 to 6.70 g/kg of kernel and oil content varied from 2.61 to 4.36 g/10g of kernel (Sridharan and Venugopal 1998). These studies indicate a negative influence of total rainy days during fruiting season (April-August) on the azadirachtin content and significant positive influence of sunshine hours during off-season (September-March) on the oil content of the seeds.

Eeswara *et al.* (1997) reported that azadirachtin content in seeds of neem trees from 3 sites in the dry zone and one site in the intermediate zones of Sri Lanka, varied from 2000 to 6500 $\mu g/g^{-1}$. In Malaysia samples from Jalan Sungai, Batu Pahat, Perlis, and northern Peninsular Malaysia contained the highest percentage of neem oil (63.8%), while that from Tanjung Rhu, Langkawi contained the highest content of azadirachtin (0.4%) (Kadir *et al.* (1998).

Kumar *et al.* (2000) evaluated the insecticidal property of 38 neem trees, sampled from six locations in Karnataka, through laboratory bioassays of neem seed

Table 5. Average azadirachtin and oil contents of neem seed kernel samples from different Asian countries

Country	No. of samples	Azadirachtin (mg/g±S.E.)	Oil (%±S.E.)
Iran	4	2.75 ± 1.65	45.4 ± 1.2
India	9	5.14 ± 1.80	47.6 ± 5.5
Myanmar (Burma)	3	6.10 ± 0.70	48.8 ± 5.0
Sri Lanka	3	3.40 ± 0.34	50.1 ± 5.0
Thailand	6	5.20 ± 1.10	45.0 ± 5.0
Yemen	7	4.44 ± 0.90	49.7 ± 2.5

Source: Ermel (1995)

Table 6. Variations in azadirachtin content in neem from different states of India

State	Place	Azadirachtin content	
		mg/g	Per cent
Karnataka	Bijapur-1	5.69	0.57
	Bijapur-2	3.59	0.36
	Bijapur-3	4.16	0.42
	Dharwad	2.19	0.22
Punjab	Central zone	1.65	0.16
Orissa	Bhubaneswar	4.52	0.45
Rajasthan	Chitor	2.72	0.27
Tamil Nadu	Aundipatti	2.65	0.26
	Irukkangudi	1.32	0.13
Uttar Pradesh	Kanpur	2.12	0.21

Source: Kondal *et al.* (2002)

kernel extract against the second winter larvae of *Plutella xylostella*. The assays revealed a four-fold difference between trees for LC_{50} values, which may be attributed to both qualitative and quantitative differences in neem seed chemicals.

From the point of view of the chemistry of this tree, in general, extracts of neem fruit, seeds, seed kernels, twigs, stem bark and root bark have been studied and scores of compounds isolated worldwide (see recent review by Kraus, 2002). However, a new triterpenoid, 1 α, 7 α-diacetoxyapotirucall-14-ene-3 α, 21, 22, 24, 25-pentaol was recently isolated in China from a methanolic extract of the seed kernels of neem collected from Myanmar (Luo-Xiao *et al.*, 2000), otherwise all literature on Asian studies have been discussed in the review by Kraus.

4. USES OF NEEM

Neem and its derivatives have found use in agriculture, public health, medicine, toiletries, cosmetics and livestock production and health. Neem has now been universally accepted as a wonder tree due to its multifarious uses (Koul *et al.*, 1990, 1996). From a tree for solving the problems of developing third world countries, neem is now being sought after by the elite nations of the world (Schmutterer, 1995a). On the basis of the uses of neem, the ancient Indian names for the neem tree include *Prabhadra* (very useful), *paribhadrak* (spreading its utility over large distances), *sarvobhadrak* (useful in every way) and *rajbhadrak* (best among all the useful trees), all of which point to its immense usefulness in Indian way of life (Dube, 1996). The exploitation of neem for various purposes like pest management, fertilizer management, medicinal use, population regulation, agroforestry and reforestation and biomass production are well documented in different Asian countries and are discussed in the present section.

4.1 Pest Management

Neem has had a long history of use primarily against household and storage pests and to a limited extent against crop pests in the Asian continent. However, a breakthrough in the insecticidal application of neem was made by Pradhan *et al.* (1962) who successfully protected the standing crops by spraying them with 0.1 per cent neem seed kernel suspension during a locust invasion. Till date neem products have been evaluated against 450-500 species of insects in different countries around the world and 413 of these are reportedly susceptible at different concentrations (Schmutterer and Singh, 1995). In India alone, neem has been evaluated against 103 species of insects, 12 nematodes and many pathogenic fungi (Singh and Kataria, 1991; Arora and Dhaliwal, 1994, Suresh *et al.*, present volume). Some of the recent reviews on potential of neem in pest management include Singh (1996, 2000), Singh and Raheja (1996), Naqvi (1996), Abdul Kareem *et al.* (1998), Saxena (1998) and Dhaliwal and Arora (2001).

A major advantage of using neem-based pesticides in IPM is that in addition to causing mortality, neem products exhibit a wide range of physiological and behavioural effects on the target insects. The repellent and antifeedant effects of neem have been reported against a wide range of insect pests including desert locust, *Schistocerca gregaria*; migratory locust, *Locusta migratoria*; rice plant hoppers, *Nilaparvata lugens* and *Sogetella furcifera*; the leaf folder, *Cnaphalocrocis medinalis*; the ear cutting caterpillar, *Mythimna separata*; etc. (Ketkar, 1976; Saxena, 1989). Even starved *N. lugens* avoided feeding on plants sprayed with neem oil and spent most of their time searching for suitable sites. A concentration of 0.1 per cent of NSKE deterred feeding by all the larval instars of *Epilachna dodecastigma* on ribbed gourd leaves (Islam and Islam, 1988). Concentrations ranging from 0.001 to 0.4 per cent of various neem seed kernel (NSK) extracts have generally been found to deter the feeding of most of the insects evaluated so far (Singh, 2000). The growth inhibitory effects of neem derivatives result in various developmental defects and even mortality. The larvae of various lepidopterous and coleopterous pests like *P. xylostella*, *Spodoptera frugiperda*, *Helicoverpa zea*, *Pectinophora gossypiella*, *Epilachna varivestis* and *Ephestia kuehniella* studied in various developing countries in Asia show impaired development on neem-treated diet (Saxena, 1993). Neem products also affect insect vigour, longevity and fecundity. Neem compounds sterilized females of *E. varivestis* and *Leptinotarsa decemlineata*, while reproductive maturation was inhibited in *N. lugens* males. At higher concentrations, most females did not emit normal male eliciting signals (Saxena, 1993). Neem products have also been found to act as ovipositional deterrents for *Dacus cucurbitae*, *Helicoverpa armigera*, *Spodoptera litura*, *Callosobruchus* spp., etc. (Parmar and Singh, 1993; Chari and Ramaprasad, 1993).

A significant reduction in fecundity and egg hatchability was observed when *E. dodecastigma* beetles were fed on leaves of ribbed gourd treated with different concentrations (0.01-0.5%) of NSKE (Islam and Islam, 1998). Ovicidal activity of neem products has also been reported in other insect species including *Corcyra cephalonica*, *Earias vittella* and *S. litura* (Arora and Dhaliwal, 1994).

A number of studies have been carried out in several countries of Asia to evaluate neem alone or in combination/alternation with conventional insecticides and other approaches against insect pests of agricultural crops. In the Philippines, plots treated with a 2:10 neem cake-urea mixture applied at 120 kg/ha had lower incidence of ragged stunt, grassy stunt and tungro viruses and yielded significantly more than control plots in both dry and wet seasons. Also, weekly spraying of 50 per cent neem oil-custard apple mixture in 4:1 proportion (v/v) at 8 litres/ha from seedling to the maximum tillering stages significantly reduced tungro incidence and increased grain yield (Abdul Kareem *et al.*, 1987). In field trials conducted in India, neem treatments were found effective against green leafhopper, yellow stem borer, rice gall midge, rice leaf folder and grasshopper (Dhaliwal *et al.*, 1996; Nanda *et al.*, 1996).

Sprays of NeemAzal 5% @ 1.0 and 0.5 ml/l were as effective as monocrotophos 5.6ml/l against rice leaf folder, *Cnaphalocrocis medinalis* and yellow stem borer, *Scirpophaga incertulas*, respectively (Dhaliwal *et al.*, 2002). However, field evaluation of neem oil applied to rice in Bangladesh, China, India, Philippines and Thailand has not provided consistent results (Lim and Bottrell, 1994).

The neem oil at a concentration of 150 ppm was effective against cotton bollworms, *Pectinophora gossypiella* and *Earias insulana* in Egypt and caused reduction in infestation very near to that of chemical insecticides applied (Dimetry, 1996). However, in field trials conducted in India, neem products alone or with *Bacillus thuringiensis* or conventional synthetic insecticides failed to suppress the cotton pest complex. However, the pests were controlled and cotton yield and quality improved when neem products were applied in combination with synthetic pyrethroids (Gupta, 1996; Gupta *et al.*, 1999). An azadirachtin-rich insecticide, RD-9 Repelin, controlled the bollworm complex on cotton in Punjab (Dhawan and Simwat, 1996) and in Andhra Pradesh on par with quinalphos (Rosaiah and Reddy, 1996).

Alternate application of neem products and conventional insecticides made at the economic threshold level of 6-8 proved quite effective against *Bemisia tabaci* adults (Mann *et al.*, 2001). NeemAzal and Rakshakgold @ 2l/ha when alternated with endosulfan and chlorpyriphos received only 4 sprays during 50 days of economic damage period and persisted for 11-18 and 10-17 days, respectively, to maintain B. *tabaci* below economic threshold level (Table 7).

Neem treatments in cabbage intercropped with other vegetables in Mauritius controlled *P. xylostella* infestation and the combined effect was more effective than cartap hydrochloride, the recommended insecticide (Facknath, 1996). Neem products provided effective control of *Lipaphis erysimi*, *Spodoptera litura* and *Pieris brassicae* on cabbage in India, though the control was less than that provided by endosulfan. However, neem formulations were safer to various parasitoids (see chapter 8 for details). Moreover, feeding efficiency of *Coccinella septempunctata* (Linnaeus) on *L. erysimi*, treated with neem products, was higher as compared to that on the aphids treated with endosulfan (Dhaliwal *et al.*, 1998). In field trials conducted in Karnataka, neem seed kernel powder extract (4%) was found to be effective against *P. xylostella* and *Crocidolomia binotalis* (Moorthy and Kumar, 2000). A mixture of aqueous extracts (1%) of neem seed kernel and *Bacillus thuringiensis* subsp. *Kurstaki* was synergistic to *P. xylostella* (Rajamohan, 2002). In Sri Lanka, low-volume spray of neem derivatives was found to be effective against second instar larvae of *P. xylostella* (Ganesalingam, 1993). Neem products were found as effective as the recommended rates of the commercial insecticides for controlling leaf-eating caterpillar complex of cabbage and were safe to the larval parasitoid, *Apanteles plutellae* (Bandara and Kudagamage, 1996). In Thailand, all Chinese kale plants treated with Thai neem seed extracts withstood *P. xylostella* and *S. litura* and showed better resistance to *Hellula undalis* (Sombatsiri, 1996; Sombatsiri *et al.*, 1995).

A study conducted for two consecutive years in Assam (India) revealed that infestation of top shoot borer, plassey borers and termite was 4.82, 9.62 and 10.44 per cent, respectively in neem (Econeem)-treated plots as compared to 23.99, 25.04 and 28.96 per cent, respectively in control. The yield recorded from the neem-treated plot also increased to 93.69t/ha as compared to 70.36t/ha in control plot (Deka and Singh, 2001).In Thailand, a mixture of extracts of neem seed (*A. indica* var. *siamensis*) and *Hyptis suaveolens* at 5000 ppm caused 95 per cent mortality of

African red spider mite, *Eutetranychus africanus* (Tucker) on papaya, within 10.39 h after application while monocrotophos could kill within 21.30h. The mixture was also found to stimulate the growth of papaya with 45.5cm height and 93.33g stem plus root weight at 65 days after germination as compared to 25.17cm height and 13.33g stem plus root weight per plant in monocrotophos treatment (Uraisakul *et al.*, 1999). Thus, neem has bright prospects in managing insect pests of major agricultural crops in tropical Asia.

4.2 Fertilizer Management

Neem has been known to augment nitrogen use efficiency of plants under sub-tropical conditions. Bains *et al.* (1971) were the first to show under field conditions that treatment of urea with an acetone extract of dried and crushed neem kernel compared well with proven nitrification inhibitors and was superior to sulphur coated urea. Since then, a large number of studies have been conducted in Asia on nitrification inhibiting property of neem as well as effects of its application on ammonia volatilization and leaching losses of nitrogen and on the efficiency of nitrogen utilization by crops (Ketkar and Ketkar, 1995; Prasad *et al.*, 1996).

In several field studies, conservation of higher ammonium-N in soils has been encountered when neem cake treated/coated urea was applied. Subbiah and Kothandaraman (1980) recorded 8.3 ppm ammonium-N with prilled urea and 12.2 ppm with urea+neem cake when applied at 120 kg N/ha. Tiwari (1989) recorded 72 and 93 ppm ammonium N in surface (0-30 cm) soil layer in prilled urea and neem cake coated urea rice plots, respectively. Application of 100 kg N/ha to rice through neem extract-coated urea gave identical yield (6.58 t/ha) to 120 kg N/ha (6.53 t/ha) by prilled urea, resulting in net saving of 20 kg fertilizer N/ha (Bhandari *et al.*, 1996).

A two-year study with Java citronella (*Cymbopogon winterianus* Jowitt.) showed that neem-coated urea significantly increased the uptake of N, P and K by 17, 15 and 25 per cent, respectively over ordinary urea (Prakasa Rao, 1996). It increased apparent recoveries by 90 and 45 per cent over ordinary urea at 300 and 400 kg N/ha/year, respectively, and reduced NH_3 volatilization losses by 31 per cent over ordinary urea. However, under submergence and field capacity conditions, the ammonia loss from neem oil-coated urea was 1.7 and 6.7 per cent, which was markedly less than that of 6.5 and 21.3 per cent from prilled urea, respectively (Singh and Kakkar 1996).

Shah and Faheem (2000) evaluated seed cakes of neem, bakain (*Melia azedarach*) and arend (*Ricinus communis*) for their nitrification inhibition properties in three soils in Peshawar valley in Pakistan. The extent of nitrification inhibition was highest for neem followed by bakain and arend treatments.

In studies conducted in Sri Lanka, neem cake amended urea at 20 and 30 per cent conserved ammonium ions and reduced nitrate ions compared to urea alone up to 8 weeks in reddish brown latosolic (RBL-Ultisol) soil and up to 12 weeks in reddish brown earth (RBE-Alfisol) soil. In contrast, in red yellow podzolic (RYP-Ultisol) soil, all neem treatments increased nitrate content and reduced ammonium content up to 6 weeks of incubation (Gnanavelraja and Kumaragamage, 1998).

Table 7. Effect of economic threshold level (ETL)-based alternate sprays of neem-based insecticides with conventinal insecticides on populatoin build up of *Bemisia tabaci* on cotton

Spray Schedule	ETL-Based no. of sprays	Spray interval	Whitefly adults per 15 leaves on indicated days after sowing*									
			110	128	131	136	142	146	153	156	158	160
S_1	4	11-18	123.3*	56.0 (7.53)	90.0* (9.52)	29.0 (5.46)	59.3 (7.75)	94.7* (9.76)	40.3 (6.41)	62.0 (7.92)	91.3* (9.57)	47.0 (6.90)
S_2	4	10-17	101.3*	62.7 (7.99)	98.7* (9.98)	29.7 (5.48)	67.0 (8.24)	104.0* (47.0)	47.0 (6.90)	95.0* (9.79)	43.3 (6.64)	54.0 (7.40)
S_3	5	7-15	90.3*	93.3* (9.69)	49.7 (7.10)	63.0 (7.99)	93.7* (9.71)	47.0 (6.90)	91.3* (9.60)	41.3 (6.49)	12.7 (7.92)	96.0* (9.82)
S_4	5	6-9	92.3*	111.7* (10.61)	85.7 (9.30)	118.3* (10.90)	38.7 (6.29)	60.7 (7.84)	91.0* (9.57)	82.0 (9.09)	110.0* (10.49)	58.0 (7.65)
S_5	Control		98.3	101.3 (10.11)	145.0 (12.06)	121.7 (11.06)	154.0 (12.40)	114.0 (10.68)	143.3 (12.01)	166.0 (12.91)	169.7 (13.05)	117.3 (10.85)
CD (P=0.05)			NS	(1.38)	(0.82)	(1.13)	(1.26)	(1.32)	(0.83)	(0.97)	(1.57)	(1.31)

Figure in parentheses are $\sqrt{n + 1}$ transformations
* ETL based spraying days
S_1 = NeemAzal 1% (2l/ha)/Endosulfan 35 EC (2.5l/ha)/NeemAzal 1% (2l/ha)/Chlorpyriphos 20 EC (5l/ha).
S_2 = RakshakGold 1% (2l/ha)/Endosulfan 35 EC (2.5l/ha)/RakshakGold 1% (2l/ha)/Chlorpyriphos 20 EC (5l/ha).
S_3 = ICIPE Neem 0.5% (3l/ha)/Endosulfan 35 EC (2.5l/ha)/ICIPE Neem 0.5% (3l/ha)
S_4 = Phosphamidon 85SL (187ml/ha)/Cypermethrin 10EC(0.5l/ha)/Triazophos 40EC (1.5l/ha)/Monocrotophos 36SL (1.25l/ha)/Chlorpyriphos 20 EC (5l/ha)
Source: Mann *et al.* (2001)

Similarly, neem cake treatments at 20 and 30 per cent levels, significantly reduced leaching losses of nitrate with both urea and ammonium sulphate. Application of neem cake and extract with N fertilizer gave significantly higher yield of radish compared to fertilizer application alone (Gnanavelrajah and Kumaragamage, 1999).

Thus, the efficacy of various neem-based products in increasing N use efficiency has been amply demonstrated. There is a need to develop efficient coating/incorporation techniques of neem ingredients into fertilizer urea matrix as well as on the rate of dissolution and diffusion of these products with the aim of optimally matching the nitrification rates with N uptake by plants.

4.3 Medicinal Uses

Since ancient times, the uses of neem have been documented in Indian Ayurveda and Unani systems of medicine, and millions of Asians have used neem medicinally over thousands of years (NRC, 1992; Ketkar and Ketkar, 1995). The neem twig is nature's toothbrush to over 500 million people daily in India alone. Neem fruits, seeds, oil, leaves, root and bark have such varied uses as general antiseptics, antimicrobials, treatment of urinary disorders, fever, bronchitis, diarrhoea, skin diseases, septic sores, infected burns, hypertension and inflammatory diseases (Saxena, 1993; Riar, 1996).

Singh *et al.* (1979) reported an improvement or cure in all cases of acute eczema, ringworm infection and scabies in humans with the alcoholic extract of neem leaves. Neem oil, nimbidines and the alcoholic extract of seeds have also been reported to be bactericidal (Singh and Sastry, 1988). Rao *et al.* (1986) reported a faster rate of wound contraction when neem oil was applied topically in the form of 25 per cent ointment in petroleum jelly, to laboratory animals. Nimbidin, a constituent of neem oil, has been found to reduce significantly acute paw oedema in rats induced by phlogistic agents, carrageenin and kaolin (Pillai and Santhakumari, 1981). The test drug significantly suppressed the formalin-induced arthritis of ankle joint and the fluid exudation in croton oil-induced granuloma in rats. The drug can be considered as a general anti-inflammatory agent. The crude extract of neem leaves has been reported to affect profound hypertension and a minimal negative chronotropic effect in guinea pigs and rats, which increased at higher doses. In one rabbit, 200 mg of extract/kg body weight decreased heart rate from 280 to 150 beats/min (Riar, 1996).

Neem bark has been used as astringent and in the treatment of malarial fevers in India. Aqueous extract of the leaves is used as remedy for malaria in Thailand. Because of its schizontocidal action, it has been pointed out that indigenous Asian people use this extract to treat malaria over a period of time and it may be breaking the alternate day cycle of fever (Sharma, 1996). Ethanol extracts of neem leaves and seeds have been found to be effective against chloroquine-sensitive and chloroquine-resistant strains of *Plasmodium falciparum* (Badam *et al.*, 1987). Since malaria is creeping back in many Asian countries and there is a growing problem of resistance to the conventional treatments, there is a need for further exploration of role of neem in malaria control.

Recently, a herbal drug preparation containing neem oil and Karanja oil was shown to provide complete recovery to cattle from dermatitis within 5-8 days of treatment, with no recurrence of the condition observed for one year (Kulkarni and Bansod, 2001).

4.4 Human Population Regulation

The Indian scientists have carried out detailed investigations on the antifertility properties of neem products (Jacobson, 1995). Neem oil has been reported to possess strong spermicidal action against rhesus monkey and human spermatozoa *in vitro* and killed spermatozoa within 30 seconds of mixing with semen (Riar, 1996). When used intravaginally, a dose of 1ml in rhesus monkeys and human beings, and 20 μl in rats, before sexual intercourse was 100 per cent effective in preventing pregnancy. Moreover, there were no side effects on repeated application as confirmed by histopathological studies on reproductive organs or other tissues (Sinha *et al.*, 1984). Upadhyay *et al.* (1990) observed that injection of neem oil (single dose 100 μl) into the uterine horn of rats, created an immunological response and prevented pregnancy for nearly five months. A novel use of neem oil, based on its ability to stimulate locally at the site of application, the cell-mediated immune reactions has been developed by Talwar *et al.* (1996). A single administration of purified neem extract 'Praneem' into the uterus causes a long lasting effect on fertility in rats and monkeys without any disturbance of ovulation and sex steroid hormone production.

The male antifertility effect of neem has been studied in mice, rats, rabbits and guinea pigs by daily oral feeding of a cold-water extract of fresh green leaves (Sadre *et al.*, 1983). The infertility effect was seen in treated rats as there was 66.7 per cent reduction in fertility after 6 weeks, 80 per cent after 9 weeks and 100 per cent after 11 weeks. There was no inhibition of spermatogenesis but the motility of spermatozoa markedly decreased.

Implications of neem for human fertility control are being studied. This could be a major breakthrough because of the future population increase in Asian countries where neem could be easily grown.

4.5 Agroforestry and Reforestation

Neem, in several regions of Asia, is the most suitable tree component in many agroforestry species. Under semi-arid conditions at Jhansi (Uttar Pradesh, India), neem along with other tree species increased the productivity of a silvipastoral system up to 8.5 t/ha (Gill and Deb Roy, 1996). It has been reported that the forage production can be increased from 0.05 to 3.6 t/ha in the arid zones of Thar desert by growing suitable grasses and legumes along with neem and other trees (Singh and Rai, 1989). However, Hazra and Tripathi (1989) reported that forage yield was 74 per cent under neem as compared to open at Jhansi.

Growing trees with arable crops has been practiced in many Asian countries for a very long time. However, depending on the crop geometry and tree species, large variations have been reported. Ramshe (1989) reported that reduction in grain production in chickpea, pearl millet and sunflower under *A. indica* during second and third year varied from 13 to 33 per cent while this reduction in *Leucaena leucocephala* varied from 46 to 90 per cent for second year and 66 to 99 per cent for third year. The paddy and wheat grown under 10 tree species indicated that paddy yield was better under *A. indica* and *Albizia lebbek* (5t/ha), and wheat yield under *Dalbergia sissoo* (4.4t/ha) (Anonymous, 1992). Under semi-arid conditions at Hyderabad (India) grain yield of sorghum was the maximum with *A. indica* (1,248 kg/ha) and the least under *Acacia nilotica* (831 kg/ha) (Nimbole and Dass, 1990). Thus, the compatibility of neem as an intercrop needs to be thoroughly investigated. Recent advances in biotechnology should be employed to select phenotypes with desirable height and stature for use in agroforestry.

Being a hardy tree, neem is ideal for reforestation programmes and for rehabilitating degraded, semi-arid and arid lands and coastal areas (Maramorosch, 1999). Neem exhibited the best performance both in growth parameters and survival under saline conditions in Gujarat, India (Hegde *et al.*, 1990). Singh *et al.* (1990) recommended the planting of neem with intercrop for the management of salt effected soils. About 50,000 trees of neem were planted in the plains of Arafat, Saudi Arabia, to provide shade to 2-3 million Muslim pilgrims visiting Mecca. This plantation is irrigated with saline water and the plants seem to be tolerant to the ill effects of saline waters because the earliest plantings represent more than 15 years old healthy plants now (Ahmed *et al.*, 1989). Neem has been found very effective as wind breaks in drier areas, particularly on sandy soil where sand blasting and desiccation can affect crop establishment (Gill and Deb Roy, 1996).

4.6 Biomass Production

Neem after 7-8 years of planting is known to give 1000-kg/ha biomass (Srivastava and Rama Mohan Rao, 1989). The tree has been recommended for fuelwood plantations in Maharashtra and Kandi area of Punjab in India. Studies on fuel and timber production of 10 different tree species planted at 1mx1m distance in medium textured soil under scarce water regime indicated that performance of neem was comparable with other species (Ramshe, 1989). Studies conducted at Hyderabad revealed that fuel yield from pruning at 18 months after transplanting was highest in *Acacia nilotica* (3.8 t/ha) and next in order were *A. indica* (1.7 t/ha) and eucalyptus (1.6 t/ha) (Gill and Deb Roy, 1996).

5. COMMERCIALIZATION

Neem has been exploited in many Asian countries particularly India for commercial production of a large number of products such as pesticides and allied agrochemicals, plant nutrients, animal feed, medicines, toiletries, cosmetics, etc. (Parmar and Ketkar, 1996). In India, about 100 products have either been marketed or are awaiting commercialization as pesticides (Gahukar, 1998). Majority of the formulations contain 300 or 1500 ppm azadirachtin (Table 8). However, recently formulations having 10,000 and even 50,000 ppm of azadirachtin have been developed. The Central Insecticides Board approved the guidelines and data requirements for registration of neem pesticides for domestic and export purposes in 1991. Currently, more than 40 name-based products have been registered in India either provisionally or with full registration (Walia *et al.*, 2002). Registration status in other Asian countries is either provisional or unknown (see chapter 1)

A number of neem-based medicinal products have been commercialized in India (Ketkar and Ketkar, 1995). Between these two medicinal aspects of neem, viz. use as a contraceptive and in the management of secondary hyperglycemia have been exploited commercially. A commercial preparation 'Sensal' containing neem oil (98% w/v) has been developed which is a safe, reliable, intravaginal, pre-coital, spermicidal contraceptive. Neem oil has been formulated in gelatin capsules 'Nimbola', which is an oral treatment for diabetes control and is stated to be devoid of any side effects. Several skin care products have also been developed, the important being 'Clean 'N' Care' (pimples), 'Curoline' (antiseptic skin cream), 'Neemcure' (antiseptic product against skin diseases, piles, burns, wounds and injury) and 'Greneem' (blood purifier, useful in acne, skin disorders and bacterial and viral infections) (Parmar and Ketkar, 1996).

Other neem based products, which have been marketed in India for various purposes such as cosmetics (Neemtulsi, Neemal, Licika, etc.), soap (Feu Drop, Homacol, Kutir Neem Sandal Soap, Parashais Limda Soap), shampoo (Margosa Neem), toothpaste and tooth powder (Neem, ORA Neem Gel) are quite popular. In Karachi, Pakistan, M/S Hamdard Co is marketing a tooth powder 'Nimodent'. In addition, several products are being manufactured as manures and fertilizers (Neem Manure, Humi-Gold, Wellgro, Godrej NESU, Nimin, Neemax, etc.) and cattle and poultry feed (Pasutone).

6. ECONOMIC EFFICIENCY

As vast area in Asia is under marginal lands, plantation of neem on a portion of these lands can make a significant contribution in boosting general and agricultural economy, besides helping in maintaining environmental balance. Mruthyunjaya and Jha (1996) carried out a detailed analysis to evaluate the economic feasibility of neem plantation. The economic felling cycle for neem was fixed at 23 years and the discount rate selected for the study was 12 per cent per annum (Table 9). It may be seen from the table that the net present worth of Rs.40,838 implies a return in excess of the value of the capital invested plus the specified rate of return (12%) on that capital. Thus, the individual investor can expect to receive a net income of Rs.5289 per ha per year from raising the neem tree. Since net present worth (NPW) is a positive sum (Rs.40, 838), B-C ratio is greater than unity (3.59) and internal rate of return (IRR) is much above the discount rate (45.88%), the investment in neem plantation is a worthy preposition. An economic appraisal of 12 multipurpose

Table 8. Selected commercial formulations of neem-based pesticides in India

S. No.	Formulation	Azadirachtin content (%)	Manufacturer/ Formulator
1	Achook	0.03	Godrej Agrovet Ltd., Mumbai
2	Bioneem	0.03	Zuari Industries Ltd., Margoa, Goa
3	Econeem	0.03	Margo Biocontrols Pvt. Ltd., Bangalore
4	Fortune Aza	0.15	Fortune Biotech Ltd., Secunderabad
5	Gronim-T	10.00	National Tree Growers Co-operative Federation Ltd., Anand
6	Kranti	0.15	Pragti Glyxal Pvt. Ltd., Musore
7	Margocide	0.03, 0.15	Monofix Agro Products Ltd., Hubhi
8	Margosom	0.03,0.15	Som Phytopharma (India) Ltd., Hyderabad
9	Multineem	0.03	Karnataka Agro Chemicals, Bangalore
10	Multiplex	0.03	Multiplex Fertilizers Pvt. Ltd., Bangalore
11	Neemactin	0.15	Wockhardt (Biostadt Agrisciences), Mumbai
12	Neemazal	1,5	EID Parry (India) Ltd., Chennai
13	Neemarin	0.15	Biotech International Ltd., New Delhi
14	Neemark	0.03	West Coast Herbochem Ltd., Mumbai
15	NeemGold	0.15, 10.00	Southern Petrochemical Industries Corporation Ltd., Chennai
16	Neemguard	0.03	Gharda Chemicals Ltd., Mumbai
17	Neemitaf	0.15	Rallis India Ltd., Pune
18	Neemnath	0.03	Nath Seeds Ltd., Aurangabad
19	Neemol	0.03	Ramson Agrotech Pvt. Ltd., Vijaywada
20	Neemolin	0.03, 0.15	Khatau Agrotech Ltd., Mumbai
21	Neemstar	0.03	Universal Pesticides & Chemicals Industries, Coimbatore
22	Nimbasol	0.15	Nimba Foods & Chemicals Pvt. Ltd., New Delhi
23	Nimbecidine	0.03	T. Stanes & Co., Coimbatore
24	Peekrakshak	0.03	Yawalkar Pesticides Pvt. Ltd., Nagpur
25	Rakshak	0.15	Murkumbi Bioagro Pvt. Ltd., Belgaum
26	RakshakGold	1.00	Murkumbi Bioagro Pvt. Ltd., Belgaum
27	RD-9 Repelin	0.03	Indian Tobacco Co. Ltd., Hyderabad
28	Reconeem	0.15	Ramson Agrotech Pvt. Ltd., Vijaywada
29	Sukrina	0.15	Conster Chemicals Pvt. Ltd., Chennai
30	Uttamneem	0.03	Chambal Fertilizers & Chemicals Ltd., New Delhi

Source: Gahukar (1998)

tree species grown uniformly for 8 years on denuded shallow black soils in Bijapur, Karnataka, showed that neem ranked first in terms of the values of these appraisal tools, viz. net present value (18061.75), benefit cost ratio (7.50) and internal rate of return (13.02%) (Poddar *et al.*, 2000).

Ahmed (1995) cited several examples of economic analysis of neem products for pest control, soap production, fertilizer, etc. However, Kandaswamy and Raveendran (1988)'s economic analysis of 7 field trials is most interesting. While neem products increased crop yield by 150-1000 kg/ha and gave a benefit: cost ratio (BCR) of about 5, monocrotophos increased paddy yield by 1000-2000 kg/ha and gave a higher BCR of 10-31. In one experiment, NSKE5% applied by low-volume sprayer gave a more attractive BCR of 44 against 31 obtained with monocrotophos and in another experiment; both products gave the same BCR of 25. Thus, for limited-resource farmers of Asia, there is need to promote the use of homemade products. There is also need to provide incentives to promote propagation, processing and use of the ecofriendly neem-based products to protect our environment.

7. FARMERS' OUTLOOK

Majority of the farmers in Asia are limited-resource farmers and they cannot afford to purchase high cost synthetic pesticides (Ahmed, 1995). It is, therefore, essential to demonstrate and convince these farmers about the possible benefits of using neem in agriculture. A survey of 300 farmers located in Coimbatore, Madurai and Aduthurai regions of Tamil Nadu, revealed that farmers using neem products along with chemicals obtained comparatively higher yields than those using only chemicals (Palanisamy, 1992; Lim and Bottrell, 1994). The results of several studies have revealed that though most farmers are aware of the neem products, their adoption rate was comparatively low (Jayaraj, 1993). Considering the long-term benefits in terms of health and environment, neem should be encouraged in a cost effective manner. The studies in farmers' fields at 4 different locations in Tamil Nadu, India, revealed that rice yield was 20.5 per cent more and income was US$119.77/ha more under IPM treatments using neem than under farmers' practice treatment (Table 10). Yields and increased incomes were only slightly less under IPM treatment using synthetic insecticides (Lim and Bottrell, 1994). In a study conducted in two areas of Andhra Pradesh to know the farmers' perception about natural pest control products (neem seed kernel extract and nucleopolyhedrovirus), it was revealed that when farmers adopt these products, they use criteria of cost and efficiency, rather than health and environmental considerations (Tripp and Ali, 2001). It was suggested that allowing farmers more opportunities to experiment with them and supporting the commercial manufacture of these products would encourage widespread use of alternative pest control products.

Table 9. Annual and present value of costs and gross and net returns (Rs) from raising one ha of neem over a felling cycle of 23 years

Year	Annual Costs	Annual returns	Present worth of annual costs	Present worth of annual returns	Discount factor at 12%	Cash flow	Present worth of cash flow
1	3650	--	3260	--	0.893	-3650	-3259
2	2350	--	1873	--	0.797	-2350	-1873
3	850	--	605	--	0.712	-850	-605
4	850	--	541	--	0.636	-850	-541
5	1650	9400	936	3629	0.567	7750	4394
6	1050	9400	532	3245	0.507	8350	4233
7	1050	9400	475	2893	0.452	8350	3774
8	1050	9400	424	2586	0.404	8350	3373
9	1050	9400	379	2310	0.361	8350	3014
10	1650	9400	531	2061	0.322	7750	2496
11	1250	12400	359	3559	0.287	11150	3200
12	1250	12400	321	3187	0.257	11150	2866
13	1250	12400	286	2840	0.229	11150	2553
14	1250	12400	256	2542	0.205	11150	2286
15	1850	12400	339	2269	0.183	10550	1930
16	1700	15400	277	2510	0.163	13700	2233
17	1700	15400	248	2248	0.146	13700	2000
18	1700	15400	221	2002	0.130	13700	1781
19	1700	15400	197	1786	0.116	13700	1589
20	2300	15400	239	1602	0.104	13100	1362
21	1700	15400	158	1432	0.093	13700	1274
22	1700	15400	141	1278	0.083	13700	1137
23	2100	24000	155	1776	0.074	21900	1621
Total			12753	45755	7.721		40838

Annuity (Rs/ha) = 40838 = 5289; (ii) B.C. ratio (Rs) = 45755 = 3.59; (iii) NPW (Rs/ha) = 40838; (iv) IRR (%) = 45.88 12753

Source: Mruthyunjaya and Jha (1996)

Table 10. Comparison of average rice yields and incomes from neem-incorporated IPM (NIPM), synthetic insecticide-based IPM (SIPM) and farmers' practices (FP) at different locations in Tamil Nadu, India

Location	Yield (kg/ha)			Yield increase over FP (%)		Increase in net income over FP	
	FP	NIPM	SIPM	NIPM	SIPM	NIPM	SIPM
Aduthurai	7092	7820	7971	10.3	12.4	49.18	66.28
Coimbatore (Gobichettipalayam)	4446	4964	5000	11.7	14.7	32.43	24.86
Thanjavur (Rajendran)	7316	8730	9064	19.3	23.9	146.63	177.02
(Vaidyanathampatti)	5010	7242	7428	44.6	48.3	250.86	262.43
Mean	5966	7189	7366	20.5	23.5	119.77	132.65

Source: Lim and Bottrell (1994)

8. PROSPECTS AND POSSIBILITIES FOR FUTURE

Neem is a versatile tree with immense potential to protect the environment and developing sustainable agriculture in tropical Asia. There is, thus, an urgent need to popularise its cultivation on marginal lands and also bringing about awareness about the benefits and economic advantages of neem cultivation. It has been estimated that 150-250 million neem trees are required in South Asia to meet the requirements of limited resource farmers, which would need one million hectares of land to be brought under neem (Ahmed, 1995). Taking into consideration millions of hectares of semi-arid lands lying under-utilized in South Asia, this target does not appear to be difficult to achieve. As the demand for neem products increases further, more land would need to be brought under neem in South Asia and elsewhere. There is also need to tap the potential of already existing trees as in India less than 20 per cent of the seed crop is harvested due to unorganized scattered plantations (Walia *et al.*, 2002). Under the optimistic assumption of an azadirachtin content of 5g per kg of dried kernels, there should be a reservoir of about 150 tonnes of azadirachtin per year but also a yield of less than 50g per tree from 15-25 million trees in India (Rembold, 1996).

A number of studies have revealed the vast variations in azadirachtin content in neem in different regions of Asia. There are indications that the variations in azadirachtin content are not only due to ecological conditions but also due to genetical variations. The maximum variation is likely to be in its natural distribution range, i.e. Shivalik hills (India) to Myanmar. Moreover, the presence of neem tree in Myanmar and Thailand, and its subspecies, *A. indica siamensis* and intermediate forms is indicative of the presence of various ecotypes. There is thus a strong need to conduct thorough survey for elite trees for possible variations. A number of neem-based products have become available in market in India and other Asian countries, particularly for pest management. However, there are many reports of inconsistency in the field performance of these products. Moreover, these products have generally low persistence. Hence, there is need for stabilization of neem products against photo, thermal and microbial degradation. Simple formulation technology will have to be developed so that ready-to-use pesticides can be produced at the local level. Quality control of neem-based products is also a major problem. There is a large variation in the quality and quantity of extractives obtained from a plant due to variation in ecotypes, environmental factors, etc. Such variations affect the performance and shelf life of formulated products. There is an urgent need to develop and prescribe suitable standards for registration of neem products in various Asian countries.

Cost-benefit ratios and partial budgets that estimate farmer profitability are required to show whether neem is more cost-effective than synthetic pesticides. Such estimates should be derived from long term trials in farmers' fields in representative areas, taking environmental perspectives into consideration. The analysis should also include the social factors influencing farmers' acceptance.

REFERENCES

Abdul Kareem, A., Gunasekaran, K. and Anbalagan, G. (1996), Botanical pesticides in integrated pest management, in *Ecological Agriculture and Sustainable Development*, Vol. 1, eds. G.S. Dhaliwal, N.S. Randhawa, R. Arora and A.K. Dhawan , Indian Ecological Society and Centre for Research in Rural and Industrial Development, Chandigarh, India, pp. 146-161.

Abdul Kareem, A., Saxena, R.C. and Justo, Jr., H.D. (1987), Cost comparison of neem oil and insecticide in rice tungro virus (RTV). *Inernational Rice Research Newsletter*, **12,** 28-29.

Ahmed, S. (1995), Economic, Socio-economic and policy considerations on neem's current use for pest management, in *The Neem Tree Azadirachta indica A. Juss. and Other Meliaceous Plants*, ed. H.Schmutterer, VCH, Weinheim, Germany, pp. 559-578.

Ahmed, S., Bamofleh, S. and Munshi, M. (1989), Cultivation of neem (*Azadirachta indica*, Meliaceae) in Saudi Arabia. *Economic Botany*, **43,** 34-38.

Anonymous (1992), *Annual Progress Report (1990-91)*, All India Coordinated Research Project on Agroforestry, New Delhi.

Arora, R. and Dhaliwal, G.S. (1994), Botanical pesticides in insect pest management: Ecological perspective, in *Management of Agricultural Pollution in India*, eds. G.S. Dhaliwal and B.D. Kansal, Commonwealth Publishers, New Delhi, pp. 213-247.

Badam, L., Deolankar, R.P., Kulkarni, M.M., Nagsampgi, B.A. and Wagh, U.V. (1987), *In vitro* antimalarial activity of neem (*Azadirachta indica* A. Juss) leaf and seed extracts. *Indian Journal of Malariology*, **24**, 111-117.

Bains, S.S., Prasad, R. and Bhatia, P.C. (1971), Use of idigenous materials to enhance the efficiency of fertilizer nitrogen for rice, *Fertilizer News*, **16**, 30-32.

Bandara, K.A.N.P. and Kudagamage, C. (1996), Potential of neem for controlling leaf-eating caterpillar complex on cabbage, in *Neem and Environment*, Vol.1, eds. R.P. Singh, M.S. Chari, A.K. Raheja and W. Kraus, Oxford & IBH Publishing Co. Pvt. Ltd., New Delhi, pp. 547-557.

Bhandari, A.L., Sharma, K.N., Brar, R.S. and Rana, D.S. (1996), Relative agronomic efficiency of neem extract-coated urea versus pulled urea with rice, in *Neem and Environment*, Vol.2, R.P. Singh, M.S. Chari, A.K. Raheja and W. Kraus, Oxford & IBH Publishing Co. Pvt. Ltd., New Delhi, pp. 821-825.

Chari, M.S. and Ramaprasad, G. (1993), Neem as an anti-feedant and ovpositional repellent for *Spodoptera litura* F, in *Souvenir World Neem Conf.*, Feb.24-28, 1993, Bangalore, pp. 81-90.

Chiu, S.-F. and Zhang, Y.-G. (1989), Successful introduction of the neem tree (*Azadirachta indica* A. Juss) in China, *Neem Newsletter,* **6**, 16-17.

Deka, M.K. and Singh, S.N. (2001), Neem formulations in the management of sugarcane insects and pests, in *Proc. 63rd Annual Convention of the Sugar Technologists' Association of India,* August 25-27, 2001, Jaipur, pp. A33-A38.

Dhaliwal, G.S. and Arora, R. (2001), Role of phytochemicals in integrated pest management, in *Phytochemical Biopesticides*, eds. O. Koul and G.S. Dhaliwal, Harwood Academic Publishers, Amsterdam, The Netherlands, pp. 97-117.

Dhaliwal, G.S., Gill, R.S. and Dilawari, V.K. (1998), Management of insect pest complex of cabbage with neem based insecticides, in *Ecological Agriculture and Sustainable Development*, Vol. 2, eds. G.S. Dhaliwal, N.S. Randhawa, R. Arora and A.K. Dhawan, Indian Ecological Society and Centre for Research in Rural and Industrial Development, Chandigarh, India, pp. 306-314.

Dhaliwal, G.S., Singh, J. and Dilawari, V.K. (1996), Potential of neem in insect pest management in rice, in *Neem and Environment*, Vol. 1, eds. R.P. Singh, M.S. Chari, A.K. Raheja and W. Kraus, Oxford & IBH Publishing Co. Pvt. Ltd., New Delhi, pp. 425-431.

Dhaliwal, G.S., Multani, J.S., Singh, S., Kaur, G., Dilawari, V.K. and Singh, J. (2002), Field evaluation of azadirachtin-rich neem formulations against *Cnathalocrocis medinalis* (Guenee) and *Scirpophaga incertulas* (Walker) on rice. *Pesticide Research Journal*, **14**, 77-82.

Dhawan, A.K. and Simwat, G.S. (1996), Field evaluation of some botanical insecticides alone or in combination with other insecticides for management of bollworm complex on cotton, in *Neem and Environment*, Vol. 1, eds. R.P. Singh, M.S. Chari, A.K. Raheja and W. Kraus, Oxford & IBH Publishing Co. Pvt. Ltd., New Delhi, pp. 485-491.

Dimetry, N.Z. (1996), Neem as a safe trend for pest control in the future, in *Neem and Environment*, Vol. 1, eds. R.P. Singh, M.S. Chari, A.K. Raheja and W. Kraus, Oxford & IBH Publishing Co. Pvt. Ltd., New Delhi, pp. 51-60.

Dube, R. (1996), *Neem Kranti*, Shubham Prakashan, Lucknow.

Eeswara, J.P., Mordue, A.J., Stuchbury, T. and Allan, E.J. (1997), Azadirachtin, salannin and nimbin contents of seeds and callus of neem (*Azadirachta indica*

A. Juss), in *Practice Oriented Results on Use and Production of Neem-Ingredients and Pheromones*, eds. H. Kleeberg and C.P.W. Zebitz, Proceedings of the 5th Workshop, Wetzlar, Germany, pp. 251-261.

Ermel, K. (1995), Azadirachtin content of neem seed kernels from different regions of the world, in *The Neem Tree Azadirachta indica A. Juss. and Other Meliaceous Plants: Sources of Unique Natural Products for Integrated Pest Management, Medicine, Industry and Other Purposes*, ed. H. Schmutterer, VCH Weinheim, Germany, pp. 89-92.

Ermel, K., Pahlich, E. and Schmutterer, H. (1984), Comparison of azadirachtin content of neem seeds from ecotypes of Asian and African origin, in *Proc. 2nd Int. Neem Conf.,* 1983, eds. H. Schmutterer and K.R.S. Ascher, GTZ, Rauischholzhausen, Germany, pp. 91-93.

Ermel, K., Pahlich, E. and Schmutterer, H. (1987), Azadirachtin content of neem kernels from different geographical locations, in *Proc. 3rd Int. Neem Conf.*, 1986, eds. H. Schmutterer and K.R.S. Ascher, GTZ, Eschborn, Germany, pp. 171-185.

Facknath, S. (1996), Application of neem and intercropping for control of important cabbage pests. Paper presented at Int. Neem Conf., Gatton College, Australia, February 1996 (Abstract).

Gahukar, R.T. (1998), Neem pesticides, *Pestology*, **22**, 15-41.

Ganesalingam, V.K. (1993), Effect of low-volume spray of neem derivatives on the larvae of *Plutella xylostella* (L.) In: *Abstract World Neem Conference*, February 24-28, 1993, Bangalore, p.8.

Gill, A.S. and Deb Roy, R. (1996), Agroforestry uses, in *Neem*, eds. N.S. Randhawa and B.S. Parmar, New Age International (P) Ltd., New Delhi, pp. 49-58.

Gnanavelrajah, N. and Kumaragamage, D. (1998), Effect of neem (*Azadirachta indica* A. Juss) materials on nitrification of applied urea in three selected soils of Sri Lanka, *Tropical Agricultural Research*, **10**, 61-73.

Gnanavelrajah, N. and Kumaragamage, D. (1999), Nitrogen leaching losses and plant response to nitrogen fertilizers as influenced by application of neem (*Azadirachta indica* A. Juss) materials, *Tropical Agricultural Research*, **11**, 29-41.

Gupta, G.P. (1996), Use of neem in management of pest complex in cotton, in *Neem and Environment*, Vol.1, eds. R.P.Singh, M.S. Chari, A.K. Raheja and W. Kraus, Oxford & IBH Publishing Co. Pvt. Ltd., New Delhi, pp. 469-483.

Gupta, G.P., Katiyar, K.N., Sharma, K. and Sharma, K. (1999), Neem in the management strategies of insect pests of cotton, in *Azadirachta indica* A. Juss., eds. R.P.Singh and R.C. Saxena, Science Publishers Inc., Enfield, USA, pp. 177-189.

Gupta, P.K. and Prabhu, V.V. (1997), On the high performance liquid chromatographic determination and monitoring of azadirachtin in neem ecotypes, *Indian Forester*, **123**, 1067-1071.

Hazra, C.R. and Tripathi, S.B. (1989), Performance of oat genotypes under different tree covers and open, *Agriculture Science Digest*, **9**, 146-148.

Hegde, N.G. (1996), Improving the productivity of neem trees, in *Neem and Environment,* Vol. 1, eds. R.P. Singh, M.S. Chari, A.K. Raheja and W. Kraus, Oxford & IBH Publishing Co. Pvt. Ltd., New Delhi, pp. 69-79.

Hegde, N.G., Relwani, L.L., Gandhe, A.M., Joglekar, S.R. and Dinesh Kumar, K.T. (1990), Evaluation of MPTS for agroforestry in the agricultural ecozones of Gujarat, Maharashtra and Karnataka, in *MPTS for Agroforestry Systems*, eds. P.S. Pathak, R. Deb Roy and P. Singh, RMSI, Jhansi, India, pp. 53-60.

Islam, K.S. and Islam, B.N. (1988), Effects of methanol extracts of neem *A. indica* seed extracts on *Epilachna* beetle, *Bangladesh Journal of Agricultural Sciences*, 15, 27-32.

Jacobson, M. (1995), Antifertility effects and population control agents, in *The Neem Tree Azadirachta indica A. Juss. and Other Meliaceous Plants*, ed. H. Schmutterer, VCH, Weinheim, Germany, pp. 518-525.

Jayaraj, S. (1993), Neem in pest control: Progress and perspectives, in *Souvenir. World Neem Conference*, February 24-28, 1993, Bangalore, pp. 37-43.

Kadir, A.A., Ariffin, M., Shaari, K. and Heng, C.K. (1998), The distribution and azadirachtin content of Malaysian neem trees. *Journal of Tropical Forest Products*, **4**, 17-23.

Kandaswamy, A. and Raveendaran, N. (1988), Economic analysis of field experiments using neem products as plant protection agents. *Final Workshop of the IRRI-ADB-EWC Project on Botanical Pest Control in Rice-based Cropping Systems*. IRRI, Los Banos, Philippines.

Ketkar, A.Y. and Ketkar, C.M. (1995), Medicinal uses including pharmacology in Asia, in *The Neem Tree Azadirachta indica* A. Juss *and Other Meliaceous Plants*, ed. H. Schmutterer, VCH, Weinheim, Germany, pp. 518-525.

Ketkar, C.M. (1976), *Utilization of Neem (Azadirachta indica A. Juss) and Its Byproducts,* Khadi and Village Industries Commission, Bombay.

Ketkar, C.M. and Ketkar, M.S. (1993), Azadirachtin contents of neem and its byproducts from different parts of India, in *Abstracts World Neem Conference*, February 24-28, 1993, Bangalore, India, p.54.

Ketkar, C.M. and Ketkar, M.S. (1995), Neem seed crush and deoiled cake as manure and as nitrification inhibitors, in T*he Neem Tree Azadirachta indica* A. Juss. *and other Meliaceous Plants*, ed. H. Schmutterer, VCH, Weinheim, Germany, pp. 531-540.

Koundal, R., Khajuria, H.N., Dilawari, V.K., Chauhan, S. and Gera, V.K. (2003), Variation in the magnitude of azadirachtin content in different provenances of neem (*Azadirachta indica* A. Juss), Abstract *National Symposium Agroforestry in 21st Century*, PAU, Ludhiana, India, p. 70

Koul, O. (1996) Neem research and development : present and future scenario, in *Supplement to Cultivation and Utilization of Medicinal Plants*, eds. S. S. Handa and M. K. Koul , PID, CSIR, New Delhi, pp. 583-611.

Koul, O., Isman, M.B. and Ketkar, C.M. (1990), Properties and uses of neem, *Azadirachta indica* A. Juss, *Canadian Journal of Botany*, **68**, 1-11.

Kraus, W. (2002), Biologically active ingredients, in *The Neem Tree, 2nd Edition*, ed. H. Schmutterer, The Neem Foundation, Mumbai, India, pp. 39-111.

Kulkarni, M.D. and Bansod, R.S. (2001), Studies on herbal preparation against dermatitis in crossbred cattle, *Indian Veternary Journal*, **78**, 64-65.

Kumar, A.R.V., Jayappa, J. and Chandrashekara, K. (2000), Relative insecticidal value: An index for identifying neem trees with high insecticidal yield. *Current Science*, **79**, 1474-1478.

Lim, G.S. and Bottrell, D.G. (1994), *Neem Pesticides in Rice : Potential and Limitations*, International Rice Research Institute, Los Banos, Philippines.

Luo-Xiao, Dong, Wu-Shao, Hua, Ma-Yun, Bao, Wu-Da, Gang, Luo-X.D., Wu, -S.H., Ma, -Y.B. and Wu, -D.G. (2000), A new triterpenoid from *Azadirachta indica*, *Fitoterapia*, **71**, 668-672.

Mann, G.S., Dhaliwal, G.S. and Dhawan, A.K. (2001), Effect of alternate application of neem products and insecticides on population of *Bemisia tabaci* (Gennadius) and its impact on bollworm damage in upland cotton, *Annals of Plant Protectin Science*, **9**, 22-25.

Maramorosch, K. (1999), Neem for reforestation, in *Azadirachta indica* A. Juss., eds. R.P. Singh and R.C. Saxena, Science Publishers Inc., Enfield, USA, pp. 11-20.

Moorthy, P.N.K. and Kumar, N.K.K. (2000), Efficacy of neem seed kernel powder extracts on cabbage pests, *Pest Management and Horticultural Ecosystm*, **6**, 27-31.

Mruthyunjaya and Jha, D. (1996), Economics and policy issues, in *Neem*, eds. N.S. Randhawa and B.S. Parmar, New Age International (P) Ltd., New Delhi, pp. 311-317.

Nanda, U.K., Parija, B., Pradhan, W.C., Nanda, B. and Dash, D.D. (1996), Bioefficacy of neem derivatives against insect pest complex of rice, in *Neem and Environment*, Vol. 1, eds. R.P. Singh, M.S. Chari, A.K. Raheja and W. Kraus, Oxford & IBH Publishing Co. Pvt. Ltd., New Delhi, pp. 517-527.

Naqvi, S.N.H. (1996), Prospects and development of a neem based pesticide in Pakistan. *Proceedings Pakistan Congress Zoology*, **16**, 325-338.

Nimbole, N.N. and Dass, S.K. (1990), *Annual Report (1989-90)*, Central Research Institute for Dryland Agriculture, Hyderabad.

NRC (1992), *Neem : A Tree for Solving Global Problems*, National Academy Press, Washington, DC, USA.

Oo, L.T. (1987), *Neem Tree Research*, Terminal Report, Burmese-German Plant. Protection and Rodent Control Project, Rangoon, Tana Press, Bangkok, Thailand.

Palanisami (1992), *Economic Analysis of Botanical Pest Control in Rice-based Cropping Systems*, Social Sciences Division, International Rice Research Institute, Los Banos, Philippines.

Parmar, B.S. and Ketkar, C.M. (1996), Commercialization, in *Neem*, eds. N.S. Randhawa and B.S. Parmar, New Age International (P) Ltd., New Delhi, pp. 318-332.

Parmar, B.S. and Singh, R.P. (1993), *Neem in Agriculture*, Indian Agricultural Research Institute, New Delhi.

Pillai, N.R. and Santhakumari, G. (1981), Hypoglycemic activity of *Melia azadirachta* Linn. (Neem), *Indian Journa of Pharmacology*, **13,** 91.

Poddar, R.S., Devarnadagi, S.B. and Kiresur, V.R. (2000), Economic potential of growing multi-purpose tree species on denuded soils in semi-arid tropics, *Karnataka Journal of Agricultural Sciences*, **13**, 642-645.

Pradhan, S., Jotwani, M.G. and Rai, B.K. (1962), The neem seed deterrent to locusts, *Indian Farming*, **12**, 7-11.

Prakasa Rao, E.V.S. (1996), Potential use of urea coated with neem cake to improve nutrient uptake and N recoveries and to reduce N losses in a perennial aromatic grass, Java citronella (*Cymbopogon winterianus* Jowitt.), in *Neem and Environment*, Vol. 2, eds. R.P. Singh, M.S. Chari, A.K. Raheja and W. Kraus, Oxford & IBH Publishing Co. Pvt. Ltd., New Delhi, pp. 815-819.

Prasad, R., Devakumar, C. and Shivay, Y.S. (1996), Significance in increasing fertilizer nitrogen efficiency, in *Neem*, eds. N.S. Randhawa and B.S. Parmar, New Age International (P) Ltd., New Delhi, pp. 121-132.

Rajasekaran, R. (1991), *An Analysis of Neem Seed Industry*, Department of Agricultural Economics, Tamil Nadu Agricultural University, Coimbatore, India.

Rajamohan, H. (2002), Joint action of neem seed extarcts (*Azadirachtin indica* A. Juss.) and *Bacillus thuringiensis* subsp. *Kurstaki* against *Plutella xylostella* (Linn.), Ph.D. Thesis, Post-Graduate School, Indian Agricultural Research Institute, New Delhi, 62 pp.

Ramshe, D.G. (1989), Techniques for agroforestry in semi-arid tropics, *Indian Farming*, **39**, 26-28.

Rao, D.V.K., Singh, I., Chopra, P., Chhabra, P.C. and Ramanujala, G. (1986), *In vitro* anti-bacterial activity of neem oil. *Indian Journal of Medical Research*, **84**, 314-316.

Rembold, H. (1996), Neem and its natural development for pest control. in *Neem and Environment*, Vol. 1, eds. R.P. Singh, M.S. Chari, A.K. Raheja and W. Kraus, Oxford & IBH Publishing Co. Pvt. Ltd., New Delhi, pp. 1-10.

Rengasamy, S., Kaushik, N., Kumar, J., Koul, O. and Parmar, B.S. (1996), Azadirachtin content and bioactivity of some neem ecotypes of India, in *Neem and Environment*, Vol. 1, eds. R.P. Singh, M.S. Chari, A.K. Raheja and W. Kraus, Oxford & IBH Publishing Co. Pvt. Ltd., New Delhi, pp. 207-217.

Riar, S.S. (1996), Antifertility and other medical applications, in *Neem*, eds. N.S. Randhawa and B.S. Parmar, New Age International (P) Ltd., New Delhi, pp. 268-274.

Rosaiah, B. and Reddy, A.S. (1996), Utility of botanicals and biocides as additives to insecticides in *Helicoverpa armigera* Hubner management on cotton, in *Neem and Environment*, Vol. 2, eds. R.P. Singh, M.S. Chari, A.K. Raheja and W. Kraus, Oxford & IBH Publishing Co. Pvt. Ltd., New Delhi, pp. 1101-1107.

Sadre, N.L., Deshpande, V.Y., Mendulkar, K.N. and Nandal, D.H. (1983), Male antifertility activity of *Azadirachta indica* in different species, in *Natural Pesticides from the Neem Tree (Azadirachta indica A. Juss) and Other Tropical Plants*, eds. H. Schmutterer and K.R.S Ascher, GTZ, Eschborn, pp.473-482.

Saxena, R.C. (1989), Insecticides from neem, in *Insecticides of Plant Origin*, eds. J.T. Arnason, B.J.R. Philogene and P. Morand, American Chemical Society, Washington, DC, pp. 110-135.

Saxena, R.C. (1993), Scope of neem for developing countries. In: *Souvenir World Neem Conference*, February 24-28, 1993, Bangalore, pp. 30-36.

Saxena, R.C. (1998), "Green revolutions" without blues : Botanicals for pest management, in *Ecological Agriculture and Sustainable Development*, Vol. 2, eds. G.S.Dhaliwal, N.S. Randhawa, R. Arora and A.K. Dhawan, Indian Ecological Society and Centre for Research in Rural and Industrial Development, Chandigarh, India, pp. 111-127.

Schmutterer, H. (1995a), *The Neem Tree Azadirachta indica A. Juss. and other Meliaceous Plants : Sources of Unique Natural Products for Integrated Pest Management, Medicine, Industry and Other Purposes.* VCH Weinheim, Germany.

Schmutterer, H. (1995b), The tree and its characteristics, in *The Neem Tree Azadirachta indica A. Juss. and other Meliaceous Plants : Sources of Unique Natural Products for Integrated Pest Management, Medicine, Industry and Other Purposes*, ed. H. Schmutterer , VCH Weinheim, Germany, pp. 1-34.

Schmutterer, H. and Singh, R.P. (1995), List of insect pests susceptible to neem products, in *The Neem Tree Azadirachta indica A. Juss. and other Meliaceous Plants : Sources of Unique Natural Products for Integrated Pest Management, Medicine, Industry and Other Purposes*, ed. H. Schmutterer, VCH Weinheim, Germany, 325-326.

Shah, Z. and Faheem, A. (2000), Evaluation of indigenous plant materials for the nitrification inhibition property in soil. *Sarhad Journal of Agriculture,* **16**, 69-75.

Sharma, V.P. (1996), Malaria control, in *Neem*, eds. N.S. Randhawa and B.S. Parmar, New Age International (P) Ltd., New Delhi, pp. 283-289.

Singh, G., Gill, H.S., Abrol, I.P. and Singh, N.T. (1990), Agroforestry for improvement and management of salt affected soils, in *Proc. Int. Symp. Water Erosion, Sedimentation, Resources Conservation,* October 9-10, 1990, Dehra Dun, India, pp.420-428.

Singh, M. and Takkar, P.N. (1996), Influence of neem oil coating on prilled urea on volatilization loss from swell-shrink soil amended with organic matter under different moisture regimes, in *Neem and Environmen*, Vol. 2, eds. R.P. Singh, M.S. Chari, A.K. Raheja and W. Kraus, Oxford & IBH Publishing Co. Pvt. Ltd., New Delhi, pp. 851-854.

Singh, N., Misra, N., Singh, S.P. and Kohli, R.P. (1979), *Melia azadirachta* in some common disorders (a clinical evaluation), *Antiseptic*, **76**. 677-679.

Singh, N.K. and Sastry, M.S. (1988), Anti-microbial activity of neem oil, *Indian Journal of Pharmacology*, **13**, 102.

Singh, P. and Rai, P. (1989), Herbage production from grasslands under stressed environments in India. Paper presented at the *International Grassland Congress*, France.

Singh, R.P. (1987), Comparison of antifeedant efficacy and extract yields from different parts and ecotypes of neem (*Azadirachta indica* A. Juss) tree, in *Natural Pesticides from the Neem Tree (Azadirachta indica A. Juss) and Other Tropical Plants*, eds. H. Schmutterer and K.R.S Ascher, GTZ, Eschborn, pp. 185-194.

Singh, R.P. (1996), Bioactivity against insect pests, in *Neem*, eds. N.S. Randhawa and B.S. Parmar, New Age International (P) Ltd., New Delhi, pp. 146-159.

Singh, R.P. (2000), Botanicals in pest management. An ecological perspective, in *Pesticides and Environment*, eds. G.S. Dhaliwal and B. Singh, Commonwealth Publishers, New Delhi, pp. 279-343.

Singh, R.P. and Kataria, P.K. (1991), Insects, nematodes and fungi evaluated with neem (*Azadirachta indica* A. Juss). *Neem Newsletter*, **8**, 3-10.

Singh, R.P. and Raheja, A.K. (1996), Strategies in management of insect pests with neem (*Azadirachta indica* A. Juss), in *Neem and Environment,* Vol. 1, R.P. Singh, M.S. Chari, A.K. Raheja and W. Kraus, Oxford & IBH Publishing Co. Pvt. Ltd., New Delhi, pp. 103-120.

Singh, R.P., Singh, S. and Wahab, S. (1998), Biodiversity and importance of botanical pesticides, in *Ecological Agriculture and Sustainable Development.* Vol. 2, eds. G.S. Dhaliwal, N.S. Randhawa, R. Arora and A.K. Dhawan, Indian

Ecological Society and Centre for Research in Rural and Industrial Development, Chandigarh, pp.128-145.

Sinha, K.C., Riar, S.S., Tiwary, R.S., Dhawan, A.K., Bardhan, J., Thomas, P., Kain, A.K. and Jain, R.K. (1984), Neem oil as a vaginal contraceptive. *Indian Journal of Medical Research*, 79, 131-136.

Sombatsiri, K. (1996), Effects of the extracted substances from Thai neem seed, *Azadirachta indica* var. *siamenis*, on the insect pests of Chinese kale. In *Neem and Environment*, Vol. 1, eds. R.P. Singh, M.S. Chari, A.K. Raheja and W. Kraus, Oxford & IBH Publishing Co. Pvt. Ltd., New Delhi, pp. 565-569.

Sombatsiri, K., Ermel, K. and Schmutterer, H. (1995), Other meliaceous plants containing ingredients for integrated pest management and further purposes. *Azadirachta* spp. The Thai neem tree: *Azadirachta siamensis* (Val.), in T*he Neem Tree Azadirachta indica A. Juss. and Other Meliaceous Plants : Sources of Unique Natural Products for Integrated Pest Management, Medicine, Industry and Other Purposes*, ed. H. Schmutterer, VCH, Weinheim, Germany, pp. 585-597.

Sridharan, S. and Venugopal, M.S. (1998), Effect of environmental conditions on the yield of azadirachtin and oil in neem, in *Ecological Agriculture and Sustainable Development*, Vol. 1, eds. G.S. Dhaliwal, N.S. Randhawa, R. Arora and A.K. Dhawan, Indian Ecological Society and Centre for Research in Rural and Industrial Development, Chandigarh, pp. 510-518.

Srivastava, A.K. and Rama Mohan Rao, M.S. (1989), Agroforestry for sustained agricultural production. *Indian Farmimg*, October (1989), 33-35.

Subbiah, S. and Kothandaraman, G.V. (1980), Effect of neem cake blending of urea on ammonium and nitrate N in soils under field conditions. *Journal of Indian Society of Soil Sciences,* **28**, 136-137.

Talwar, G.P., Upadhyay, S., Garg, S., Kaushic, C., Kaur, R. and Dhawan, S. (1996), Induction of cell mediated immunity in genital tract, in *Neem*, eds. N.S. Randhawa and B.S. Parmar, New Age International (P) Ltd., New Delhi, pp. 275-282.

Tiwari, S.C. (1989), Effect of neem cake blending of urea on ammoniacal and nitrate content in vertisol, *Journal of Indian Society of Soil Sciences,* **37**, 830-831.

Tripp, R. and Ali, A. (2001), *Farmers' Access to Natural Pest Control Products : Experience from an IPM Project in India.* Agricultural Research and Extension Network Paper, Overseas Development Institute, London.

Upadhyay, S.N., Kaushic, C. and Talwar, G.P. (1990), Antifertility effects of neem (*Azadirachta indica*) oil by single intrauterine administration : A novel method for contraception. *Proceedings Royal Society London*, **B242**, 175-179.

Uraisakul-K., Kanok-Uraisakul and Oates, C.G. (1999), Effectiveness of some botanical extracts on vegetative growth of papaya and egg hatching inhibition of African red spider mite [*Eutetranychus africanus* (Tucker)], in *The 37th Kasetsart University Annual Conference*, February 3-5, 1999, pp.11-18.

Walia, S., Kumar, J. and Parmar, B.S. (2002), Ecologically sound botanical pesticides : Development and use, in *Biopesticides and Pest Management*, Vol. 1, eds. O. Koul, G.S. Dhaliwal, S.S. Marwaha and J.K. Arora, Campus Books International, New Delhi, pp. 56-63.

Chapter 6

RISK ASSESSMENT OF NEEM PRODUCTS: ENVIRONMENTAL IMPLICATIONS

JOHN D. STARK
Washington State University
Puyallup Research and Extension Center
Puyallup, Washington 98371, USA

1. INTRODUCTION

Pesticides from the neem tree have often been considered environmentally friendly, especially when compared to the synthetic neurotoxic insecticides that are widely used today. Even the U.S. Environmental Protection Agency designates neem pesticides as reduced risk products. However, does this mean that pesticides derived from the neem tree present no risk to the environment?

First of all, although azadirachtin is the chemical most often associated with the neem tree, there are many biologically active components found in the leaves and seed kernels of neem (Jacobson, 1989; Koul, 1992; Mordue (Luntz) and Blackwell, 1993; Schmutterer 1990, 1995, 2002; Stark and Walter, 1995). This is important when considering the potential risk of neem pesticides to ecosystems because the type and concentration of active ingredients can vary greatly among different neem formulations and this in turn can significantly alter toxicity (Stark and Walter, 1995). Risk assessment of neem pesticides is, therefore, more complex than that of a pesticide containing only one active ingredient. We are in fact dealing with a mixture of pesticides not a single active ingredient at least with some of the commercial formulations of neem pesticides.

Many studies have been published on the effects of pesticides derived from neem on nontarget organisms. It is not my intention to review the literature on the

O. Koul and S. Wahab (eds.), Neem: Today and in the New Millennium, 97–107.

toxicity of neem products to various organisms as this has been done several times in the past. For a comprehensive review of the side effects of neem pesticides on nontarget organisms see Schmutterer (1995). Instead, I will examine the risk that neem insecticides might pose to ecosystems after application to control pests.

Much of the emphasis on the effects of neem products on nontarget organisms has been placed on biological controls of pest species and for the most part, neem pesticides appear to be safer than synthetic neurotoxic insecticides to biocontrols (Schmutterer, 1990; Stark, 1991; Stark *et al.*, 1992; Lowery and Isman, 1994; Stark, 1997). However, even with biological controls, some studies indicate that neem insecticides can have negative effects (Lowrey and Isman, 1994; Banken and Stark, 1997,1998). In addition, the universal molting hormone of arthropods is ecdysone and because azadirachtin interferes with this hormone, effects on non-target arthropods such as crustaceans, spiders and predatory mites is a concern. Much less work has been done on aquatic organisms compared to terrestrial invertebrates.

2. TOXICITY OF NEEM IS DICTATED BY THE ACTIVE INGREDIENTS AND THE FORMULATION

As mentioned above, neem pesticides are mixtures of active ingredients even though azadirachtin is the major active ingredient in most of the commercial neem insecticides produced today. Even minor amounts of neem oil and limonoids other than azadirachtin can change the toxicity of a neem pesticide and, therefore, the potential environmental impact.

Stark and Walter (1995) examined the toxicity of three commercial neem insecticides, Margosan-O (W.R. Grace & Co., Columbia, Maryland, USA), Azatin (Agridyne, Salt Lake City, Utah, USA), RH-9999 (Rohm and Haas, Philadelphia, USA), and Neem oil (W.R. Grace & Co., Columbia, Maryland, USA) to the pea aphid. Margosan-O (MO) was found to be much more toxic than the other products and this appeared to be due to the fact that MO contained neem oil and the other products did not. Stark and Walter (1995) went on to evaluate the toxicity of MO, MO devoid of neem oil, Azatin (another neem formulation devoid of neem oil), RH-9999 (a product containing 22, 23-dihydro-azadirachtin), Azatin with 5 per cent neem oil, RH-9999 with 5 per cent neem oil and neem oil (5%). The addition of neem oil increased the efficacy of the neem insecticides that did not contain oil, while removal of neem oil from MO reduced its efficacy by 62 per cent. Extraction of neem oil with methanol and subsequently added back to MO devoid of neem oil resulted in a 30 per cent loss of efficacy. Addition of canola oil to MO devoid of neem oil gave a similar response (30% loss of efficacy compared to MO). Analysis of neem oil used in the study by Stark and Walter (1995) revealed that the following limonoids were present: nimbandiol, deacetylnimbin, 6-acetylnimbandiol, deacetylsalannin, nimbin, and salannin and two unidentified chemicals, believed to be limonoids. Stark and Walter (1995) concluded that neem oil and other oils increase the efficacy of neem insecticides, but that the small quantities of limonoids contained in neem oil also contributed to increased biological activity of neem insecticides. The importance of the findings of Stark and Walter (1995) is that the impact that neem pesticides might have on nontarget organisms and thus risk to these organisms will vary depending on the neem formulation.

Another issue concerning the impact of neem on nontarget organisms is whether components of neem formulations (surfactants) or the active ingredients found in neem formulations are responsible for toxicity, particularly to aquatic organisms (Schröder, 1992). Studies with *Daphnia magna* indicated that toxicity was primarily due to inert ingredients, not components from neem extracts (Saucke and Schmutterer, 1992). Kreutweizer *et al.* (1999) also found that the formulation contributed to some of the toxicity exhibited by Neemix. Stark (2001) found that approximately 50 per cent of the toxicity of the commercial formulation, Neemix® to *D. pulex* was due to the formulation and 50 per cent was due to neem components. So neem insecticides are toxic to *Daphnia* but the formulation is equally toxic. This means that the formulation of neem pesticides (inert ingredients) can contribute greatly to effects on nontarget organisms and thus risk.

3. PERSISTENCE OF AZADIRACHTIN

An important consideration in risk assessment is environmental persistence. The more persistent a pesticide, the more likely it is to cause damage depending on its toxicity.

Interestingly, azadirachtin is a fairly persistent chemical in water and soil compared to some of the other insecticides (Stark, 1997). Although, azadirachtin is as persistent as the carbamates and pyrethroids in water and soil, it is much less persistent than many other insecticides on plant foliage (Table 1).

4. RISK ASSESSMENT OF NEEM

Three papers dealing with risk/hazard assessment and neem have been published to date (Stark, 1997; Stark, 2001; Stark and Banks, 2001). In the earliest of these, Stark (1997) compared the risk of the commercial insecticide NeemAzal to various terrestrial and aquatic organisms using a hazard ratio method similar to the ratio method described above. In the second paper (Stark, 2001) the risk of Neemix and a Neemix formulation blank to *D. pulex* was compared. In the third paper Stark and Banks (2001) compared the risk of several new insecticides including Neemix to *D. pulex* (see Appendix 1 for a review of the risk assessment process). In the first study, the acute toxicity of a commercial neem insecticide, NeemAzal® and several other synthetic and natural insecticides to water fleas, *Daphnia* spp. were used for risk characterization. Exposure data were not readily available when this study was published and consequently the recommended field rates of pesticides were used for the exposure assessment. The justification here was that field rates reflect the amount of material that will be released relative to other products. Acute toxicity data indicated that NeemAzaL® was less toxic to *Daphnia* than all of the other classes of compounds, including other natural insecticides. By dividing the recommended field rate by the lethal concentration 50 (LC_{50}) (see Appendix 1 for a definition) a comparative measure of risk, a "Risk Index", was obtained. The higher

Table 1. Persistence (1/2 life in days) of azadirachtin and other insecticides in various environmental media.

Chemical	water	soil	foliage
Neem			
azadirachtin	<1-12 d	20d	<1 d
Organophosphates			
diazinon	180 d	14-28 d	2-14 d
malathion	21d	6 d	2 d
Carbamates			
carbaryl	1-32 d	7 d	3-10 d
methomyl	6 d	33 d	3-7 d
Pyrethroids			
esfenvalerate	4-15 d	15-45 d	ND
permethrin	2 d	21-42 d	3 d
Natural Insecticides			
rotenone	1-3 d	1-3 d	ND
avermectin	4 d	20-47 d	1.5 d

ND = data not available
Source: Stark 1997

the Risk Index number, the greater potential environmental risk of the product. Based on these criteria, NeemAzal® was found to pose the lowest risk of all of the products examined while the organophosphates had the highest risk (Stark, 1997).

In the next study, Stark (2001) evaluated the toxicity of three commercial neem insecticides, Neemix®, Azatin®, and the experimental insecticide, RH-9999 to the aquatic invertebrate, *D. pulex*. *D. pulex* was used as the study organism because it is recommended by the U.S. Environmental Protection Agency as an aquatic indicator species. Several toxicological endpoints, 48h acute mortality, 10d population growth (chronic) and No Observable Effect Concentrations (NOEC) for reproduction (chronic) were compared with a measure of environmental exposure developed for forest pest management, the estimated environmental concentration (EEC). Stark (2001) stopped short of developing an actual risk assessment. In addition, Stark (2001) and Stark and Banks (2001) evaluated different toxicological endpoints in their studies.

In the most recent study (Stark and Banks, 2001), Neemix® was compared to several other new insecticides and an older insecticide, diazinon. Risk was evaluated for each pesticide by developing extinction thresholds; mathematical projections based on chronic exposure and population growth rate (r). I have combined data from each of these studies for this paper and the combined risk assessment will be presented for the first time here (see below).

5. NEEM DOSE-RESPONSE STUDIES

Acute toxicity studies revealed that Neemix® and Azatin® were equitoxic with lethal concentration 50 estimates (LC_{50}) of 0.68 and 0.57 mg/l (ppm) while RH-9999 was significantly less toxic with an LC_{50} of 13 ppm (Stark, 2001). A population growth study (10d) was conducted for Neemix® and the formulation blank of Neemix® (Neemix® devoid of the active ingredients) to determine whether the active ingredients of Neemix® and/or components of the formulation were responsible for toxicity. Populations of *D. pulex* went to extinction after exposure to a Neemix® concentration of 0.45 mg/l azadirachtin (equivalent to the acute LC_7). The No Observable Effect Concentration (NOEC) and Lowest Observable Effect Concentration (LOEC) values for population growth were 0.045 and 0.15 ppm azadirachtin, respectively. The mean number of offspring per surviving female (*Ro*) declined in a concentration-dependent manner after exposure to Neemix® with no offspring being produced after exposure to 0.45 ppm. NOEC and LOEC values for reproduction were 0.045 and 0.15 ppm, respectively.

The formulation blank caused no mortality in the individuals used to start the population growth study but reduced reproduction and population growth accounting for 47% of the toxicity caused by Neemix® at a concentration of 0.15 ppm.

6. EXPOSURE ASSESSMENT

There is little information regarding the amounts of azadirachtin or other components of neem pesticides likely to be found in aquatic or terrestrial ecosystems. This is due to the fact that neem insecticides have not been widely used in the United States and there are no government regulations requiring monitoring for neem pesticides. Regulatory agencies that monitor water (U.S. Geological Survey) do not look for azadirachtin in their surveys. The Canadian Pest Management Regulatory Agency has developed an approach for determining the amount of a pesticide likely to end up in bodies of fresh water after application to forests. The Expected Environmental Concentration (EEC) is defined as the concentration of pesticide in 15 cm of water after a direct over spray of a forest at the maximum application rate (Anonymous, 1993). For Neemix®, the forest application rate is 50 g ai/ha (Kreutzweiser *et al.*, 1999, 2000) and the EEC is then estimated to be 0.035 mg/l. The EEC was used as the exposure component in the risk characterization below.

6.1 Risk Characterization

Dividing the estimated environmental concentration of 0.035 mg/l by the toxicological endpoints for Neemix, the following risk values were obtained:

6.1.1 Risk based on LC_{50}

Neemix Risk = 0.035 mg/l/0.68 mg/l = 0.05
Azatin Risk = 0.035 mg/l /0.57 mg/l = 0.06
RH-9999 Risk = 0.035 mg/l /13 mg/l = 0.003

6.1.2 Risk based on population growth

Neemix Risk = 0.035 mg/l /0.045 mg/l = 0.78

6.1.3 Risk based on NOEC for reproduction

Neemix Risk = 0.035 mg/l /0.045 mg/l = 0.78

6.1.4 Risk based on extinction threshold

Neemix Risk is 0.035 mg/l /0.015 mg/l = 2

6.2 Risk Assessment

All of these risk values were below 1 except the risk assessment based on extinction threshold. The majority of the evidence thus indicates that all three neem insecticides are unlikely to pose a risk to *D. pulex*. However, because azadirachtin is fairly persistent in water, caution should be used when neem pesticides are used near freshwater systems. Furthermore, only one species, *D. pulex* was evaluated in this risk assessment. Susceptibility of other organisms may be much greater or less than that of *D. pulex* this in turn would change the results of a risk assessment.

6.3 Field Studies

Although risk assessment is a valuable decision-making tool, it cannot substitute for field studies. There are several good aquatic field studies and at least one terrestrial field study on the impact of neem pesticides on non-target organisms. Some of the major findings of these studies are discussed below.

6.3.1 Aquatic field studies

Dunkel and Richards (1998) found that nontarget stream insects might be vulnerable to neem insecticides at the expected environmental concentration (EEC) of 0.035 mg/l. Furthermore, a field study conducted by Kreutweizer *et al.* (2000) indicated that a neem-based insecticide, Neemix®, caused significant changes to an aquatic community, but only at concentrations much higher than the EEC (0.035 mg/l). These conflicting results indicate that further research on the effects of neem should be conducted.

6.3.2 Terrestrial field studies

The impact of a commercial neem insecticide, Margosan-O® (MO) and the synthetic organophosphorous insecticide, chlorpyrifos had on invertebrates inhabiting a turf grass ecosystem was determined by Stark (1991). M-O had less of an impact on most of the invertebrates compared to chlorpyrifos. However, some groups of invertebrates (Oribatid mites) where actually more susceptible to neem than to chlorpyrifos. Other groups (Sminthurid and non-sminthurid collembola) were found to be less susceptible to MO than to chlorpyrifos but populations were significantly reduced compared to the control. MO had no significant effect on non-oribatid mites and spiders but chlorpyrifos significantly reduced populations of these organisms. Results of this study indicate that although MO is less detrimental in general than chlorpyrifos, it still caused damage to certain groups of terrestrial invertebrates.

7. CONCLUSIONS

The evidence to date suggests that insecticides derived from the neem tree are unlikely to cause substantial environmental damage and these products appear to be safer than synthetic neurotoxins. However, pesticides derived from neem are poisons and thus should be treated as such. Certain organisms are particularly sensitive to neem and this should be taken into consideration when contemplating their use.

REFERENCES

Anonymous (1993), *Registration Guidelines for Microbial Pest Control Agents,* Canadian Pest Management Regulatory Agency, Bulletin Pro 93-04, Ottawa, ON.

Banken, J.A.O.and Stark, J.D. (1997), Stage and age influence on the susceptibility of *Coccinella septempunctata* L. (Coleoptera: Cocinellidae) after direct exposure to the neem insecticide, Neemix, *Journal of Economic Entomology*, **90**, 1102-1105.

Banken, J.A.O. and Stark, J.D. (1998), Multiple routes of pesticide exposure and the risk of pesticides to biological controls: A study of neem and the seven-spot lady beetle, *Coccinella septempunctata* L., *Journal of Economic Entomology*, **91**, 1-6.

Dunkel, F.V. and Richards, D.C. (1998), Effects of an azadirachtin formulation on six nontarget aquatic macroinvertebrates, *Environmental Entomology*, **27**, 667-674.

Jacobson, M. (1989), *Focus on phytochemical pesticides*, Vol. 1, *The neem tree*, CRC Press, Boca Raton, Florida.

Koul, O. (1992), Neem allelochemicals and insect control, in *Allelopathy, Basic and Applied*, eds. S.J.H. Rizvi and V. Rizvi, Chapman & Hall, London, pp. 389-412.

Kreutzweiser, D.P., Capell, S.S. and Scarr, T.A. (1999), Acute lethal and sublethal effects of a neem-based insecticide on nontarget aquatic insects in stream channels, *Bulletin of Environmental Contamination and Toxicology*, **63**, 365-371.

Kreutzweiser, D.P., Capell, S.S.and Scarr, T.A. (2000), Community-level responses by stream insects to neem products containing azadirachtin, *Environmental Toxicology and Chemistry*, **19**, 855-861.

Lowery, D.T. and Isman, M.B. (1994), Effects of neem and azadirachtin on aphids and their natural enemies, in *Bioregulators for Crop Protection and Pest Control*, ed. P.A. Hedin, ACS Symposium Ser. 557, American Chemical Society, Washington, DC, pp. 78-91.

Mordue (Luntz), A.J. and Blackwell, A. (1993), Azadirachtin: An update. *Insect Physiology*, **39**, 903-924.

National Research Council (1983), *Risk assessment in the federal government: managing the process*, National Academy Press, Washington, D.C.

Saucke, V.H. and Schmutterer, H. (1992), Untersuchungen über Nebenwirkungen verschiedener Niemprodukte bei *Daphnia magna* Strauss (Crustacea: Cladocera), *Anz Schädlingskde Pflanzenschutz Umweltschutz,* **65,** 121-126.

Schmutterer, H. (1990), Properties and potential of natural pesticides from the neem tree, *Azadirachta indica*, *Annual Review of Entomology*, **35**, 271-297.

Schmutterer, H. 1995. *The neem tree : Azadirachta indica A. Juss. and other meliaceous plants: sources of unique natural products for integrated pest management, medicine, industry and other purposes*, VCH, Weinheim, Germany.

Schmutterer, H. (2002), *The Neem Tree, 2nd Edition*, The Neem Foundation, Mumbai, India.

Schröder, P. (1992), Neem Azal-F in the aquatic environment. in *Practice Oriented Results on the Use and Production of Neem Ingredients*, ed. H. Kleeburg, Druck & Graphic, Giessen, Germany, pp 109-121.

Stark, J.D. (1991), Comparison of the impact of Margosan-O and Chlorpyrifos on non-target invertebrates inhabiting turf grass, *Pesticide Science*, **36**, 293-299.

Stark, J.D. (1997), Risk assessment of neem insecticides: Persistence in the environment and potential impact on nontarget organisms, in *Biopesticides: Toxicity, Safety, Development and Proper Use*, Proceedings First International Symposium on Biopesticides, Chulalongkorn, University Press, Bangkok, Thailand, pp. 69-74.

Stark, J.D. (2001), Population-level effects of the neem insecticide, Neemix on *Daphnia pulex*, *Journal of Environmental Science and Health*, Part B , **36**, 457-465.

Stark, J.D. and Banks, J.E. (2001), "Selective" Pesticides": Are They Less Hazardous to the Environment? *BioScience,* **51**: 980-982.

Stark, J.D. and Walter, J.F. (1995), Persistence of azadirachtin A and B in soil: Effects of temperature and microbial activity, *Journal of Environmental Science and Health*, Part B, **30**, 685-698.

Stark, J.D., Wong, T.T.Y., Vargas, R.I. and Thalman, R.K. (1992), Survival, longevity, and reproduction of tephritid fruit fly parasitoids (Hymenoptera: Braconidae) reared from fruit flies exposed to azadirachtin, *Journal of Economic entomology*, **85**, 1125-1129.

APPENDIX 1

THE RISK ASSESSMENT PROCESS

According to the United States National Academy of Sciences, Risk Assessment is "the determination of the probability that an adverse effect will result from a defined exposure" (NRC 1983). Risk assessment is a combination of science and expert judgment whereby a risk assessor makes a decision based on the best available data. There are two major approaches to risk assessment commonly used today, deterministic and probabilistic risk assessment. Risk assessment based on a point estimate is called a deterministic risk assessment. The ratio method described below is an example of deterministic risk assessment. Deterministic risk assessments are based on a single estimate of exposure (usually the worst case scenario). As such they do not incorporate information about variability and uncertainty that may associate with a risk. However, a tiered decision-making progress is often utilized in deterministic risk assessment whereby a series of decisions are made based on the outcome of a previous result.

An assessment based on the probability of occurrence is called a probabilistic risk assessment. This method gives a measure of risk and the associated probabilities of their occurrence. Mathematical models are used to develop probabilistic risk assessment.

The process of risk assessment can vary among different agencies within a county and among countries. In the United States the same basic principles are usually followed and consist of four steps.

1. Hazard Identification

The first part of any risk assessment is to ascertain whether a chemical is hazardous. This is accomplished by gathering basic information about its toxicity.

2. Dose Response Assessment

The relationship between the dose or concentration and the incidence of adverse effects in exposed populations is developed in dose response assessment. The most widely used measure of toxicity is the Lethal Dose50 (LD50) or Lethal Concentration 50 (LC50). The LD50 is statistically derived measure of the dose-response relationship and is an estimate of the dose that causes 50% mortality (death, reproduction etc.) of a population of organisms. The difference between the LD50 and LC50 is that lethal doses are based on a known amount of toxicant per amount of body weight (mg toxicant/kg bodyweight) or amount of toxicant per animal (mg toxicant/animal). In both cases the amount of toxicant the organism receives is known. On the other hand, the LC50 is based on the amount of toxicant in an environmental medium such as water, soil or air (mg toxicant/liter of water) and the amount of toxicant that enters the organism is not known.

The No Observable Effect Concentration (NOEC) or Level (NOEL) is the highest concentration or dose where no effect is observed. This measure is often used in risk assessment and the endpoint of choice is reproduction. However, by selecting toxicological endpoints the question can be asked that what do we evaluate? Either acute and chronic exposure data are used in risk assessment. Acute data is generated for a short time period (24 hours or less) and after a single exposure. Continuous exposure to a chemical over many days to a lifetime results in chronic toxicity data. Mortality is often the endpoint of interest in acute studies while life span, reproduction, or weight-gain are often the endpoints evaluated in chronic toxicological studies.

3. Exposure Assessment

The amount of exposure and the duration of exposure are estimated in the exposure assessment part of risk assessment. The various routes of exposure are considered in exposure assessment. For example, application of a pesticide may result in exposure through contact with contaminated water, soil, air and/or food.

4. Risk Characterization

Risk characterization combines toxicity data and exposure assessment to arrive at probabilities of effects occurring. There are several approaches to risk characterization the simplest of which is the ratio method.

$$\textbf{Risk} = \frac{\textbf{estimated environmental concentration}}{\textbf{toxicological endpoint concentration}}$$

An example of the ratio method is: the LC_{50} for a particular species is 10 ppm and a pesticide is present at 20 ppm, then the most simple risk assessment would be determined as follows:

Risk = exposure/susceptibility - 20/10 = 2
Numbers greater than 1 indicate an unacceptable risk.

Chapter 7

NEEM BASED PRODUCTS: REGISTRATION REQUIREMENTS, REGULATORY PROCESSES AND GLOBAL IMPLICATIONS

HUBERTUS KLEEBERG
Trifolio-M GmbH, Sonnenstrasse 22, D-35633 Lahnau, Germany

1. INTRODUCTION

For the development of new pesticides from naturally occurring active ingredients from various biological sources have been of great importance for a long time now. Researchers have tried to obtain number of lead compounds or structures. Also due to the competitive situation of the large pesticide producing companies it seems inevitable and indispensable to patent and produce their own compounds from these natural substances. This could be achieved by modifying them in order to obtain higher/lower persistence, higher biological efficacy, lesser side effects, or just an advantage in the production and/or marketing of the compound.

The natural substances as such have not come into large use till date with a few exceptions (like pyrethrins, neem products, bacterial products, or some fungal products). Some locally used extracts, like rotenone or nicotine, are not widely used due to their disadvantageous toxicological and/or eco-toxicological properties.

However, the production process of the chemical pesticides as well as the handling of their precursors, intermediates and by-products is a problem in itself. Additionally residues of these substances frequently lead to discussions in the public due to neglect of the required waiting periods. These problems seem to be major reasons for the public call for "organic or biological plant protection" in many countries.

O. Koul and S. Wahab (eds.), Neem: Today and in the New Millennium, 109–123.

The loss of confidence by public in agricultural produce has resulted from different crisis concerning food safety (eg: BSE, Dioxine, FMD, GMOs, Nitrofen). The public in most countries does not accept the argument that intensive use of chemical pesticides is necessary for feeding an increasing world population. The new strategy of multinational companies to achieve this goal is through genetically modified organisms (GMOs). However, there is substantiated public suspicion in the broad scale agricultural application of genetically modified organisms too, since they may exhibit unexpected negative side effects.

The intrinsic dilemma of the multinational chemical as well as GMO-producing companies is the cost of the development of new products. Reliable data for the developmental cost of chemical pesticides had been published for Europe for example in 1996 (see Fig. 1). It is obvious that these data vary considerably from one final product to the other. But they show a general trend of increasing developmental cost over the years. Nowadays the developmental cost is of the order of 150 million Euro for one single product. This tremendous developmental investment is only affordable for very large markets, that means for the large cash crops like cotton, maize, wheat, rice, and potato. For small and minor crops specific and toxicologically as well as eco-toxicologically acceptable pesticides are already lacking in many cases.

This situation has lead to increased research activities in many countries for alternative biological products. In this case the term "biological" has a two-fold meaning:

- The product has to be of biological origin (i.e. a naturally occurring micro-organism or an extract from plants, micro-organisms or soil, or a mineral product which occurs usually in nature in abundance).

- The product has to be acceptable to organic or biological farming principles as laid down in the guidelines of the respective organisations (as for example in the EEC Guideline Nr. 2092/91 of 24. June 1991).

In USA the Environmental Protection Agency (EPA) has lead the responsibility for regulating the use of pesticides, under the Federal Insecticide, Fungicide and Rodenticide Act (FIFRA). As per EPA norms natural products generally fall into the category of biochemical and microbial pesticides and has specified test requirements for registration in USA in "guidelines for biorational pesticides" (subdivision M of CFR 158) (EPA, 1989). However, botanicals, although of natural origin, do not necessarily always have a non-toxic mode of action and may have to follow the stringent requirements of a conventional pesticide. The introduction of Food Quality Protection Act in 1996 in USA has not made any major impact on registration requirements for biopesticides though the act has favoured this category of pesticides under reduced risk policy and agreed to waiver some requirements and not to establish tolerances for many of the biopesticides (Neale, 2000). During the last ten years these activities have led to a number of excellent innovative and very specific products – usually for small and

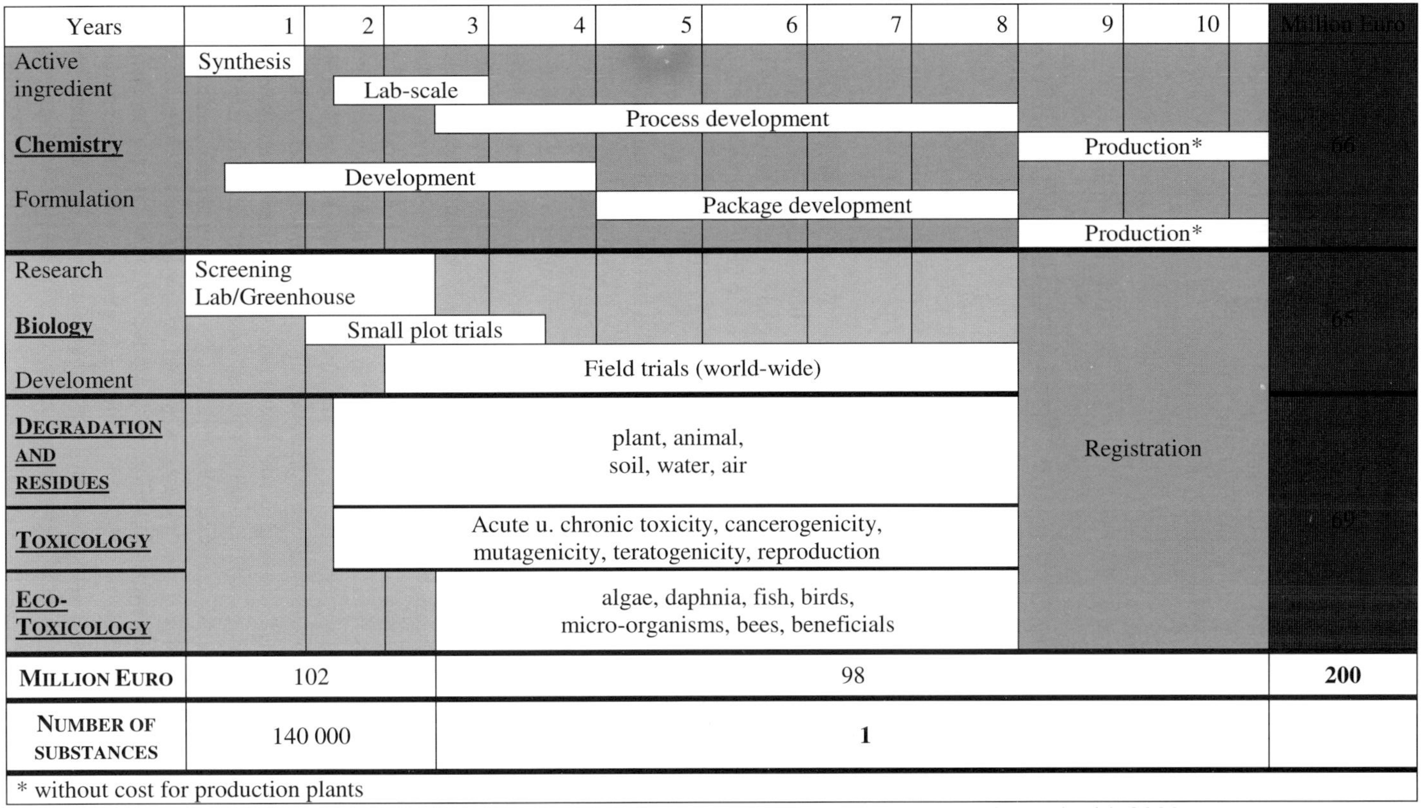

Figure 1. Developmental cost of plant protection products (After Schmid, 2003)

minor crops. This trend is understandable since the major crops are still covered commercially by a considerable number of not very expensive synthetic pesticides. As far as the economics of the development of these biological products is concerned it is clear that their developmental cost is "only" about 2 to 10 per cent of that of synthetic pesticides (i.e. of the order of several to a few ten million Euro). As it makes sense to offer and sell these products in small market, at present the developmental cost per unit volume or hectare is very considerable.

Among other possibilities the search for alternatives has led researchers to investigate azadirachtins, a group of limonoids contained in the seeds of the tropical neem tree, *Azadirachta indica* A. Juss.

2. THE NEEM TREE AND REGISTRATION REQUIREMENTS

Different parts of the tropical Neem tree *Azadirachta indica* A. Juss are used in India since times immemorial for curing many diseases (Ketkar and Ketkar, 1993; Koul, 1996)). In a holistic perception the protection of plants and animals against diseases and illness is a medical issue as well. The leaves and especially the seed kernels of the neem tree and their extracts have been used for the control of various insect pests in India. However, due to different reasons there is a demand for standardised natural products for plant protection today.

Insecticidally active neem-products are referred to internationally under the term "Azadirachtin" for the active ingredient. Since it is known that a group of limonoids called azadirachtins contribute to the biological activity of neem seed extracts this summarised term was chosen. The most abundant of these azadirachtins is azadirachtin A. The content of azadirachtin A and other azadirachtins varies with the quality of the neem seed kernels and especially with the extraction procedure of the product.

Recently the FAO has been active in documenting specification for neem-products. Since very different products are on the market the FAO has divided neem-products into two groups:

(i) Products containing neem-oil.
(ii) Azadirachtin-rich extracts free of neem-oil.

It is the aim of FAO-specifications to standardise chemical as well as biological plant protection products in order to facilitate the judgement of the authorities of all countries for their demand and quality of pesticides. In this specification process detailed information on different properties of the products as well as of the active ingredients have to be submitted, like:

- Physico-chemical properties
- Toxicology
- Ecotoxicology (including side effects on non-target organisms)
- Mode of action

- Residues (and their degradation) in agricultural produce, soil, water etc.
- Bioefficacy

Most authorities generally agree that one of the most crucial toxin of neem-products is aflatoxin. Thus it has generally been agreed that neem-products should not contain more than the permitted 4 μg total aflatoxins/kg in food. EU-countries are very strict about the requirements of the studies, which have to be submitted. However, biological products are supported and handled in different countries in different ways and there are no universal guidelines available.

In order to help companies with the registration of biological products in the UK the Pesticides Safety Directorate has a Small Business Champion - who is dedicated to helping small businesses through the UK regulatory process. What the Small Business Champion has been doing is helping the companies through the legislation and on a risk based approach and trying to remove the need to supply data packages where it is not necessary. The reason for this is that small companies in the UK who struggle with this complex area of legislation produce the majority of innovative 'green' products. In fact, small companies in the UK account for 99 per cent of UK businesses and represents 37 per cent of the UKs turnover. However, no neem-products have been registered in the UK till today.

In Italy under its interpretation of the Organic Directive 2092/91 does not require pesticides for use in organic agriculture to go through any registration. However, especially since it is unclear for how long this situation will continue most important companies register neem-products in Italy under the normal registration requirements.

In USA, according to "40 CFR ' 152.25(f))", minimum risk pesticides are exempted products. Products containing very different active ingredients are exempt from the requirements of FIFRA, alone or in combination with other substances. However, Azadirachtin does not fall into this category. Azadirachtin was exempted from tolerance according to "58 FR 8695: § 180.1119 Azadirachtin: exemption from the requirement of a tolerance". An exemption from the requirement of a tolerance has been established for the biochemical azadirachtin, which is isolated from the berries of the Neem tree (*Azadirachta indica*), when used as a pesticide at 20 grams or less per acre on all raw agircultural commodities.

In Australia the presence of negative side effects with respect to mammalian toxicity of some neem-products led to a controversial discussion of the registration authorities. As a consequence special toxicological studies and /or standardisation requirements need to be submitted for registration. Government of India approved in the year 2000 the simplified registration guidelines/data requirements for registration of biopesticides and biocontrol agents (Koul, personal communication) in consultation with the scientists, concerned associations and the prospective manufactureres.

However, in order to look into the neem based products for their registration requirements, regulatory processes and global implications, this chapter will discuss all these aspects taking NeemAzal as the product example which should give fair idea about the regulatory status required worldwide for neem based biopesticides.

3. WHAT IS NEEMAZAL®-T/S?

Our research has combined the experience of the thousands years of old Indian experience and modern demands for plant protection products. The result of our development is the formulation NeemAzal®-T/S. NeemAzal®-T/S is a formulation of the highly concentrated active ingredients of the Neem-tree, namely the Azadirachtins (Devakumar, 1993; Kraus, 2002). This concentrate – named "NeemAzal®" – contains on an average 34 per cent Azadirachtin A (Fig. 2), about 20 per cent of other Azadirachtins (Kleeberg, 2001) and 46 per cent of inert ingredients like lipids, oligosaccharides, and hydrate water. NeemAzal® has standardised composition, which is achieved by its unique extraction process. Consequently the properties of NeemAzal® and its formulations strongly depend on this composition. Thus quality control measures are a major issue for the production of formulations based on this active ingredient and its properties are not transferable to other Neem-products at all.

NeemAzal® is formulated with the help of surfactants (produced from renewable resources) and edible plant oils to obtain an emulsion concentrate (EC) containing 1 per cent Azadirachtin A. NeemAzal® is the registered trademark of Trifolio-M.

4. PHYSICO-CHEMISTRY AND DEGRADATION

The formulation has a shelf life of more than 2 years if stored below 20°C in a dark place. It forms a stable emulsion with water and spreads easily for example on leaf

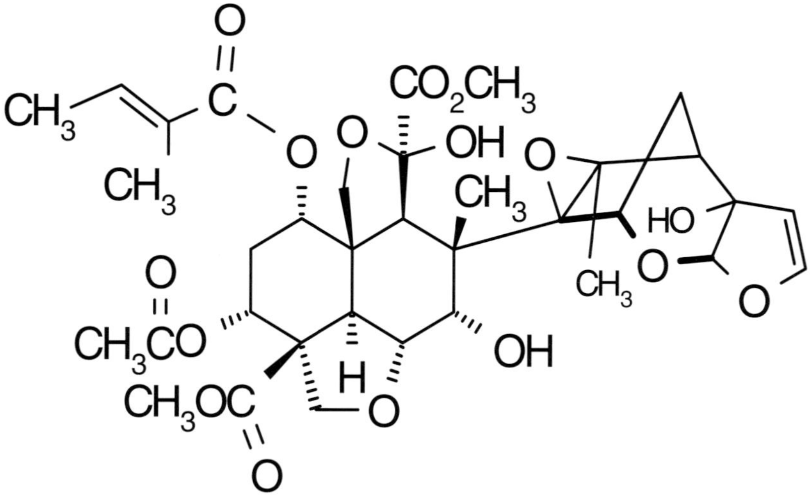

Figure 2. Structural formula of Azadirachtin A, the leading compound of NeemAzal

surfaces. An octanol-water distribution coefficient below 10 (Ruch *et al.*, 1997) indicates a low potential for accumulation in fatty tissue and hence in the food chain. Azadirachtins are not much adsorbed by soil and thus leach rapidly (Ruch *et al.* 1997). However, the degradation (especially microbiological) is very fast, so that a risk of contamination of ground water can be excluded (Ruch *et al.*, 1997). In water NeemAzal is transformed very rapidly by light (degradation half-life about 10 hours) (Ruch *et al.*, 1997; Troß *et al.*, 2000; Pussemier, 2000; Michalski, 2001).
After spray application to leaves and fruits Azadirachtin A is degraded rapidly with a half life of the order of very few days (Troß *et al.*, 2000, Ruch and Kleeberg, 2001).

5. TOXICOLOGY

NeemAzal and NeemAzal-T/S have been investigated thoroughly with respect to possible toxicological impacts to mammals. Neither acute nor subchronic or chronic studies indicate the presence of important risks for humans or mammals (Stewart, 1998, Niemann and Hilbig, 2000, Niemann, 2001) (see Table 1). This is especially established with respect to carcinogenicity, teratogenicity, reproduction etc. In this connection it is important to state that these "non-toxic" properties refer only to the concentrate NeemAzal and its standardised formulation and not to other "Neem-products" since these may have considerably different compositions.

Table 1. NeemAzal – Summary of most relevant toxicological data

Acute toxicity	
Rat LD_{50} oral	>5000 mg/kg bw
Rabbit LD_{50} dermal	>3000 mg/kg bw
Rat LC_{50} inhalation	>0.72 mg/l
Skin irritation	Very mild irritant
Eye irritation	Very mild irritant
Skin sensitization (test method used and result)	Sensitizer (M&K); **R 43**

Short term toxicity	
Target / critical effect	Liver and thyroid: organ weight↑; blood coagulation impaired; bw and food consumption↓
Lowest relevant oral NOAEL / NOEL	90 d oral, rat: 100 ppm (7.7 mg/kg bw/d)
Lowest relevant dermal NOAEL / NOEL	No study available
Lowest relevant inhalation NOAEL / NOEL	No study available

Genotoxicity	
	No evidence of a genotoxic potential

Long term toxicity and carcinogenicity	
Target/critical effect	Haematological changes; testis weight↓
Lowest relevant NOAEL / NOEL	18 mo, mouse: 100 ppm (ca. 10 mg/kg bw/d); NeemAzal-F tested
Carcinogenicity	No evidence of a carcinogenic potential

Reproductive toxicity	
Reproduction target / critical effect	No evidence of reproductive effects
Lowest relevant reproductive NOAEL / NOEL	5000 ppm (ca. 250 mg/kg bw/d) for reproduction and offspring, 200 ppm (ca. 10 mg/kg bw/d) for parental toxicity; NeemAzal-F tested
Developmental target / critical effect	No developmental toxicity or teratogenicity
Lowest relevant developmental NOAEL / NOEL	50 mg/g bw/d (rat, oral)

Neurotoxicity / Delayed neurotoxicity	
	No data since no evidence of a neurotoxic potential was found in other studies.

Medical data	
	Much human experience with other neem products; comparable NeemAzal-formulations successfully tested for scabies and head lice control in humans

Summary (proposal)	Value	Study	Safety factor
ADI	0.1 mg/kg bw	Multigeneration, rat; cancerogenicity, mouse; 90 d oral, rat	100
AOEL systemic	0.1 mg/kg bw/d	Multigeneration, rat; cancerogenicity, mouse; 90 d oral, rat	100
ArfD (acute reference dose)	No risk to consumers via acute residue exposure.		

6. ECOTOXICOLOGY

NeemAzal-T/S has been studied carefully with respect to possible side effects on the environment. Table 1 summarises results obtained for aquatic organisms. The high NOEC (No Observable Effect Concentrations) indicate an extremely low risk to aquatic organisms. This is true especially in view of the low concentrations of Azadirachtin A, which are necessary for efficient applications (i.e. of the order of 15 to 30 ppm Azadirachtin A in typical spraying solutions (see Table 2).
Obviously microbiological organisms degrade NeemAzal-T/S rapidly. This may lead to peculiar effects: for example the activation of the soil microfauna leads to an increased weight gain of earthworms after application of NeemAzal-T/S (Ruch *et al.*, 1997).
Beneficials are generally not influenced to a meaningful extent by NeemAzal-T/S applications (Forster, 2001) - with the exception of thin skinned species (like syrphids) (Hermann *et al.*, 1998).
Acute as well as reproduction studies with honey bees show (Leymann *et al.*, 2000) that no adverse effects may be considered after application of NeemAzal-T/S. Studies on chicken as well as field observations do not show any significant effects with respect to birds.

7. RESIDUES

The fast degradation of Azadirachtins on/in plants, the low amount of the active ingredient applied per hectare and the favourable toxicological properties indicated that even very short time after the application of NeemAzal-T/S residues cannot be considered a problem. Analytical investigations indicate that the concentration of residues depends on the surface area to mass ratio of the treated crops. Thus for example Azadirachtin A residues in/on leaves are significantly higher than on apple or tomato (Ruch and Kleeberg, 2001).

Table 2. Summary of the aqua toxicological results for NeemAzal and NeemAzal-T/S

Organism	**Test Substance**	**NOEC mg/l ***	**NOEC MgAzA/l****	**Time of Exposure**
Algae	NeemAzal-T/S	22	0,22	72 hours
	NeemAzal	144	49	72 hours
Daphnia				
Daphnia magna	NeemAzal	2,5***	0,74	21 days
Daphnia magna	NeemAzal-T/S	6,25 (reproductive output)	0,06	21 days
Fish:				
Rainbow trout	NeemAzal-T/S	100	1	96 hours
Freshwater carp	NeemAzal-T/S	100	1	96 hours
Zebra fish	NeemAzal	6,4***	1,9	1.5 life cycle (~ 7months)
Rainbow trout	NeemAzal-T/S	75	0,75	28 days

*NOEC: No Observable Effect Concentration
**NOEC-value of the test-substance converted to Azadirachtin A concentrations
*** highest concentration tested

8. MODE OF ACTION

After the treatment with NeemAzal-T/S larvae react with feeding and moulting inhibition and mortality; adult (beetle) show feeding inhibition, infertility and to a lesser degree mortality (Kleeberg, 1992; Otto, 1994; Hummel and Kleeberg, 1996; Hummel and Kleeberg, 1997; Schulz *et al.*, 1998).

As a result of this comparatively slow **"insectistatic"** mode of action of NeemAzal-T/S a final assessment of the treatment should be done 7-10 days after application under practical conditions. The number of dead pest insects is not necessarily a good evaluation criterion. For the assessment the following criteria are appropriate: loss of leaf mass, damage to plants, formation of honey dew, crop yield, development of the pest population, positive effects on beneficials (Kleeberg and Hummel, 1999).

Table 3. Time dependence of phenomena observed after treatment with NeemAzal-T/S

Phenomenon	**Timing**	**Description**	**Assessment**
Feeding inhibition	After hours	Reduced food consumption	Reduction of weight increase, plant damage, faeces and honey dew production
Inactivity	After days to 1-2 weeks	Over all reduction of fitness, moulting inhibition, starvation	Mortality
Fertility reduction	After weeks (next generation)	Reduction in progeny	Reduction in future population

The success of the application with NeemAzal-T/S depends on the progress of the pest infestation and adequate timing of the treatment.

In case of a temporary infestation and synchronous development of pest populations one application per generation or season is generally sufficient (under European climatic conditions). Usually one or two generations, for example: appearance of fundatrices of the Rosy Apple Aphid, *Dysaphis plantaginea,* first adults of Elder Bush Aphid, *Aphis sambuci* (Hom., Aphididae), first young larvae of Colorado Beetle, *Leptinotarsa decemlineata*, and beginning of flight of Cockchafers (*Melolontha* sp.).

In case of a permanent infestation (several generations like Aphids, Thrips, White Flies, Spider Mites etc.) repetitive applications are required. The interval between treatments is usually 7 to 14 days and depends on climatic conditions and infestation pressure.

NeemAzal-T/S is harmless to most beneficials - they are an important factor in the control of the remainder of the pest population. NeemAzal-T/S can favourably be combined with the use of beneficials in plant protection conceptions.

9. CONCLUSIONS

Currently the development of new means for plant protection has different motivations. Three major groups are apparent: synthetic chemicals, genetically modified products and biological products. The present scenario of regulatory situation in different countries is not very clear and comprehensively laid down; therefore, NeemAzal has been taken as a specific example.

An extract "NeemAzal" obtained from seed kernels of the Neem tree *Azadirachta indica* A. Juss and its formulation contains about 54 per cent azadirachtins. NeemAzal-T/S is a formulation of NeemAzal containing 1 per cent w/w of azadirachtin A. The formulation has a shelf life of more than 2 years below $20^{o}C$. An octanol-water distribution coefficient below 10 indicates a low potential for accumulation in fatty tissue and hence in the food chain. Azadirachtins are not much adsorbed by soil and thus leach rapidly. However, the degradation is very fast, so that a risk of contamination of ground water can be excluded. In water NeemAzal is transformed very rapidly by light (degradation half life about 10 hours). After spray application to leaves and fruits azadirachtin A is degraded rapidly with a half-life of the order of very few days.

NeemAzal and NeemAzal-T/S have been investigated thoroughly with respect to possible toxicological impacts to mammals. Acute, subchronic and chronic studies indicate the presence of no important risks for humans or mammals. In this connection it is important to state that these "non-toxic" properties refer only to the concentrate NeemAzal and its standardised formulation and not to other "Neem-products" since these may have considerably different compositions.

The results of studies of possible environmental impacts indicate an extremely low risk to aquatic organisms, micro-organisms, earth worms, beneficial insects, honey bees, birds etc. This is true especially in view of the low concentrations of Azadirachtin A, which are required for efficient control of pests. After the treatment with NeemAzal®-T/S larvae suffer feeding and moulting inhibition and mortality; adults show feeding inhibition, infertility and to a lesser degree, the mortality. This specific mode of action is called **"insectistatic".**

These studies with NeemAzal definitely imply that this and several other developments in neem-bsed pesticides have convinced registration authorities not only in Europe and Asia but in USA and Canada as well and Neem has been included among reduced-risk pesticides. That is why main opportunities are seen as arising from the discovery of new leads from high-throughput screening of plant extracts. It is hoped that international harmonized approach will come into force with a uniform set of rules to encourage the development of plant-based products for rational and sustainable agriculture. Of course, the lead from neem-based products now already exists and should be followed globally in order to develop safe and standardized products.

REFERENCES

Devakumar, C. and Dev, S. (1993), Chemistry, in *Neem Research and Development*, eds. N.S. Randhawa and B.S. Parmar, Society of Pesticide Science, India, New Delhi, pp. 63-96.

EPA (1989), *Subdivision M of the Pesticide Testing Guidelines: Microbial and Bio chemical Pest Control Agents*, US Environmental Protection Agency, USA.

Forster, R. (2001), Auswirkungen von niem – und pyrethrinhaltigen Pflanzen schutzmitteln auf den Naturhaushalt – Eine Übersicht über die im Rahmen des Zulassungsverfahrens verwendeten Unterlagen für die Prüfbereiche in der Terrestrik, in *Berichte aus der Biologischen Bundesanstalt für Land- und Forstwirtschaft*, Vol. 76, ed. Biologische Bundesanstalt für Land- und Forstwirtschaft, Saphir Verlag, Ribbesbüttel, Germany, pp. 62-65

Hermann, P., Zebitz, C.P.W. and Kienzle, J. (1998), Nebenwirkungen von NeemAzal-T/S auf Nützlinge, in *Biologische Pflanzenschutzverfahren im Erwerbsobstbau; Praxis, Beratung und Forschung im Gespräch*, eds. J. Kienzle und C.P.W. Zebitz, Universität Hohenheim, Germany, pp. 96-110.

Hummel, E. and Kleeberg, H. (1996), Wirkung NeemAzal-T/S auf die Wicken blattlaus *Megoura viciae* im Labor (1994), in *Practice Oriented Results on Use and Production of Neem-Ingredients and Pheromones*, eds. H. Kleeberg and V. Micheletti, Druck und Graphic, Giessen, Germany, pp. 51-56.

Hummel, E. and Kleeberg, H. (1997), Wirkung der Neem-Extraktformulierung NeemAzal-T/S auf die Grüne Erbsenblattlaus *Acyrthosiphon pisum* im Labor (1995), in *Practice Oriented Results on Use and Production of Neem-Ingredients and Pheromones*, eds. H. Kleeberg and C.P.W. Zebitz, Druck und Graphic, Giessen, Germany , pp. 33-39.

Ketkar, C.M. and Ketkar, M.S. (1993), Different Uses of Neem (*Azadirachta indica* A. Juss), in *Practice Oriented Results on Use and Production of Neem-Ingredients and Pheromones*, eds. H. Kleeberg , Druck und Graphic, Giessen, Germany, pp. 1-12.

Kleeberg, H. (1992), The NeemAzal Conception: Test of Systemic Activity, in *Practice Oriented Results on Use and Production of Neem-Ingredient*, ed. H. Kleeberg , Druck und Graphic, Giessen, Germany, pp. 5-16.

Kleeberg, H. (2001), Niem-Produkte: Herstellung, Standardisierung und Qualitätssi cherung, in *Berichte aus der Biologischen Bundesanstalt für Land- und Forstwirtschaft*, Vol. 76, ed. Biologische Bundesanstalt für Land- und Forstwirtschaft, Saphir Verlag, Ribbesbüttel, Germany , pp. 16-25.

Kleeberg, H. and Hummel, E. (1999), Experiences with NeemAzal-T/S in 1994 – 1998, *Med. Fac. Landbouw. Univ. Gent*, 64/3a, 305-310.

Koul, O. (1996), Neem research and development : present and future scenario, in *Supplement to Cultivation and Utilization of Medicinal Plants*, eds. S.S. Handa and M.K. Koul,. PID, CSIR, New Delhi, pp. 583-611.

Kraus, W. (2002), Biologically active ingredients, in *The Neem Tree*, ed. H. Schmutterer, Neem Foundation, Mumbai, India, pp. 39-111.

Leymann, B., Mühlen, W. and Edelmann, A. (2000), Auswirkungen von NeemAzal-T/S auf die Brut von Honigbienenvölkern (*Apis mellifera carnica* L., Hymenoptera, Apidae). Ergebnisse eines Halbfreilandversuches, in *Practice Oriented Results on Use and Production of Neem-Ingredients and Pheromones*, eds. H. Kleeberg and C.P.W. Zebitz, Druck und Graphic, Giessen, Germany, pp. 69-74.

Michalski B. (2001), Auswirkungen von niem- bzw. pyrethrinhaltigen Pflanzen schutzpräparaten auf den Naturhaushalt, , in *Berichte aus der Biologischen Bundesanstalt für Land- und Forstwirtschaft*, Vol. 76, ed. Biologische Bundesanstalt für Land- und Forstwirtschaft, Saphir Verlag, Ribbesbüttel, Germany, pp. 54-61.

Neale, M. (2000), The regulation of natural products as crop-protection agents, *Pest Management Science*, 56, 677-680.

Niemann, L. (2001), Toxikologische Bewertung von NeemAzal und Pyrethrum, in *Berichte aus der Biologischen Bundesanstalt für Land- und Forstwirtschaft*, Vol. 76, ed. Biologische Bundesanstalt für Land- und Forstwirtschaft, Saphir Verlag, Ribbesbüttel, Germany, pp. 68-76.

Niemann, L. and Hilbig, V. (2000), Die gesundheitliche Bewertung des Einsatzes von Naturstoffen im Pflanzenschutz am Beispiel von Neemkernextrakten, *Gesunde Pflanze*, 52, 135-141.

Otto, D. (1994), Systemic Effects of the Azadirachtin Preparation "NeemAzal-W" on Larvae and Adults of *Leptinotarsa decemlineata*, in *Practice Oriented Results on Use and Production of Neem-Ingredients and Pheromones*, ed. H. Kleeberg, Druck und Graphic, Giessen, Germany, pp. 39-53.

Pussemier, L. (2000), Environmental Behaviour and Aquatic Ecotoxicity of Azadi rachtinA, in: *Practice Oriented Results on Use and Production of Neem-Ingredients and Pheromones*, eds. H. Kleeberg and C.P.W. Zebitz, Druck und Graphic, Giessen, Germany, pp. 63-68.

Ruch, B. and Kleeberg, H. (2001), Abschätzung des Rückstandsverhaltens von NeemAzal-T/S aus Analysen der Leitsubstanz Azadirachtin A, in *Berichte aus der Biologischen Bundesanstalt für Land- und Forstwirtschaft*, Vol. 76, ed. Biologische Bundesanstalt für Land- und Forstwirtschaft, Saphir Verlag, Ribbesbüttel, Germany, pp. 84-87.

Ruch, B., Kliche-Spory, C., Schlicht, A., Schäfer, I., Kleeberg, J., Troß, R. and Kleeberg, H. (1997), Summary of Some Environmental Aspects of the Neem Ingredient NeemAzal and NeemAzal-T/S, in *Practice Oriented Results on Use and Production of Neem-Ingredients and Pheromones*, eds. H. Kleeberg and C.P.W. Zebitz, Druck und Graphic, Giessen, Germany, pp. 15-20.

Schmid, H. (2003), Forschungs- und Entwicklungskosten im Pflanzenschutz in *Industrieverband Agrar, Jahresbericht 2002/2003,* ed. Industrieverband Agrar e.V., Pressehaus Bintz-Verlag, Offenbach, Germany, pp. 21-23.

Schulz, C., Kienzle, J. and Zebitz, C.P.W. (1998), Wirkung von NeemAzal-T/S auf die Mehlige Apfelblattlaus, in *Biologische Pflanzenschutzverfahren im Erwerbsobstbau; Praxis, Beratung und Forschung im Gespräch*, eds. J. Kienzle und C.P.W. Zebitz, Universität Hohenheim, Germany, pp. 87-95.

Stewart, R.R. (1998), Toxicological Results of NeemAzal-technical and NeemAzal-formulations, in *Practice Oriented Results on Use and Production of Neem-Ingredients and Pheromones*, ed. H. Kleeberg,, Druck und Graphic, Giessen, Germany, pp. 21-26.

Troß, R., Ruch, B., Kliche-Spory, C. and Kleeberg, H. (2000), Abbauverhalten von Neem-Inhaltsstoffen in Pflanzen, Boden und Wasser, in *Practice Oriented Results on Use and Production of Neem-Ingredients and Pheromones*, eds. H. Kleeberg and C.P.W. Zebitz, Druck und Graphic, Giessen, Germany, pp. 109-112.

Chapter 8

NEEM VERSUS ENTOMOPATHOGENS AND NATURAL ENEMIES OF CROP PESTS: THE POTENTIAL IMPACT AND STRATEGIES

S. RAGURAMAN[1], N. GANAPATHY[1] AND T. VENKATESAN[2]
[1]*Tamil Nadu Agricultural University, Department of Agricultural Entomology, Agricultural College and Research Institute, Madurai 625 104, India*
[2] *Project Directorate of Biological Control, P.B. No. 2491, H.A. Farm Post, Hebbal, Bangalore-560 024, India*

1. INTRODUCTION

Agricultural crops are infested by an array of pests. Some of these pests are controllable by biological means with pathogens, parasitoids or predators. But to achieve a satisfactory level of control of pest complexes, selective use of pesticides is indispensable. In fact, biopesticides are a prerequisite in integrated pest management (IPM). In this overview an attempt has been made to find out whether neem products, home made or commercial, are selective for use in IPM. In the past, it was not well established that neem products were safer to natural enemies of pests, however, in last few years some detailed results have been published leading to a better assessment of this issue (Schmutterer, 1995,1997, 1999; Singh and Saxena, 1999; Koul *et al.*, 2002a,b)

2. NEEM AND ENTOMOPATHOGENS

2.1 Nuclear Polyhedrosis Viruses (NPV)

The compatibility of neem products with NPV on different crop pests has been studied (Table 1). Shapiro *et al.* (1994) reported that the neem seed extract (NSE) containing 2300 ppm of azadirachtin when applied in a dilution of 0.12-1.0 per cent (v/v) had little effect on the lethal infection by a polyhedrosis virus of the gypsy moth, Lymantria dispar with low LC_{50} values. Applications of a mixture of the virus

O. Koul and S. Wahab (eds.), Neem: Today and in the New Millennium, 125–182

Table 1. Compatibility of neem with NPV

	Insect	Concentration	Observation	Reference
1	*Lymantria dispar*	Neem seed extract containing 2300 ppm with NPV	Positive effect as compared to the use of virus alone	Shapiro *et al.* (1994)
2	*Spodoptera litura*	Neem bitter (0.1 %) with NPV ($5x10^5$ POBs/ml) and crude sugar (1.0 %)	Positive effect	Rabindra *et al.* (1997)
3	*S. litura*	SlNPV $2x10^4$ POBs/ml+ neem oil 1 % and neem cake extract 5%	Positive effect	Baskaran *et al.* (1999)
4	*Helicoverpa armigera*	HaNPV (500 LE/ha) + NSKE 5%	Positive effect	Sarode *et al.* (1997)
5	*Thysanoplusia* sp.	Neemark (0.25 %) with NPV (150 LE/ha)	Positive effect	Men and Thakare (1998)
6	*H. armigera*	Annona 36 EC + neem 36 EC	Positive effect	Das *et al.*, (2000)
7	*H. armigera*	Nimbecidine 2 % with NPV (250 LE/ha)	Positive effect	Reddy and Manjunath (2000)

and NSE, however, enhanced virus-caused mortality of the treated larvae compared to the use of the virus alone. Most of the virus-based studies have been conducted in India. Neem seed kernel extract (NSKE) and NPV were used for the control of H. armigera on cotton in Akola, India, during 1992-93. NSKE was applied at 3, 4, 5 and 6 per cent and the virus at 200, 300, 400 and 500 LE/ha. The treatments were applied three times at 15d intervals starting at square formation. The lowest infestation of 2.4 per cent on green fruiting bodies was recorded with 500 LE virus/ha, 7 days after treatment. NSKE (6%) showed 3.8 per cent and 400 LE virus /ha about 4.2 per cent infestation. NPV at 500 and 400 LE/ha and NSKE 6 per cent when used individually recorded minimum infestations of 6.91, 7.28 and 7.64 per cent, respectively on open bolls. Maximum yield of 8.14 q/ha was obtained from plots treated with 6 per cent NSKE , followed by 7.26 and 7.25 q/ha, respectively in viral treatments (Sarode *et al.*, 1996). In a combination study of *Bacillus thuringiensis* (Bt), neem seed extract, NPV, and half and full concentrations of the recommended insecticide against cotton crop, neem seed extract + a half concentration of insecticide/NPV also increased the seed yield (Patil and Sarode,

1996). Microbial pesticides alone neither gave adequate crop protection nor did they produce the desired yield of cotton.

Additive effects of various combinations with neem seem to be very effective. Sesame oil as synergist, diflubenzuron as insect growth regulator, NPV as biocide, neem oil as repellent and *Parthenium* leaf extract as botanical insecticide have been used against *Helicoverpa armigera* larvae (Rajasekhar *et al.*, 1996) in combination with endosulfan and cypermethrin. While sesame oil (0.5%) reduced the *H. armigera* infestation, diflubenzuron, NPV and neem oil enhanced the action of endosulfan and cypermethrin. Treatment with sesame oil as an additive gave the highest cottonseed yield. Cook *et al.* (1996) have shown that second-instar larvae of *Lymantria dispar* on semi-synthetic diet and white oak (*Quercus alba*) seedlings that had been surface-treated with azadirachtin and L. dispar NPV did affect larval development (weight gain and moulting) and survival. When consumed together, larvae died earlier as compared to those consuming only azadirachtin or virus. The combination also resulted in reduced larval survival compared to individual consumption. According to Murugan and Jeyabalan (1998), NSKE 2.5 per cent enhanced the activity of NPV at 10^2 POBs/ml and *Bacillus thuringiensis kurstaki* (Btk) at 50 mg/ml against *H. armigera* on cotton leaves, suppressing consumption, fecundity and survival. Shukla *et al.* (1998) evaluated NSKE, neem oil and Neemark along with NPV and synthetic insecticides at different concentrations against *H. armigera*. The neem products and NPV were similar in their efficacy as larval population per 10 plants ranged between 13.33 to 21.00 larvae. However, synthetic insecticides were at par where larval population of 4.3 (endosulfan, 0.07%) to 6.0 larvae (cypermethrin, 0.002%) was recorded. Significantly low pod damage was observed in neem products and was similar to synthetic insecticide treatments. Similarly, a high mortality of *Spodoptera litura* larvae treated with neem bitters (0.1%) was achieved when combined with NPV ($5x10^5$ POBs/ml) and crude sugar (1.0%). Without crude sugar, the NPV-neem bitter combination was not effective in increasing mortality. The larval weight and growth rates were significantly reduced in the NPV-neem combination (Rabindra *et al.*, 1997).

Increasing susceptibility to NPV by synergistic additives such as fluorescent brighteners, granulosis virus (GV) enhancin protein and neem extract against various lepidopterans are known. Increase in potency of up to 3.4 fold and 2.9 fold were demonstrated with *Trichoplusia ni* GV enhancin and neem extract (NeemAzal-T), respectively. Triazophos (0.05%) and methomyl (0.05%) mixed with either diflubenzuron (0.025%), S. litura NPV (SlNPV) at 250 LE/ha or neem oil (1.0%) affect predatory coccinellids of *S. litura* on groundnut (Bhanukiran *et al.*, 1997). High populations of coccinellids were observed in the untreated control, neem and NPV treatments and their combinations followed by diflubenzuron (0.025%). A mixture of conventional insecticides at half the recommended concentration with neem or NPV had little effect on predatory coccinellids.

Field studies on pigeon pea by Sarode *et al.* (1997) showed that NPV and NSKE were more effective when applied as combinations than when applied alone. HaNPV at 500 LE / ha + NSKE 5 per cent recorded the highest control at 7 and 14 days after spraying (66.5 and 55.7 %, respectively) with a highest grain yield (1110 kg/ha). Similarly, the bioefficacy of NPV in combination with neem formulations against *S. litura* was superior (96% larval kill) to the individual treatments of neem formulation, Nimba 0.1% (40% kill) or NPV (60% kill). Pigeonpea infested with *H. armigera* have also been treated with three sequential sprays of NPV (250 LE/ha), cypermethrin (0.01%) and NSKE (5%) applied at 15-day intervals starting from 50 per cent flowering (Balikai et al., 1997). The lowest pod damage of 39.8 per cent and highest seed yield of 10.2 q/ha were recorded. Sequential treatment with cypermethrin (0.01%), Bellary jali (*Prosopis juliflora*) leaf extract (5%) and NPV (250 LE/ha) did not give an acceptable level of control. Some recent studies with pigeonpea (Das *et al.*, 2000) compared the efficacy of five botanical pesticides (Annona 72 EC, Annona 36 EC + Neem 36 EC, 1500 ppm solvent based neem seed kernel extract, 300 ppm neem and 1500 ppm oil based azadirachtin) and two biocontrol agents, NPV and Btk, with endosulfan. Although, endosulfan induced maximum toxicity in *H. armigera* larvae, yet azadirachtin and Annona + neem were the best in first and second seasons, respectively. Annona + neem treatment recorded the highest grain yield (137.9 g/m^2) too.

Sonalkar *et al.* (1998) evaluated adjuvants such as Ranipal, Indigo, Sandovit, urea, neem seed extract, cotton seed extract and *Vitex negundo* leaf extract in increasing the virulence of NPV against second instar larvae of *H. armigera*. Ranipal (0.5%), indigo (0.2%) and urea (1.0%) were effective in increasing the larval mortality and reducing lethal time (LT_{50}) to 112, 124 and 149 h, respectively, compared to virus alone (175.25 h). Similarly eight treatments comprising of HaNPV, Dipel (Btk), neem seed extract and endosulfan alone or in a combination of two products against *H. armigera* in chickpea were effective in reducing larval populations, pod damage caused by this pest and besides increased grain yields compared to untreated control (Wanjari *et al.*, 1998). Btk alternated with endosulfan proved to be the best. Kumawat and Jheeba (1999) also determined the efficacy of NPV and Btk against *H. armigera* on chickpea.

In field trials in Maharashtra, India (kharif 1990-91 and 1991-92), NPV, Bt and Neemark were tested alone and in combination with endosulfan against *Thysanoplusia orichalcea* on sunflower. Consistently better results were obtained with endosulfan 0.06 per cent and combinations of a sub-lethal dose of endosulfan (0.03%) with a reduced dose of either Bt (250 ml/ha) or NPV (150 LE/ha). Bt (500 ml/ha), NPV (300 LE/ha) and Neemark (0.25%) alone or either Bt (250 ml/ha) or NPV (150 LE/ha) in combinations also yielded very significant results (Men and Thakare, 1998).

Baskaran *et al.* (1999) studied the control of *S. litura* with NPV alone at 10^8 POBs/ml and in combination at 2 x 10^4 POBs/ml (LC_{50} of SlNPV) with neem oil (NO) 1 per cent, neem seed kernel suspension (NSKS) 5 per cent and neem cake extract 5 per cent. The results indicate an increase in efficacy of NPV and the virus-induced mortality level was enhanced from 31.1 per cent to 81.5 per cent and 57.6 per cent, respectively. The combination of NPV with *Vitex negundo* 10 per cent did not increase the efficacy of NPV appreciably, while *Prosopis juliflora* and *Ipomoea carnea* had an antagonistic effect on the virus. NPV along with neem products reduced the leaf damage by larvae and LT_{50} of the virus. Suspending NPV in different concentrations (0.1-1.0 %) of neem oil and NSKS reduced the LC_{50} of NPV by 1.06 to 1.43 fold and 1.03 to 1.33-fold, respectively.

Murugan *et al.* (1999) evaluated the effects of NSKE, NO and the SlNPV on the mortality, feeding and development of *S. litura* in the laboratory. Concentration-dependent mortality occurred after NPV treatment that was relative to larval age at the time of treatment. Upon combining NPV with neem products, the mortality level was enhanced three-fold even at reduced concentrations. NSKE and NO significantly inhibited the growth and development of *S. litura* and extended the duration of the larval and pupal stages. The oviposition period, adult longevity and fecundity were significantly reduced following treatment with neem products and with a neem + NPV treatment. The efficiency of conversion of ingested food (ECI) and digested food (ECD) were also significantly decreased with NSKE, NO and NPV treatment, either alone or in combination.

Combinations of Nimbecidine 2 per cent + NPV at 250 LE/ha and Dipel 81 + NPV @250 LE/ha were the most effective treatments against *H. armigera* (Reddy and Manjunatha, 2000). The IPM components (*Trichogramma chilonis, Chrysoperla carnea*, NPV, Nimbecidine, Dipel and synthetic chemicals) were used at different intervals on the basis of pheromone trap threshold level (7 moths/trap per night) in an area of 40 ha cotton (MCU-1) fields at two locations. The results demonstrated the superiority of IPM strategy in terms of both cost versus benefit and environmental safety over that of farmers' practices where conventional control methods alone were followed. Potential of NPV (crude and formulated), endosulfan and neem products to control *H. armigera* on tomato cv. Sel-7 under field conditions revealed that treatments sprayed 15 days after flowering which included crude NPV (200 or 300 LE/ha), formulated NPV (100 LE/ha), endosulfan + crude NPV (350 g a.i./ha + 100 or 200 LE/ha), formulated NPV + Econeem (100 LE + 2.5 ml/litre), formulated NPV + Neemgold (100 LE + 2.5 ml/litre) and endosulfan (700 g a.i./ha) were very effective (Satpathy and Rai, 2000) in controlling *H. armigera* populations.

Bacteria

Neem products have also shown compatibility with *Bacillus thuringiensis* (Bt) (Table 2). Residual toxicity of 0.25-1.0 per cent suspensions of Thuricide HP (Bt) and aqueous extracts of the roots or shoots of *Eclipta erecta* and leaves of neem applied to potted rice seedlings against adults of *Nilaparvata lugens* has been recorded. The data showed that all the Thuricide HP treatments gave 100 per cent mortality with plant products after 48 h (Rao and Rao, 979).

Similarly, Hellpap and Zebitz (1986) found in a laboratory study that 4- and 10-day-old larvae of *Spodoptera frugiperda* when fed for 1-3 days on castor leaves treated with either NSKE, Dipel (Btk) or a mixture of both or when fourth instar larvae of *Aedes togoi* were kept in water treated with a neem seed kernel extract, *B. thuringiensis* var. *israeliensis* (Teknar) or a mixture of both until adult emergence; there was increased mortality in combinations compared to individual treatments. The combination also caused reduction in the pupal weights of the

surviving test insects and reduced the time required to kill 50 per cent of larvae of *S. frugiperda.* In the tests with *A. togoi*, more pupae than larvae were killed by the combination than by each component alone. The combinations had an additive effect
in most cases, but in some, a synergistic effect was observed. However, Moar and Trumble (1987) observed antagonistic effect when NSKE and Bt were applied against *Spodoptera exigua*.

Table 2. Compatibility of neem with *Bacillus thuringiensis*

	Insect	Concentration	Observation	Reference
1	*Spodoptera frugiperda*	Dipel (Btk) with NSKS	Positive effect	Hellpap and Zebitz (1986)
2	*Plutella xylostella*	NSKS with Bt	No additive effect	Krisch (1986)
3	*Leptinotarsa decemlineata*	Mixture of NeemAzal-S and low concentation of Bt Tenebrionis	Negative effect	Tillman (1992)
4	*S. litura*	NSKS (1%) + Bt 1g/l	No additive effect	Joshi *et al.*(1993)
5	*H. armigera, Earias insulana, E. vitella, Pectinophora gossypiella, Amrasca devastans, Bemisia tabaci*	Neemrich-I, Neemrich-II, Neemguard and Nimbecidine either alone or in combination with Bt or conventional synthetic pyrethroids	Positive effect	Gupta (1996)
6	*L. decemlineata*	Sublethal concentration of neem (0.45 or 0. 25 mg (a.i.)/litre with sublethal concentration of Bt (0.74 mg/a.i./litre)	Positive effect	Trisyono and Whalon (1999)

Monocrotophos (0.04%) and aqueous neem extracts (5 and 10%) and ether extracts of neem leaves and kernels (2%), extracts of *Ipomoea* and *Lantana*, other neem preparations and Bt were compared for their effectiveness against *H. armigera*.

Aqueous neem extract 10 per cent gave comparable damage reduction to quinalphos 0.04 per cent, and was recommended as an economically viable alternative (Thakare *et al.*, 1992).

Joshi *et al.* (1993) tested relative efficacy of botanicals and biocides alone and in combination against *S. litura* in tobacco nursery. The treatments were neem seed kernel suspension (NSKS) 2 per cent, B.t. (Bactospeine) (16 x 10^3 spores / mg, 2 and 1 g / l), *Pongamia* cake water (PCW) extract 1:5 (w/v), NPV 250 LE/ha, NSB (neem seed bitters) 0.25 and 0.12 per cent, NSKS 1 per cent + B.t. 1 g/l, NSKS 1 per cent + NPV 125 LE/ha, NSB (0.12%) + B.t. 1 g/l and untreated beds as control. It was observed that the two botanicals NSKS and PCW extracts were superior in protecting nursery than that of Bt and NPV. The NSKS combined well with both the biocides but mixing the two had no special advantage. Similar observations were earlier made by Krisch (1986) on *Plutella xylostella*. It was also found that there was no difference between two concentrations of NSB though higher concentrations had similar effect as NSKS, combination of NSB with Bt and combination of PCW with Bt Combined analysis revealed that botanicals alone were more effective than biocides alone. Combined treatments of NSKS with either Bt or NPV were as effective as botanicals alone. However, seven days after spraying the two botanicals continued to be more effective than Bt and NPV.

Latha *et al.* (1994) evaluated the efficacy of monocrotophos (1.0 1/ha), buprofezin (1.6 kg/ha), Btk (2.5 1/ha) and NSKE (5%) as repeated sprays either alone or in alternation against *Cnaphalocrocis medinalis* and *Marasmia patnalis*. Spray application of monocrotophos thrice at 53, 61 and 85 days after transplanting suppressed insect damage below the economic threshold of 10 per cent infestation at the vegetative growth stage and 5 per cent infestation at flowering stage. Low predation and parasitism was observed in plots treated with monocrotophos and buprofezin. Foliar application of NSKE was the safest and at par with untreated control, whereas Bt was safe to predators and parasitoids.

Kroschel and Koch (1996) tested various chemicals like fenvalerate, diflubenzuron, fenozycarb, teflubenzuron, seed extracts of chinaberry (*Melia azedarach*), neem, garlic (*Allium sativum*) and Bt (Bt mixed with fine sand and dusted) to determine their potential to control *Phthorimaea operculella*. In bioassay on tubers, products were tested either when the potatoes were inoculated with eggs of the pest after treatment, or when the larvae were already in the tubers. Fenvalerate, diflubenbzuron and Bt prevented the development of the larvae once eggs had hatched. Fenoxycarb arrested development to the adult stage, but 50 per cent of the eggs developed to the larval and pupal stages and larvae caused damage to the tubers. The effectiveness of the water extract of neem was 93.8 per cent and even sand without Bt was 94.6 percent effective. The action of sand was attributable to the high proportion of quartz, which caused damage to the very sensitive cuticle of newly hatched larvae.

The bioefficacy of neem products and their possible combinations with synthetic insecticides for the management of pest complex in cotton viz., *Amrasca devastans, Bemisia tabaci, Earias vittella, E. insulana, Pectinophora gossypiella* and *H. armigera* show that the neem products, Neemrich-I (Margocide OK),

Neemrich-II (Margocide CK), Neemguard and Nimbecidine during three year period
alone or in combination with Bt or conventional synthetic pyrethroids were very effective in managing pest complex in cotton and increasing the yield (Gupta, 1996). The seed cotton yield in such a schedule was more (1905 kg/ha and 2476 kg/ha) as compared to recommended schedule (1861 kg/ha and 2405 kg/ha). These findings show that neem products can replace conventional insecticides in a spray schedule thereby minimize the exposure of environment to toxic pesticides and will also avoid the development of resistance. Schmutterer (1990) recommended the use of neem products alternatively or mixed with other products to prevent or at least postpone the development of possible resistance.

Basedow *et al.* (1997) conducted field experiments during 1995-96 in potato fields in Germany on the biological control of *L. decemlineata*. The effect of treatments was measured by the degree of defoliation. Double treatments NeemAzal-T/S (2 l/ha) or *B. thuringiensis* subsp. *tenebrionis* (3 l/ha as Novodor) significantly increased potato yields, while NeemAzal-T/S increased crop yields in 1996. The economic output was the highest in the NeemAzal-treatment (net gain DM 4450/ha in 1996). Similarly, Trisyono and Whalon (1999) examined the toxicity of neem (Neemix, 0.25% with azadirachtin as a.i.) and combinations of neem and Bt on second instar larvae of Bt susceptible (Bt-S) and resistant (Bt-R) strains of same insect species. Using leaf-dip method, the LC_{50} values of neem determined two days post treatment to larvae of the Bt-S and Bt-R were 2.07 and 6.56 mg a.i./l, respectively. The LC_{50} values in both strains decreased significantly with increased exposure time. Cross-resistance between these toxins was not evident with a resistance ratio ranging from 1.3 to 3.2. Combinations of sublethal concentrations of neem (0.45 or 0.25 mg a.i./l) with a sublethal concentration of Bt (0.74 mg a.i./l) to larvae of the Bt-S yielded an additive effect in larval mortality. In contrast, combinations of neem (0.78 or 0.43 mg a.i./l) and Bt (319.8 mg a.i./l) resulted in a synergistic effect against larvae of the Bt-R strain. Sublethal concentrations of neem or Bt applied separately or in combinations decreased the mean larval weight and retarded the larval growth of both strains. These results suggest that neem is a potential insect growth regulator for larvae of the Bt-S and Bt-R strains and azadirachtin is possibly Bt resistance-breaking compound. Surprisingly, on the contrary mixtures of NeemAzal-S and low concentrations of Btt (B. thuringiensis var. tenebrionis) gave antagonistic effects in laboratory trials of Tillmann (1992) and in field trials of Schrod *et al.* (1996). Btt in the mixture was phagodeterrent, preventing the ingestion of lethal concentration of NeemAzal-S by grubs of the Colorado potato beetle, *Leptinotarsa decemlineata*. When NeemAzal-S was applied first and Btt two days afterwards, an additive effect was observed.

Nagesh and Verma (1997) conducted a field trial during the year 1995 and 1996 in cabbage variety Goldenacre to determine the comparative efficacy of certain eco-friendly pesticides viz., neem, Bt, diflubenzuron, lufenuron and cartap and synthetic organic insecticides against *P. xylostella*. Cartap at 0.05% concentration was found to be the most effective in controlling P. xylostella. The percent larval reduction over the untreated check was 93.3, 98.2 and 64.8, respectively after 1, 7 and 14 days of each spray. It was closely followed by lufenuron (0.006%), Bt

(0.2%) and then NeemAzal-T/S (0.002%). However, diflubenzuron (0.025%) was least effective. Hence, sequential spraying with these chemicals with different modes of action could be recommended to solve the problem of development of resistance by *P. xylostella*. Field trials with *S. litura, P. xylostella* and *Crocidolomia binotalis* on cauliflowers using individual treatments of cartap and quinalphos and their combinations with neem oil (1%) and Btk (0.15%) were highly effective and required fewer (2 to 3) sprays at much longer intervals of 12 to 15 days (Babu and Krishnayya, 1998). Other recent study with same species (Malathi and Sriramulu, 2000) using Btk formulations Dipel, Delfin, Biobit, Biolep and Bioasp at 0.075 per cent, Neemgold at 0.03 per cent, the nematode formulation Green-commandos @ 25 sponges/acre and endosulfan at 0.07 percent against second instar larvae were also very effective killing 86.66-100 per cent larvae in 72 hours in various combinations.

There are, however, some results obtained in various studies that implicate negative impact of neem on bacterial products. Tomar (1998) conducted field studies in the kharif season of 1995 and 1996 at Bilaspur, Madhya Pradesh, India to determine the efficacy of Btk (Dipel) mixed with lower concentrations of Multineem (neem extract) and chemical insecticides for the control of *E. vittella* infesting okra. Dipel+endosulfan and Dipel+fenvalerate were very effective in containing the shoot and fruit borer infestation. The maximum yield of healthy fruits was obtained using Dipel+fenvalerate, which also gave the highest profit and cost-benefit ratio. However, Dipel+Multineem and Dipel alone were least effective in reducing the shoot and fruit borer infestation by number and also by weight. Maximum yield (2168.75 kg /ha) of healthy fruits was obtained from the plot treated with Dipel+endosulfan. The Dipel and Multineem combination recorded low yield (1465.58 kg/ ha), which was at par with untreated control (999.75 kg / ha). Similarly, Gupta and Sharma (1998) tested eight different spray schedules against bollworms in cotton system during 1993 and 1994 and their impact was studied on the build up of whitefly population. Neem alone or when used alternately with Bt or with conventional synthetic insecticides failed to check bollworms, however, no build-up in whitefly population was noted. Alternate use of conventional insecticides with synthetic pyrethroids proved effective against bollworms but whitefly resurgence was evident. However, spray schedule in which neem was used alternately with Bt and atleast one spray of synthetic pyrethroid proved effective in managing bollworm complex without resurgence of whitefly. Murli Krishna *et al.* (1998) found that the application of Btk and neem alone was ineffective against the three major insect pests viz., *Amrasca biguttula biguttula, Bemisia tabaci* and *Leucinodes orbonalis* of brinjal. But, Awad *et al.* (1998) reported enhanced detrimental effects of neem derivatives by the presence of Bt.

Gupta *et al.* (1999) evaluated the bioefficacy of neem products against cotton bollworms, viz., *Earias* spp., *Pectinophora gossypiella* and *H. armigera* and their impact on *B. tabaci*. Neem products were applied alone and in combination with synthetic insecticides against bollworms. Application of neem alone or in combination with Bt formulations or broad-spectrum conventional insecticides failed to check the incidence of bollworms. However, neem in combination with one spray of synthetic pyrethroid gave significant control of bollworms as minimum incidence (10.8%, 18.2%) and maximum yields of seed cotton (1970, 1819 kg/ha) were recorded during 1994. In general, the high population of whitefly was recorded in synthetic pyrethroid-treated plots; the maximum population (1248 adults/30 leaves) was noted in plots treated with bifenthrin, at any stage of crop growth. There was no build up of whitefly population in neem, as only 627 adults/30 leaves were recorded, thus reducing the infestation by about 50 per cent. Thus management strategy based on neem, Bt formulations and 84 per cent reduced rate of synthetic pyrethroids proved effective against bollworms under field conditions and also safe for the environment with no resurgence of whitefly.

Toxicity of several biorational pesticides and chemicals to *H. armigera* and *H. punctigera* as well as major predators on cotton in field condition at Dalby, Queensland, Australia have been recorded (Ma et al., 2000). Moderate rate-dependent control was obtained in plots treated with neem seed extract - azadirachtin (AZA) at rates of 30, 60 and 90 g/ha. Plots treated with Talstar EC (bifenthrin) formulations achieved the best results, followed by treatment with alternation of chemicals (methomyl, bifenthrin, thiodicarb and endosulfan) and biorational insecticides (neem oil, azadirachtin and Btk). Predators, including coccinellids, chrysopids, araneae and hemipterans were insensitive to AZA, toosendanin (Tsdn) and Bt applications. In contrast, chemicals were very toxic to predators. All treatments provided some protection from infestation by *H. armigera* and *H. punctigera*. The effect of AZA on *Helicoverpa* spp. was reflected in a relatively higher yield of seed cotton harvested from AZA-treated plots compared to control, but chemical control achieved significantly higher yields than any other treatments.

Similarly neem as a component in the management of *L. orbonalis* involving eco-friendly methods has been reported (Sasikala *et al.*, 1999) using NSKE (5%), neem oil (0.2%), Btk (0.15%), lufenuron (0.02%), carbaryl (0.15%), and their combinations. Neem oil (0.2%) resulted in very good control of shoot and fruit borer compared to control. Plots treated with neem oil (0.2%), neem oil (0.1%) + Bt (0.075%), neem oil (0.1%) + lufenuron (0.01%), and neem oil (0.1%) + carbaryl (0.075%) gave higher fruit yield (40.76, 33.80, 31.35 and 29.07 kg/plot, respectively, compared to 17.5 kg/plot obtained from control plots).

2.3 Fungi

The compatibility of neem with fungi is also known (Table 3). Aguda *et al.* (1986) reported from laboratory trials that neem oil 5% significantly reduced the production of conidia by the entomopathogenic fungus *Metarhizium anisopliae*. Wilps *et al.* (1992) investigated in cage experiments, the effects of teflubenzuron, extracts of *A. indica* and *Melia volkensii* and spores of *M. flavoviride* and *Beauveria bassiana* on adults of *Schistocerca gregaria*. Teflubenzuron caused about 70 per cent mortality. The other treatments caused 40-100 per cent mortality. Of the survivors in various treatments, flight rate was reduced by 40-70 per cent and the provision of energy for musculature was reduced to 30-50 per cent of controls. Jani *et al.* (1992) found that *M. anisopliae* var. *anisopliae* and *B. bronginartii* tolerated the combination of Neemark.

In vitro application of nicotine sulfate, RD-9 Repelin (extracts of *A. indica*, *Pongamia glabra* (*P. pinnata*) and *Madhuca indica* (*M. longifolia*) and indiara (diallyl disulfide and allyl propyl disulfide) inhibited the growth of *B. bassiana* and *M. anisopliae* (Vyas *et al.*, 1992). Phytoalexin (an unspecified herbal extract with a biostimulant) had greater inhibitory effect. Neemark (azadirachtin) did not inhibit the growth of either fungus.

Table 3. Compatibility of neem with entomopathogenic fungi (EPF)

	EPF	Concentration	Observation	Reference
1	*Metarhizium anisopliae*	Neem oil 5 %	Negative effect	Aguda *et al.* (1986)
2	*Verticilium lecanii*	Aqueous suspension of oil seed cakes of neem	Positive effect	Rao *et al.* (1996)
3	*Beauveria bassiana*	Easmar 0.5, 2.5 and 5 %	Positive effect	Rodriguez *et al.* (1997)

Nasseh *et al.* (1992) conducted field trials in cages in northern Niger, during 1990-91, to test various spray and bait treatments against *S. gregaria*. The growth regulator teflubenzuron (50 g a.i./ha) was compared with a standard dieldrin treatment (20 g a.i./ha) by direct application to adults and nymphs of *S. gregaria* and adults of *Anacaridiumi wernerellum* and to the food plant *Schouwia thebaica*. Sprays of pure and enriched neem oil (10 l/ha) and oil formulations of *B. bassiana* (1 x 10^{13} conidia/kg applied at 320 and 960 g/ha) and baits containing *B. bassiana* and *Nosema locustae* (2.5 x 10^{9}/kg bran/ha) were also tested. Teflubenzuron caused low adult mortality than dieldrin but prevented development of 3rd- and 4th-instar nymphs. Neem oil caused no mortality by topical application but had strong antifeedant effects. However, severe burning of *S. thebaica* foliage by neem oil was also observed. None of the treatments involving spraying with *B. bassiana* or baiting with *B. bassiana* or *N. locustae* caused any mortality.

Various oils, including the neem oil, studied for viability of freshly harvested conidia of *M. flavoviride* stored at 25^{0}C clearly showed the reduction in the viability of the candidate fungus to less than 40 per cent (Stathers *et al.*, 1993). However, Devi and Prashad (1996) tested seed kernel extracts from *A. indica, M. azedarach* and *P. pinnata*, whole plant extracts from *Tephrosia purpurea, Parthenium hysterophorus* and *Cleome viscosa*, and vegetable oils from sunflowers, safflower, groundnut, rapeseed, sesame, coconut and cotton seed for their compatibility with the entomogenous fungus, *Nomuraea rileyi* applied to larvae of *S. litura*. None of the oils was detrimental to the fungus. Ranaivo *et al.* (1996) studied the effect of low doses of *Metarhizium flavoviride* and neem on walking activity, food consumption and weight of *Locusta migratoria*. Blastospore suspension was applied at 1 x 10^{4}, 5 x 10^{4} and 1 x 10^{5} spores/ml (LC_{10}, LC_{40} and LC_{50}, respectively). Neem oil was applied at 5 and 12.5 per cent, corresponding to LC_{10} and LC_{50}. Differences in walking activity were recorded for nymphs treated with increasing blastospore concentrations. Walking activity was reduced to 24 and 7 per cent of controls, 10 days after treatment with 5 and 12.5 per cent neem oil. Food intake was reduced for all blastospore and neem oil treatments. Rao *et al.* (1996) tested the efficacy of aqueous suspensions of oilseed cakes of neem (*A. indica*), castor (*Ricinus communis*) and pongamia (*Pongamia pinnata*) to support the growth of *Verticillium lecanii*. Results showed that autoclaved suspensions of neem and castor cake increased the growth of fungal mycelium and sporulation.

Evaluation of *M. anisopliae* as part of an IPM program against the coffee berry borer (*Hypothenemus hampei*) with the commercial dose (CD) and half commercial doses (1/2 CD) of various substances including neem extract showed that neem preparations had lower inhibitory effect (<30%). The fungistatic effect of the pesticides on the fungus decreased 48 h after germination.

Field trials conducted for four years (1989-93) in different rice seasons in various areas endemic to rice hispa, *Dicladispa armigera* in Assam, India to assess the efficacy of a mycopathogen, *B. bassiana* for controlling the insect (Hazarika and Puzari, 1997) have shown that among *B. bassiana* (10 million spores/ml dilution), neem-seed oil 1 per cent and a conventional insecticide (monocrotophos 0.072%) treatments, the mycopesticide was superior to neem-seed oil but at par with monocrotophos in controlling the rice hispa, leading to an increase in crop yield. Rodriguez *et al.* (1997) carried out a trial for evaluation of *B. bassiana* with aqueous extracts of *A. indica* seeds (Easmar), to observe their compatibility. Doses of Easmar of 0.5, 2.5 and 5.0 per cent did not inhibit mycelial growth or spore viability of the entomopathogen.

Bajan *et al.* (1998) Cultured the isolates of *B. bassiana* from dead larvae of *Achroia grisella* from pine forests in zones of different degrees of sulphur dioxide and nitric oxide pollution in the laboratory on pure substrate or substrate contaminated with field equivalent doses of Fastak (alpha-cypermethrin) or BioNEEM (extract of *A. indica*). Addition of Fastak to the substrate considerably inhibited the growth of an isolate and also significantly reduced the pathogenicity of the isolate from the heavily contaminated zone. Addition of BioNEEM caused substantial inhibition of colony growth, regardless of the ecotype. The substrate with BioNEEM or Fastak affects the isolate by altering their biological properties to some extent, however, it does not change the role of the fungus in the habitat. It was suggested that preparations of BioNEEM or Fastak and appropriate *B. bassiana* isolates could be applied simultaneously in integrated pest control.

The pathogenecity of three isolates of *B. bassiana* and one isolate of *M. anisopliae* in the laboratory have been compared and assessed against adults of *Leptoglossus zonatus* and *Pachycoris klugii*, the two most frequent pest species in physic nut (*Jatropha curcas*) plantations in Nicaragua. In a dipping bioassay, the median lethal concentration (LC_{50}) of the most efficient strain, *M. anisopliae* NB, was determined as 4.34 x 10^6 conidia/ml for adult *P. klugii*. In a field trial, a scheduled high-volume spray regime using *B. bassiana* increased fruit yield by 28 per cent, and was more effective than Malathion or an aqueous extract of ground neem seeds. The effectiveness of *M. anisopliae* was further tested in field cages covering entire trees and containing a predetermined number of insects. Mineral oil based ultra-low volume controlled droplet applications of *M. anisopliae* at a rate of

1×10^{10} conidia / tree were made using hand-held Micron ULVA + sprayers. The corrected mortalities ranged from 65 percent in *P. klugii* to 94 per cent in *L. zonatus* (Grimm and Guharay, 1998). However, it was concluded that effect of *B. bassiana* was better than neem products.

2.4 Nematodes

The effect of neem on entomopathogenic nematodes is shown in Table 4. Rovesti and Deseo (1991) studied the influence of an aqueous NSKE on steinernematid and rhabditid nematodes such as *Steinernema carpocapsae, S. glaseri, S. feltiae, Heterorhabditis bacteriophora* and *H. heliothidis*. In a viability/movement test using third instar larvae of *Heterorhabditis* spp. inconsistent results were obtained but a high mortality among the juvenile nematodes was caused by the highest concentration of NSKE (2% w/v). *Steinernema* spp. were less susceptible but the highest concentration tested (2%) caused a reduction in the activity (fitness) of J3 instar juveniles. The mortality of the treated nematodes reached 20-30%. Lower concentrations were not effective. Neem–treated larvae of *Steinernema* spp. killed all *Galleria mellonella* larvae offered to them but *Heterorhabditis* sp. was unable to parasitize the hosts. Based on these results the authors did not recommend the use of neem products and entomoparasitic nematodes, especially *Heterorhabditis* spp. at the same time. However, it is expected that parasitic nematodes, while in the soil, be exposed to lower concentrations of neem products than in the above discussed laboratory tests.

Gill and Raupp (1994) observed bagworm, *Thyridopteryx ephemeraeformis*, damaging a wide range of evergreen and deciduous plants in the USA. Trials were conducted in Maryland during 1992-93 to control mid-to late-instar bagworm larvae on arborvitae (*Thuja occidentalis*) grown in large containers.

Table 4. Effect of neem with entomopathogenic nematodes (EPN)

	EPN	Concentration	Observation	Reference
1	*Heterorhabditis bacteriophora*	2% w/v at ½, ¼, 1/8, 1/16 and 1/32 of NSKE	Negative effect specially at higher concentrations	Rovesti and Deseo (1991)
2	*Steinernema feltiae*	Neem seed shell extract 5g/100 ml water	Negative effect	Pezowicz *et al.* (1997)
3	*S. carpocapsae, S. feltiae and S. glaseri*	Margosan-O	Negative effect at much higher concentartions than recommended rate of Azadirahtin 20 mg/l of water	Stark (1997)
4	*Steinernema sp. and H. indica*	Neem Suraksha (EC 0.05 µl/ml)	Positive effect	Hussaini *et al.* (2001a)

Treatments included two formulations of neem, carbaryl, acephate, cyfluthrin, formulations of *S. carpocapsae* and *S. feltiae*, and Btk. Neem gave the least control (36-56% reduction) while the nematodes, either alone or with oil or antidesiccant, gave 91-100 per cent control of larvae. The synthetic pyrethroid cyfluthrin gave 100 per cent control; carbaryl and acephate also gave acceptable control (83 and 86%, respectively). In 1993, cyfluthrin recorded the maximum control of both mid-and late-instar larvae (95-97%). The biological control agents, *Bacillus* and the nematodes, provided intermediate levels of control. Both species of *Steinernema* were effective.

Pezowicz *et al.* (1997) observed high mortality levels (>50%) of eco-friendly juveniles of *S. feltiae* after five days of exposure to neem seed shell extract (5g /100 ml water), but not after exposure to seed kernel extracts. The botanical insecticide, Neemix (azadirachtin), had a slight effect on the survival of infective juveniles as well. Bancol 50 WP 0.75 per cent (bensultap) resulted in 100 percent mortality after three days. The exposure of infective nematodes to aqueous extracts did not affect their ability to infect larvae of *G. mellonella*, but limited their establishment in the insect. The mean number of established nematodes per *G. mellonella* decreased with increasing Neemix concentrations. Stark (1997) studied the effect of Margosan-O, a commercial neem based insecticide, *S. carpocapsae*, *S. feltiae*, and *S. glaseri*. Both acute and chronic toxicity were estimated along with effects on nematode infectivity after incubation with the insecticide. Margosan-O was toxic to all the three species at much higher concentrations than the recommended field rate of 20 mg azadirachtin/l water. The acute LC_{50} for *S. carpocapsae* exposed to Margosan-O was 425 mg azadirachtin/l. Chronic toxicity was not significantly different from acute toxicity until the 15th day of incubation and chronic toxicity values were not significantly different until day 15. The acute LC_{50} for *S. feltiae* was 380 mg azadirachtin/l. The susceptibility of *S. feltiae* increased substantially between 12 and 15 days of incubation with Margosan-O. The acute LC_{30} for *S. glaseri* was 351 mg azadirachtin /l. Unlike the other 2 species, no difference was detected in susceptibility among incubation times for *S. glaseri*. When three species were compared for susceptibility to Margosan-O at LC_{50} levels, incubation time interval was not significantly different until 15 days; however, *S. feltiae* was significantly more susceptible than the other species after exposure to Margosan-O for 15 days. Infectivity measured as the ability to kill Greater waxmoth larvae, *G. mellonella* it was reduced in all three species at concentrations > 200 mg azadirachtin/l. When Margosan-O and nematodes were applied together at field rates to soil, infectivity was not reduced.

The feasibility of using EPN as effective bioinsecticides, tolerence studies were carried out with commonly used synthetic and one botanical pesticide viz., fenvalerate, quinalphos, endosulfan and neem (Hussaini *et al.*, 2001a). Significant differences existed in the levels of tolerance between *Steinernema* and *Heterorhabditis indica* isolates. Neem Suraksha (EC 0.05 μl/ml) was found to be safe with survival percentages ranging from 88.8 –99.2 % followed by endosulfan (68-97.6) and fenvalerate (66.4-98.4). Quinalphos was found to be deleterious to some isolates, as survival and infectivity were impaired. Use of these EPN isolates along with pesticides other than quinalphos was suggested. In addition, it was found that mancozeb and neem (1.5 ppm) were safe to all nematode populations while the latter alone was deleterious to *H. indica*. Fifteen out of twenty combinations tested were compatible and were recommended in IPM schedule (Hussaini *et al.*, 2001b).

3. NEEM AND PARASITOIDS

The biological activity of neem products on parasitoids is summarised in Table 5 and the activity has been studied against egg, larval and pupal parasitoids.

3.1 Egg Parasitoids

Joshi *et al.* (1982) studied the side effects of neem products on egg-parasitoids in India. NSKE 2 per cent was applied on the egg masses of *S. litura*. The egg parasitoid *Telenomus remus* was not repelled from egg laying. When the treatment was carried out before egg laying of the parasitoid, the emergence of adult parasitoids was normal but their duration of life was shorter than that of controls. On the other hand, spraying NSKE after oviposition, *T. remus* increased the fecundity of the wasps developed in treated eggs and prolonged their life as compared to that of untreated controls. Li *et al.* (1986) tested 29 insecticides including Bt and NO in order to study their side-effects on *Trichogramma japonicum* in the laboratory and concluded from the results that NO and Bt were the safest pesticides for the parasitoid.

Velayuddan *et al.* (1988) studied the effect of host species on the sex ratio and parasitism rate of *Anastatus ramakrishnanae* in the laboratory and field in Tamil Nadu, India with the pentatomid *Halys dentata* (on *Cassia marginata, Azadirachta indica* and *Casuarina equisetifolia*) and the coreid *Homoeocerus prominulus* (on *Cassia marginata, Prosopis specigera* and *Acacia leucophloea*) as hosts. The order of parasitism was *C. marginata* > *A. indica* > *C. equisetifolia* with *H. dentata*, and *C. marginata* > *P. spicigera* > *A. leucophoea* with *H. prominulus*. Higher rates of parasitism were recorded with *H. dentata* than *H. prominulus*, throughout the year.
Fernandez *et al.* (1992) conducted experiments with the eggs of yellow stem borer of rice, *Tryporyza incertulas* by dipping for 30 seconds in Bordan, neem products and water. The eggs were exposed for 40 h of parasitization by the parasitoid, *Telenomus rowani*. The data revealed highest mean number of parasitoids emergence of 65.8 per cent in water treatment, 59.90 per cent in 5 percent aqueous NSKE, 31.4 percent in 3 per cent neem oil and 38.2 per cent in Bordan treatment.
Klemm and Schmutterer (1993) applied NSKE (2.5% and 3%) against *Trichogramma* spp., egg-parasitoids of *Plutella xylostella*. *T. principium* accepted neem treated eggs in the laboratory and *T. pretiosum* in the field but two treatments prevented the eclosion of adult parasitoids from treated *P. xylostella* eggs completely. Spraying of eggs with NO 0.2 per cent reduced the number of eggs parasitized per female wasp by 13.3 per cent. NO also reduced the emergence of *T. principium* from treated eggs by 45.1 per cent. However, neem seed kernel suspension (5%) and neem oil 50 EC (3%) were safe to the parasitoid *T. japonicum* in cotton ecosystem (Jayaraj *et al.*, 1993).

Table 5. Compatibility of neem with parasitoids

	Parasitoid	Concentration	Observation	Reference
1	*Telenomus remus*	Aqueous NSKE (2%)	Positive effect	Joshi *et al.* (1982)
2	*Trichogramma pretiosum*	NSKE (2.5 % and 3 %)	Negative effect	Klemm and Schmutterer (1993)
3	*Trichogramma chilonis*	Aqueous, ethanolic and hexane extracts of NSK @ 0.3 to 5.0%	Both positive and Negative effects	Raguraman (1993)
4	*T. pretiosum*	NSK based extract NIM-20	Positive effect	Cano and Gladstone (1994)
5	*T. chilonis*	NeemAzal-F 5%(1ml/lit), NeemAzal-T/S, NSKE (5 %) (50 g/lit)	Positive effect	Saikia and Parameshwaran (2001)
6	*Diaraetiella rapae, Aphidius cerasieola*	NSKE (5%)	Positive effect	Schauer (1985)
7	*Bracon hebetor*	Neem oil (5 %)	Positive effect	Jhansi and Sundarababu (1987)
8	*B. hebetor*	Repelin and Neemguard	Positive effect at lower concentration	Srinivasa Babu *et al.* (1996)
9	*B. hebetor*	Aqueous, ethanolic and hexane extracts of NSK @ 0.3, 0.6, 1.2, 2.5 & 5.0%	Both positive and Negative effects	Raguraman (1993)
10	*Diadegma semiclausum*	NSKE (0.1-5 %)	Positive effect	Schneider and Madel (1991)
11	*Cotesia glomerata*	Higher concentration (40 ppm) of azadirahtin and of azadirahtin free fraction and 50 and 100 ppm of enriched product	Negative effect	Schmutterer, (1992)
12	*Tetrastichus israeli*	Neem seed kernel suspension (5 %) and neem oil 50 EC(3 %)	Positive effect	Jayaraj *et al.* (1993)

Neem-treated eggs of *Ephestia kuehniella* in shell vials when offered to single females of *T. minutum* for parasitization by fixing the eggs with adhesive strips and held until all parasitoids had emerged from them (Lyons *et al.*, 1996); variable results were obtained. Azatin, Neem EC (4.6% AZA) and pure AZA were tested at concentrations of 50g and 500 g /ha. At 50 g/ha no significant effect was observed, at 500 g /ha Azatin and Neem EC reduced the female survival by 64 and 40 per cent, respectively whereas pure AZA showed no effect. Likewise, at 500 g/ha the number of parasitized eggs was reduced by 89 per cent by Azatin, 29 per cent by Neem EC and no reduction by AZA was recorded. The parasitoid development was reduced by all treatments.

Cano and Gladstone (1994) studied the influence of the NSK-based extract NIM-20 on parasitization of eggs of *Helicoverpa zea* in a melon field in Nicaragua. Mass-reared *T. pretiosum* were released at six weekly intervals 1,2,6 and 24h after application of NIM-20 at 2.5g/l. No negative effect was observed as up to 84 per cent of the eggs of the pest were parasitized. Oswald (1989) treated the eggs of the coconut bug, *Pseudotheraptus wayi*, with aqueous NSKE (5% w/v). The eggs were offered to the parasitoid *Oencyrtus albicrus* (Encyrtidae). There was a significant reduction in the number of wasps emerging from the treated eggs in comparison with controls. Srinivasa Babu *et al.* (1996) studied the effects of neem-based commercial insecticides such as Repelin and Neemguard on *T. australicum* in laboratory and field conditions. They reported that both the insecticides were relatively safe at lower concentrations but higher concentrations adversely affected the parasitoids both in laboratory and in field.

Effects of insecticides on the emergence of *T. japonicum* from eggs of *Corcyra cephalonica* on the third or sixth day after parasitization using chlorpyrifos, quinalphos, monocrotophos, cypermethrin, dimethoate, phosphamidon, fenvalerate, Biolep and Bioasp (both Btk products) and NeemAzal-F and Fortune Aza (both neem-based products) clearly indicate that Bt and neem products had the least effect on the emergence of parasitoids. Of the other insecticides, fenvalerate and monocrotophos had the least effect while quinalphos had the most. Adult emergence was relatively less when eggs were sprayed on the sixth day after parasitization compared to third day after parasitization (Borah and Basit, 1996). Similar results were obtained against *T. japonicum* using Econeem and NeemAzal-T/S (0.1-1.0 %) (Lakshmi *et al.*, 1998a), other neem-based pesticides in an IPM approach for rice pest management (Garg and Baranwal, 1998), and egg parasitoid, *Tetrastichus pyrillae* (Deepak and Choudhary, 1998). On the whole it has been assessed that neem products were fairly safe to *Trichogramma* spp. (Sreenivasa and Patil, 1998; Sarode and Sonalkar, 1996b).

However, some neem formulations such as Nimbecidine (0.25-4.0%), Neemgold (2.0-4.0%) and Rakshak (1.0%) are reported to possess adverse effects on parasitism (Laksmi *et al.*, 1998a). Raguraman and Singh (1999) tested in detail the neem seed oil at concentrations of 5.0, 2.5, 1.2, 0.6 and 0.3% for oviposition deterrence, feeding deterrence, toxicity, sterility and insect growth regulator effects against *Trichogramma chilonis.* Neem seed oil at 0.3% deterred oviposition (parasitization) by the parasitoid but the sensitivity varied considerably both under choice and no-choice conditions. Neem seed oil also deterred feeding at or above

1.2% concentration both in choice and no-choice tests. In feeding toxicity tests, neem seed oil at 5% concentration caused < 50% mortality to both males and females but in contact toxicity tests, females were affected sparing males. No sterility effect was observed when the parasitoid was fed with neem seed oil treated honey. Both pre-and post-treatment of host eggs revealed no adverse effects on the development of the parasitoid. Thakur and Pawar (2000) tested two neem-based insecticides (3 g Achook/litre and 2 ml Neemactin/litre), two biopesticides [1 g Halt (cypermethrin)/litre] and 1 ml Dipel (Btk)/litre], and endosulfan (1.5 ml/litre) in the laboratory for their relative toxicity to newly emerged adults of *T. chilonis*. Results revealed that neem-based pesticides and biopesticides were harmless while endosulfan was slightly toxic to egg parasitoid. These observations also get support from the studies on different groups of chemicals viz., insecticides, moult inhibitors and biopesticides against rice leaffolder, *C. medinalis* and its parasitoid *T. chilonis*. Sprays of monocrotophos 36 WSC applied at 2.0 ml/l caused 100 per cent larval mortality to *C. medinalis* seven days after treatment followed by buprofezin 25 WP applied at 3.2 g/l with 66.66 per cent larval mortality and neem seed kernal extract (NSKE) 5 per cent (50 g/l) with 63.33 per cent larval mortality. Application of *Bacillus thuringiensis* subsp. *galleriae* (Btg) at 5.0 ml/l and NeemAzal-F 5 per cent at 1g/l recorded 56.66 and 53.33 per cent larval mortality, respectively. More than 90 per cent emergence of *T. chilonis* was recorded from eggs treated with Btg and NeemAzal-F and NeemAzal T/S followed by NSKE (89.80%) and buprofezin (82.60%). Only 73.80 per cent of adult *T. chilonis* emerged from monocrotophos treated host eggs after parasitization (Saikia and Parameswaran, 2001). Similarly in Thailand Asian corn stem borer, *Ostrinia furnacalis* was controlled by neem preparations and it was observed that such treatments had no side effects on the parasitoid, *T. plasseyensis* wasps (Breithaupt *et al.*, 1999)

3.2 Larval Parasitoids

Saxena *et al.* (1981) sprayed 50 per cent neem oil as low volume application against the rice folder *C. medinalis*. The pest larvae were parasitised by ichneumonids, braconids and encyrtid groups of parasitoids in the field. Surprisingly the parasitization of the leaf folder larvae in neem oil treated plots was double than control. This is due to the fact that most of larvae could not spin the leaves together due to high toxicity of neem oil, thereby gave enough opportunity for parasitization. However, the neem oil had no side effects on parasitoids. Such an increase in parasitoid population was also observed for the parasitoid *Diadegma semiclausum* than in control after the treatment of neem based product 'Biosol' in cabbage plots (Chandra Mohan and Nanjan, 1990)). Similarly, various endoparasitic hymenoptera pupated and emerged normally from parasitized 4th and 5th instar *C. medinalis* larvae that were reared on rice leaves treated with neem fractions or extracts (Schmutterer *et al.*, 1983). Schauer (1985) found that aphid mummies containing larvae or pupae of braconid parasitoids, *Diaraetiella rapae* and *Aphidius cerasicola* were unaffected by 5 per cent neem seed kernel suspension. Neem seed oil was also quite safe for the natural enemies like *Lycosa pseudoannulata* and *Apanteles cypris*

(Wu, 1986), ichneumonid parasitoid, *Campoletis chlorideae* of *H. armigera* (Prasad *et al.*, 1987) and external larval parasitoid, *Bracon hebetor* of pod borer, *Maruca testulalis* (Jhansi and Sundara Babu, 1987). Other studies with *B. hebetor* also support the fact that neem is safer for this parasitoid as aqueous suspension and an ethanolic extract of neem seed kernel (NSK) at 0.3, 0.6, 1.2, 2.5 and 5.0 per cent administered via food or by contact had no influence on the *B. hebetor* oviposition (parasitization) on *C. cephalonica*. Parasitoid eggs and pupae were also unaffected by the extracts tested. The parasitoid larvae, however, were killed by feeding on contaminated host larvae and also through contact with neem extracts. Thus, use of a minimum safety period is suggested for inundative release of *B. hebetor* in integrated pest management (Raguraman and Singh (1998). In order to determine the toxicity of oil/extracts to *Chelonus blackburni* to explore the possibility of using parasitoid along with oils/extracts in integrated control programme of potato tuber moth; it was observed that all the vegetable oils including neem oil were safe to *C. blackburni*, an egg-larval parasitoid of *P. operculella* (Shilke *et al.*, 1990).

Schneider and Madel (1992) reported that there was no adverse effect on adults of the braconid *Diadegma semiclausum* after exposure for 3 days or during their lifetime in cages to residues of an aqueous NSKE (0.1-5%). The longevity of the wasps exposed to neem residues was even prolonged but the difference between treated and untreated individuals was statistically not significant. Females of the braconid, derived from larvae developed in neem–treated larvae of *P. xylostella*, showed no reduced fecundity or activity as compared to controls. Fresh extracts showed no repellent effect. The influence of AZA on *Diadegma terebrans*, parasitoid of the European corn borer, *Ostrinia nubilalis*, was investigated in the laboratory by McCloskey *et al.* (1993). These authors added sub lethal doses (0.1 ppm and 0.3 ppm) of AZA or ethanol (carrier solvent) to diets of second instar larvae of the pyralid. Both AZA concentrations showed no significant difference of the parasitization percentage; host acceptance by the parasitoids was also not influenced. However, significantly higher mortality of parasitoids was observed in AZA-treated groups compared to untreated groups, especially after emergence from the hosts. The duration of the larval instars in the hosts were prolonged and the weight of pupae and adults from treated groups was reduced. Lowery and Isman (1996) tested the effects of extracts from neem on aphids and their natural enemies. In field trials, populations of aphid natural enemies (predators and parasitoids) were not affected by application of neem insecticides, suggesting compatibility of neem with biological control agents.

Safety of natural enemies after neem application is also shown by the studies of Mani and Krishnamoorthy (1996) where encyrtid *Tetracnemoidea indica*, a dominant parasitoid of the pseudococcid, *Planococcus lilacinus* on acid lime were exposed to acid lime leaves treated with 34 pesticides at field recommended doses. Fenvalerate (0.01%) and NSKE (2%) were nontoxic to the adult parasitoids. Similarly there was no adverse effect of neem seed kernel water extract (NSKWE) (25 g/l) on the adult of *A. plutellae*, a parasitoid of leaf eating caterpillar complex of cabbage (Bandara and Kudagamage, 1996). It was also observed that after the spraying of the NSKWE and the two insecticides on the cocoons, there was no significant reduction in the adult emergence. Thus, NSKWE (25 g/l) had no adverse effect on *A. plutellae*. Similar results were obtained for *A. africanus* and *Telenomus remus* (Chari *et al*., 1997). Dobelin (1997) studied the side effects of NeemAzal-T/S (1.0% azadirachtin) against two parasitoids of aphids viz. *Aphidius colemani* and *Aphidoletes aphidimyza*. It was observed that neem products had no effects on these natural enemies. Michelakis and Vacante (1997) advocated Neemark as a safe device for the control of *Phyllocnistis citrella* that would not affect the parasitoid *Pnigalia* sp., which was very abundant. A biological control programme was implemented using *Ageniaspis citricola, Citrostichus phyllcnistoides* and *Semielacher petiolatus*, which were all introduced during 1996.

Stansly and Liu (1997) found that neem extract, insecticidal soap and sugar esters had little or no effect on *Encarsia pergandiella* the most abundant parasitoid of *Bemisia argentifolii* in south Florida vegetable fields and can contribute significantly to natural biological control of this and other whitefly species. Of the 10 species of leaf-mining Lepidoptera collected in apple orchards in south-western Germany in 1996, the most abundant were *Phyllonorycter blancardella, Lyonetia clerkella* and *Stigmella malella* and a mining curculionid, *Rhamphus oxyacanthae*. Total parasitism by chalcidoidea and ichneumonoidea ranged from 10 to 29 per cent. Use of a neem preparation for pest control had no effect on the rate of parasitism (Olivella and Vogt, 1997). Sharma *et al.* (1999) also reported that the extracts from neem and custard apple kernels were effective against the spotted stem borer, *Chilo partellus*, Oriental armyworm, *Mythimna separata*, head bugs, *Calocoris angustatus*, and the yellow sugarcane aphid, *Melanaphis sacchari* in sorghum, but neem extract was non-toxic to the parasitoids and predators of the sorghum midge, but reduced the parasitism to some extent.

However, certain negative effects have also been recorded. Sharma *et al.* (1984) reported that an active neem fraction of NSK had adverse effect on larval parasitoid, *Apanteles ruficrus* of Oriental armyworm, *M. separata*. Injection of 2.5 to 10 µg of azadirachtin to newly ecdysed fourth and fifth instar larvae of host either partially inhibited or totally suppressed the first larval ecdysis of braconid, *Cotesia congregata* an internal larval parasitoid of tobacco hornworm, *Manduca sexta* (Beckage *et al*., 1988). They also reported that the parasitoid growth was arrested, while the host larvae survived for two weeks or longer, following injection of azadirachtin but their parasitoids never recovered and died encased within exuvial cuticle. Lamb and Saxena (1988) gave topical treatment to the females of ectoparasite *Goniozus triangulifer* at doses from 5-50 µg/l solution of neem seed bitters. The results indicated decreased fecundity at 50 µg per female dose. When the rice plants were sprayed with 1000 ppm neem seed bitters, very few larvae of leaf folder *Marasmia patnalis* sustained the development of *G. triangulifer* up to pupation stage. However, when 1000 ppm neem seed bitters were sprayed three times, there was negative influence on parasitization by *G. triangulifer*. The studies further reported that when *Tetrastichus howardi* parasitized pupae of *M. patnalis* were dipped in 1000 ppm of neem seed bitters, the adult emergence decreased significantly. However, topical application of 1000-10000 ppm of the neem seed bitters had no effect on *T. howardi*.

Topical treatment of cocoons of *C. plutellae* treated with neem oil in the laboratory inhibited adult eclosion significantly at 2.5 per cent oil concentration and no adult emergence was observed at 10 per cent level (Loke *et al.*, 1992). Treated cocoons produced adults with reduced longevity but no morphological deformities. However, Osman and Bradley (1993) reported high mortality of larvae and morphogenetic defects of adult parasitoid, *C. glomerata* developed from hosts treated with NSKE. Neem products though did not affect adult parasitoids even after spraying with higher concentration, i.e. AZT-VR-K, 2000 ppm. Srivastava *et al.* (1997) reported that alcohol and hexane extracts of 17 neem ecotypes in India were found to be toxic to the egg, larval and pupal stages of the *B. brevicornis*. In general, the hexane extracts showed higher toxicity against the egg and pupal stages, whereas the alcohol extracts were more toxic against the larvae. Azadirachtin content of the neem ecotypes revealed no apparent correlation with the observed toxicity against different stages of the parasitoid.

It seems that there are mixed responses of parasitoids to various neem preparations. For instance, Hoelmer *et al.* (1990) did experiments with parasitoids of *B. tabaci* and *Aphis gossypii* with the neem product Margosan-O. It was found that the aphid parasitoids namely *Lysiphlebus testaceipes* and *Aphelinus asychis* were more sensitive to neem treated surface whereas the survival of the aphid parasitoid, *Eretmocerus californicus* was same on treated and untreated *Hibiscus* foliage. The *E. californicus* pairs in sealed petridishes with treated and untreated foliage survived for five days. It was also observed that dipping of aphid mummies parasitized by *L. testaceipes* and also dipping the parasitized puparia of *B. tabaci* by *Encarsia formosa* and *E. transversa*, did not affect the emergence of the parasitoids. However, when *E. californicus* parasitized white fly puparia was dipped, the emergence of parasitoids was reduced by more than 50 per cent. Similarly, Stark *et al.* (1990) on the other hand found that a highly purified and concentrated neem extract prevented adult of fruit flies emergence from puparia. However, the parasitoid *Opius* sp. emerged freely.

Schmutterer (1992) found in laboratory experiments that concentrations of 10 and 20 ppm of azadirachtin of an azadirachtin-free fraction and of an enriched formulated seed kernel extract of *A. indica* were only slightly harmful to *Cotesia glomerata*, provided they were applied against the 5th instar of *Pieris brassicae*. Under these circumstances, numerous larvae of *C. glomerata* emerged from their hosts, pupated and hatched as normal adults. However, higher concentration (40 ppm) of azadirachtin and of the azadirachtin-free fraction as well as 50 and 100 ppm of the enriched product reduced the number of parasitoids considerably. The parasitoids were mainly killed by lack of food and died within their hosts. Larvae of *P. brassicae* under the influence of metamorphosis disturbance by neem products did not die immediately after uptake of active principles, but there was reduced food uptake, leading to increased intraspecific competition among the gregarious grubs of *C. glomerata*. Direct growth regulation effects of neem products against *C. glomerata* were not observed. Application of neem products against young (1st–3rd) larval instars of *P. brassicae*, led to the death of the caterpillars together with the grubs of the parasitoid.

Thus it is obvious from the studies available so far that neem and its various products/formulations do have some side effects against some natural enemies. We have some specific reports of such side effects available. Beitzen and Hofmann (1992) studied the side effects of neem product AZT-VR-NR on the endoparasitic tachnid fly, *Drino inconspicua* when the fly was exposed for 7 days with residues of neem product (45 g ai/ha), it did not harm the adult flies but fecundity was reduced by 18.5 per cent in comparison with control. Similarly, Serra (1992) observed no or only side effects of neem products on the parasitoids of the genera *Ganaspidium, Desorygma* and *Opius*, which emerged from the tomato leaf miner *Liriomyza sativae*. He also obtained similar results on the genera *Pseudapanteles* and *Glyptapanteles* that emerged from tomato pinworm, *Keiferia lycopersicella*. Moser (1994) though observed no side effects of aqueous NSE (2.5 and 5.0%) among natural enemies (coccinellids, syrphids, chrysopids and braconids) of *Aphis gossypii* on okra in Dominican Republic fields, but at the same slight harmful effects viz., morphogenetic defects, delay of larval and pupal development were recorded in the laboratory experiments. Mineo *et al.* (2000) tested the side effects of azadirachtin mixed with mineral paraffin oil and a surfactant against the natural parasitoids of *P. citrella*. The observations revealed 16.67 per cent parasitoid larvae showing teratological symptoms.

Stark *et al.* (1992) studied the survival, longevity and reproduction of the three braconid parasitoids namely *Psystallia incisi* and *Diachasmimorpha longicaudata* from *Bactrocera dorsalis* and *Diachasmimorpha tryoni* from *Ceratitis capitata*. They also studied the effect of azadirachtin concentration on these three parasitoids. Two experiments were conducted, the first test, in which azadirachtin concentrations were used in a range of low fly emergence (1:0, 5 and 10 ppm azadirachtin) and in the second test with higher concentrations where no fly emergence would occur (2:20, 50 and 100 ppm azadirachtin). Results of the first test were in conformity with Stark *et al.* (1990). All larvae that were exposed to sand treated with azadirachtin, pupated. Adult eclosion was concentration-dependent in both fly species, with little or no fly eclosion at 10 ppm. However, *P. incisi* and *D. longicaudata* successfully eclosed from pupae treated with $\leq$ 10ppm azadirachtin. In all the cases after the exposure of azadirachtin, the adult eclosion was inhibited. Even life spans of parasitoids that emerged form treated flies were not significantly different from controls. The azadirachtin had no effect on the longevity of parasitoid species tested in this study, indicating that the parasitoids were less sensitive to this chemical than were their hosts. The reproduction of *P. incisi* that developed in flies exposed to azadiracthin concentration of > 20 ppm was reduced by 63.88 per cent. The reproduction of *D. longicaudatus* and *D. tryoni* was unaffected. This implies that neem-based products are safer at lower concentrations but induce adverse affects at higer levels of treatment, also obvious from the studies on neem products like Repelin and Neemguard that were tested on *Bracon hebetor* in laboratory and field conditions by Srinivasa Babu *et al.* (1996) to reveal their safety at lower concentrations against larval parasitoids. But at higher concentrations both preparations adversely affected the development.

It is, however, also possible that parasitoids may be adversely affected due to lack of appropriate food. This is clear from the results of Jakob and Dickler

(1996) where adults of the ectoparasitic, gregarious eulophid, *Colpoclypeus florus*, an important parasitoid of the tortricid, *Adoxophyes orana* were not adversely affected by application of NeemAzal-S (25 ppm and 100 ppm) in the laboratory and in the field, but 100 per cent of the larvae died, apparently due to lack of appropriate food on the neem-treated decaying larvae of the host. Schmutterer (1996) also described the varying sensitivity of bioagents to neem products, like eggs of predators such as coccinellids and chrysopids are not sensitive but ectoparasitic gregarious larvae of *Bracon* sp. and *Colpoclypeus* sp. showed high mortality after contact with neem. Endoparasitic solitary or gregarious hymenopteran larvae were less endangered as their hosts protected them. Schmutterer suggested that often, lack of food in neem treated hosts resulted in the death of parasitoids due to starvation. Other aspect of interest is the IPM compatibility of neem with other products vis-à-vis the safety of natural enemies. The relative toxicity of pesticides to *Phyllocnistis citrella* and its parasitoid *Ageniaspis citricola* was compared by several bioassay methods. Azadirachtin (Neemix) + oil, diflubenzuron (Micromite) + oil, fenoxycarb (Eclipse) + oil, and oil alone (FC 435-66) were classified as IPM-compatible insecticides. Sprays of azadirachtin (Align)+oil, neem oil (Neemgard), and drenched imidacloprid (Admire) were ranked as semi-compatible insecticides. The fungicide copper hydroxide (Kocide 101) and a fish oil-based foliar fertilizer (Zapata HFE) were considered compatible. Avermectin (Agri-Mek) + oil, ethion (Ethion), and imidacloprid (Provado) applied as a spray were IPM-incompatible insecticides (Villanueva and Hoy, 1998).

Teggelli *et al.* (1998) studied the effects of Nimbecidine (5 ml/litre) and Achook (5 ml/l), NPV (1.5 ml/l) and some recommended insecticides on the emergence of *Compoletes chlorideae* from host larvae 3, 5, 8 and 11 days after parasitization (DAP). Among insecticides, Achook resulted in the highest adult emergence (42.33%) at eight DAP, while fenvalerate, methomyl, malathion, chloropyriphos and monocrotophos completely inhibited emergence. At 11 DAP, the biopesticides namely Nimbecidine, Achook and NPV recorded the highest percentage of emergence (58.66, 56.33 and 53.33%, respectively), while monocrotophos was most toxic (8.66% adult emergence). The toxicity of all insecticides was lower on cocoons. Nimbecidine and NPV did not cause mortality 24 h after treatment. Similarly, *Hypomecis* sp. caused severe damage to *Azadirachta indica* in Akola, India, during October 1988. *Apanteles fabiae* and *Aleides* [*Aleiodes*] sp. were observed parasitizing *Hypomecis* sp. (Men, 1999).

Facknath (1999) while developing IPM strategy for the control of *P. xylostella* found that neem did not affect the population of *C. plutellae* an introduced parasitoid of *P. xylostella*. Reddy and Guerrero (2000) evaluated biorational and regular insecticide applications for management of the diamondback moth, *P. xylostella* in cabbage. The IPM programme, based on the pheromone trap catch threshold of eight moths per trap per night, included utilization of *C. plutellae* (250000 adults/ha), *Chrysoperla carnea* (2500 eggs/ha), nimbecidine (625 ml/ha), Bt (500 ml/ha), and phosalone (2.8 l/ha). The IPM programme induced a reduction of trap catches, egg and larval populations and, therefore, a low level of damage to the crop. All neem concentrations gave poor to very slight control of *Myzus persicae* when applied as contact action foliar sprays, with Pirimor R providing the greatest contact kill. Neem at 180 ppm, when applied as a soil drench, gave total aphid control within 24 h, apparently through systemic action. Aphid parasitoids and other beneficial insects were not affected by neem treatments, whereas Pirimor R treatments reduced beneficial insect numbers. Although Pirimor R would be the preferred choice for immediate aphid control through contact action in commercial crop production, neem still has a place in the control of aphids in situations such as organic crop production, or in crops where resistance to other chemicals by aphids has resulted. Other uses may be in indoor and outdoor landscape situations where human health is of major concern and a long lasting systemic method of aphid control desirable. In these cases neem could be applied as a soil drench at concentrations of 180 ppm, possibly through existing irrigation systems (Holmes *et al.*, 1999).

Perera et al., (2000) studied the effect of three feeding deterrents: denatonium benzoate (5, 50 and 250 mg/l), azatin [azadirachtin] EC (0.01, 0.1 and 1 ml/l) and Pestistat^R (0.1, 1 and 2 ml/l) on the fourth instar larvae of important cabbage pests, *Chrysodeixis eriosoma* and *P. xylostella* and on the parasitoid, *Cotesia plutellae*. Results suggested that the three antifeedants were effective in managing cabbage pests, *C. eriosoma* and *P. xylostella* and could be used in integrated pest management programmes. Denatonium benzoate was comparatively safer to the parasitoids *C. plutellae*.

3.3 Pupal Parasitoids

Several studies show mixed impact of neem preparation on pupal parasitoids. Tewari and Moorthy (1985) studied the effect of neem oil+acetone+Triton X-100 on the degree of parasitization by eulophid parasitoid, *Pediobius foveolatus* of the phytophagous coccinellid on *Epilachna vigintioctopunctata*. When the pest larvae were exposed to neem oil (0.075 and 0.05%) spray and then exposed to parasitoids, the rate of parasitization was very low but exposure one day after treatment had no reduction in the rate of parasitization.

Stark *et al.* (1992) studied under laboratory conditions the influence of AZA on survival, longevity and reproduction of parasitoids of tephritid flies. The braconids *Psytallia incisi* and *Biosteres longicaudatus* developed in and eclosed from the tephritid, *Bactrocera dorsalis* exposed to a diet with AZA concentrations that inhibited adult eclosion. *Diachismomorpha tryoni* also eclosed from *Ceratitis capitata*, exposed to concentrations of AZA that prevented eclosion of adult fruitflies. The longevity of parasitoids that emerged from treated flies did not differ significantly from controls but reproduction of *P. incisi*, developed in flies exposed to 20 ppm AZA was reduced by 63.88 per cent. The reproduction of other braconid species was not adversely affected. Neem seed kernel suspension (5%) and neem oil 50 EC (3%) were safe to *Tetrastichus israeli*, a pupal parasitoid of coconut black headed caterpillar, *Opisina arenosella* (Jayaraj *et al.*, 1993). In laboratory trials, Feldhege and Schmutterer (1993) used Margosan-O as pesticide and *E. formosa*, parasitoid of *Trialeurodes vaporariorum*, as target insect. The parasitized puparia of the whitefly were dipped in Margosan-O solution containing 10 or 20 ppm AZA.

The lower concentrations showed little effect on the parasitoid emergence from the puparia and on longevity, but higher concentrations caused reduction of the walking activity of the wasps. The parasitization capacity of the females decreased by 60-70 per cent.

Lafleur (1994) tested the effects of carbofuran, isofenophos and neem on the principal pests of rice and the two parasitoids of *Orseolia oryzivora* namely *Platygaster* sp. and *Tetrastichus* sp. It was observed that under the combined action of heavy pest pressure and washing neem subsequently proved ineffective. It was also observed that the synthetic insecticides reduced natural parasitism of *O. oryzivora* by the two parasites by 15-30 percent. In field trials in Orissa, India some neem derivatives (alone or in combination with synthetic organic insecticides) produced no effective control of the rice gall midge *O. oryzae*. Attack by the parasitoid, *Platygaster oryzae* was not adversely affected by neem derivatives alone or in combination with monocrotophos or chlorpyriphos. There was no linear relationship between the percentage of silver shoots and the extent of parasitism. Maximum parasitism was observed in an untreated control plot (53.2%), followed by a plot treated with 3 per cent neem oil spray (51%). The percentage parasitism was lower in insecticide-treated plots (Dash *et al.*, 1994).

4. NEEM AND PREDATORS

The bioactivity of neem on predatory insects, mites and spiders has been studied in detail (Table 6).

4.1 Predatory Insects

4.1.1 Earwigs

Sauphanor *et al.* (1995) tested the effect of NeemAzal-F against the European earwig, *Forficula auricularia*. This polyphagous insect is a crop pest and a predator at the same time. In peach and apricot orchards, for instance, it may cause serious damage by feeding on ripening fruit, whereas in applied orchards it can be useful by reduction of harmful populations of various aphid species. Adults of *F. auricularia*, exposed to 50 ppm AZA on glassplates (standarized method of IOBC/WPRS) in the laboratory, did not show increased mortality or reduced ingestion of food; fecundity was also not adversely influenced. On the other hand second instar nymphs treated with 25, 50 or 250 ppm of NeemAzal-F could not complete their metamorphosis and died. They also exhibited reduced food intake and extended stadia. Neither repellent nor phagodeterrent effects were observed. Under field conditions in a peach orchard, the nymphal population of the earwig was reduced by 70 per cent by spraying of NeemAzal-F at a concentration of 50 ppm. Hence, NeemAzal-F could be applied in peach or apricot orchards when a reduction of the nymphal population of the earwig is required but avoided if high numbers of the predator are desirable, for instances in apple orchards to obtain a significant reduction of aphids. Earlier, Schauer (1985) and Eisenlohr *et al.* (1992) observed that *F. auricularia* had no side effects by neem products.

4.1.2 Crickets

The cricket, *Metioche vittaticollis* preying upon eggs of rice leaffolders in Asia, was not affected by spraying neem seed bitters (containing AZA and other active ingredients) at 10000 ppm (8 l/ha ULV) in field trials in the Philippines (Lamb and Saxena, 1988).

Table 6. Compatibility of neem with predators

	Predator	Concentration	Observation	Reference
1	*Cyrtorhinus lividipennis*	Neem oil (3 %), Aqueous NSKE (5 %)	Positive effect	Fernandez *et al.* (1992)
2	*Orius majusculus*	NeemAzal-T/S (1:200, 1:100, 1:50)	Positive effect at lower concentration	Drescher and Madel (1995)
3	*Coccinella undecimpunctata*	Neem oil (1%)	Positive effect	Lowery and Isman (1995)
4	*Cryptolaemus montrouzieri*	NSKE (5 %)	Positive effect	Mani *et al.* (1997)
5	*Cheilomenes sexmaculata*	Neem oil (20-200ppm)	Positive effect	Oi *et al.* (2001)
6	*Chrysopa scelestes*	NSKE (2 %)	Positive effect	Joshi *et al.* (1982)
7	*Chrysoperla carnea*	AZT-VR-K (1000 ppm) with mixture of neem oil (250-30,000 ppm)	Positive effect	Kaethner (1990, 1991)
8	*Typhlodromus athiasae, Amblyseius barkeri, A. zakuri*	Margosan –O, Azatin, Repelin	Positive effect	Mansour *et al.* (1993)
9	*Lycosa pseudoannulata*	NSKE (5%), Neem oil (3 %) or aqueous extract of neem cake (10 %)	Positive effect	Raguraman (1987)

4.1.3 True bugs

Sharma *et al.* (1984) observed that *Orius* sp., a predator of sorghum midge, *Contarinia sorghicola* was unaffected by an active neem fraction. Chelliah and Rajendran (1984) tested the toxicity of seven insecticides against *Cyrtorhinus lividipennis*. The least toxic of the sprays was 0.07 per cent endosulfan, which was effective against rice hoppers, followed by 5 per cent neem oil, the corrected mortality percentages were 34.72 – 36.90 mortality on one day after spraying, 7.40-38.11 on the second day and 20.01-20.03 on third day. The other sprays tested (0.075% quinalphos, 0.04% chlorpyrifos, 0.07% phosalone, 0.08% monocrotophos and 1.0% carbaryl) were highly toxic to the bugs giving 36.07-53.96 per cent mortality on the third day.

Neem oil was slightly harmful to the mirid bug, *C. lividipennis* (Saxena *et al.*, 1984). Fernandez *et al.* (1992) conducted trial in green house against *C. lividipennis* using four treatments viz. neem oil 3 per cent, aqueous NSKE 5 per cent, endosulfan and water. They observed no mortality in the case of neem oil and aqueous NSKE while endosulfan induced cent percent mortality. However, in laboratory tests malformed symptoms of few nymphs of the predator *Nesidiocoris* sp. after the spraying of neem seed water extract 4 per cent and 2 per cent neem oil were observed (Serra, 1992). It was observed that these neem products had no significant effect on the field population of this bug. The toxicity of all the sprays diminished five days after spraying. Similarly a delayed moulting and morphogenetic defects after spraying of Margosan-O on third instar nymphs of the pentatomid predator, *Perillus bioculatus* of the Colorado potato beetle in the USA have been recorded (Hough and Keil, 1991). Krishnaiah and Kalode (1992) reported that the LD_{50} for NO 50 per cent for the black mirid bug, *Tytthus parviceps*, was 2.88 per cent, whereas for its prey, the green rice leafhopper, *Nephotettix virescens*, it was only 1.39 per cent.

Increasing mortality in test populations of the anthocorid, *Orius majusculus* was observed after NeemAzal-T/S treatment (concentrations 1:50 to 1:200) and after oral intake, the rate of emergence of first instar nymphs was reduced by 3 per cent (Drescher and Madel, 1995). No repellent or phagodeterrent effect was observed when treated eggs of *Sitotroga cerealella* served as food for the bugs. According to the guidelines of IOBC/WPRS for standardized tests on side effects of pesticides NeemAzal-T/S at 1:50 was 'slightly harmful' under laboratory conditions. No negative effect on fecundity, sex ratio, rate of emergence or behaviour was observed. Kareem *et al.* (1988) reported the effect of neem seed kernel extract on population of predatory mirid and Araneae in rice and compared with those of monocrotophos (0.75 kg a.i./ha). It was revealed that the populations of mirids and Araneae were also lower in plots treated with monocrotophos than in plots treated with neem, 48 days after treatment. Safety of neem formulations and insecticides to *Microvelia douglasi atrolineata*, studied for a predator of planthopper in rice ecosystem revealed that Neemix (2 and 4%) and Rakshak (0.2 and 0.5%) were the safest neem formulations where as phorate and carbofuran (1kg a.i./ha) granular application and quinalphos spray at 0.5 per cent were the least toxic to the predator (Lakshmi *et al.*, 1998b). Also neem formulations vis-a-vis insecticides were safe to *C. lividipennis* after the application of Neemgold at 0.5 per cent and Neemax at 2.0 per cent even after 72h exposure (Lakshmi *et al.*, 1998c) though chlorpyrifos and monocrotophos recommended for rice caused 100 per cent mortality within 24 h of exposure. Ghelani *et al.* (2000) tested various synthetic and botanical pesticides for their contact toxicity to the eggs and nymphs of *R. fuscipes* .The data on mortality of eggs and nymphs revealed that all the synthetic insecticides were more toxic than botanical insecticides. Among synthetic insecticides, quinalphos was highly toxic, while endosulfan was least toxic. However, among botanical insecticides, nicotine sulfate was least toxic to the eggs and nymphs of *R. fuscipes*.

Toxic effects of leaf extracts of *Azadirachta indica*, *Vitex negundo*, *Pongamia glabra* and *Calotropis gigantea* on different life stages of reduviid predator, *Rhynocoris marginatus* determined by contact and stomach toxicity clearly indicated that adults were more sensitive than nymphal instars. No mortality was observed in adults and nymphal instars of *R. marginatus* following contact toxicity studies (Sahayaraj and Paulraj, 1999).

Tedeschi *et al.* (2001) studied the side effects of three neem formulations (Neem-Amin EC, Stardoor and B.P. 20/S) on the mirid predator, *M. caliginosus* in the laboratory. Direct toxicity tests on first instar nymphs exposed to fresh dry residues on glass plates at different doses demonstrated that all the products were harmful to the insects with LD_{50} values much lower than the maximum recommended rate (1.217, 0.264, 1.083 mg a.i./l instead of 15, 31.5 and 80 mg a.i./l for Neem-Amin EC, Stardoor and B.P. 20/S, respectively). Moreover a reduction of fecundity of the surviving females was assessed with Neem-Amin EC and B.P. 20/S. High mortality was recorded when the insects were introduced onto the plants just after the treatment, but no significant differences compared with the controls were observed five days after the treatment. The experiments showed that azadirachtin being biodegradable, thus having short persistence, makes this active ingredient a promising component in integrated pest management programmes, if time gap is guaranteed between the treatment and the introduction of the predator.

4.1.4 Ants

Hellpap (1985) tested the neem product (AZT-VR-K) fed larvae of *Spodoptera frugiperda* to the colonies of the ant, *Ectatomma vuridum*. The ants accepted the neem-treated larvae. The predatory earwigs, *Doru laeniatum* were also exposed to the armyworm larvae. It was observed that after seven days of exposure, there was significant difference in mortality among earwigs fed with treated larvae of armyworm. Schmidt and Pesel (1987) reported that worker ants were resistant when sprayed with neem products. On the other hand, feeding of AZT-VR-K and MTB/H_2O-K-NR to the red forest ant, *Formica polyctena* led to a stimulation of egg production when low concentrations were used. In contrast, higher concentrations reduced the number of eggs drastically, some times down to zero after a few weeks. This effect could be reversed if feeding of neem products was stopped and untreated food supplied instead. Use of neem-based products with predatory ants, *Oecophylla smaragdina* gave excellent control of fruit flies, *Bactrocera cucurbitae* in organic agriculture system but it was not sufficiently active to manage *Aulacophora* spp (Rohan, 2000).

4.1.5 Beetles

In laboratory experiments, adult *C. septempunctata* kept on NO-treated glass-plates according to IOBC/WPRS guidelines, did not show increased mortality or reduction of fecundity compared to untreated control, but the metamorphosis of the larvae was interrupted (Schmutterer, 1981). The same insect species when treated in laboratory and semi-field trials with AZT-VR-K (1000 ppm) and a combination of it with NO (250-30000 ppm), there was no affect on the emergence of first instar larvae from treated eggs (Kaethner, 1990). Spraying on adults had no adverse effects on fecundity and activity (fitness), whereas the same treatment on fourth instar larvae under laboratory conditions induced mortality, especially of pupae that developed from treated larvae. Numerous adults that emerged from surviving pupae exhibited morphogenetic defects of their wings. In contrast, spraying on two coccinellid species including *C. septempunctata* in field cages did not result in any side effects. In laboratory studies of Lowery and Isman (1995) topical treatment of early second instar larvae of *C. undecimpunctata*, using 1 per cent NO, did not result in reduced pupation or emergence of adults as compared to controls.

Neem product Margosan-O had no harmful effect on *Delphastus pusillus* predating on *Bemisia tabaci* and *Scymnus* sp. predating on *Aphis gossypii* and *Myzus persicae* (Hoelmer *et al.*, 1990). Margosan-O also did not show any adverse affect against predatory carabid beetle, *Platynus dorsalis* when their soil habitat was treated with the neem product (Forster, 1991). Saleem and Matter (1991) observed that the neem oil acted as temporary repellent against the predatory staphylinid beetle, *Paederus alfierii*, the coccinellid, *C. undecimpunctata* and the lacewing, *Chrysoperla carnea* in cotton but otherwise neem oil had no adverse effect on these predators of *Spodoptera littoralis*. That neem oil had no adverse effect on predators is also obvious from the studies of Kaethner (1991), as it was found harmless to the eggs, larvae or adults of *Chrysoperla carnea* and *C. septempunctata*. It is also obvious from the findings of Mohapatra *et al.* (1991) where even 24 per cent concentration of NO had no significant adverse effect on the coleopteran predators in rice and Matter *et al.* (1993) demonstrated that although neem oil had residual activity for up to 6 days, yet it had no effect on survival or behaviour of larvae of *C. undecimpunctata* except for a prolongation of the fourth instar larva. Consumption of the aphids by this predator was unaffected. Eisenlohr *et al.* (1992) reported that NeemAzal-F had no effect on oviposition of coccinellids in peach orchards, though residual toxicity of some insecticides and neem seed kernel extracts against the predatory beetle, *Brumoides suturalis* has been recorded (Chandrababu *et al.*, 1997). It was found that NSKE extract and endosulfan exhibited low toxicity to *B. suturalis* larvae and adults.

An interesting study of Patel and Yadav (1993) on the toxicity of some botanical and chemical insecticides to *Cheilomenes sexmaculata*, and its hyperparasite, *Tetrastichus coccinellae* shows that among the botanicals, nicotine sulphate (0.05, 0.04 and 0.03%), Repelin (0.5, 0.75 and 1.0%) and Neemark (0.05, 0.2 and 0.4%) were highly toxic to adults of *T. coccinellae*, whereas they were absolutely safe to *C. sexmaculata*. In a detailed study of mortality and predation efficiency of *Coleomegila maculata* following applications of neem extracts, it was observed that the toxicity of the neem extracts to *C. maculata* was almost 100 per cent when both neem formulations were used at 10 per cent concentrations. The azadirachtin contents in neem oil (v/v) and neem seed kernels (w/v) were 13.7 and 91.0 ppm, respectively. Malathion was also tested at the field rate of 2.85 g a.i./l (Roger *et al.*, 1995). Adult mortality rate of the coccinellids after 72 h was 100 per cent following Malathion treatments. No toxicity was observed after the treatments with the aqueous suspension of ground neem seeds. The predation efficiency of *C. maculata* was also evaluated after topical application of these three insecticides at sublethal doses. Fifteen minutes after treatments, adult coccinellids were provided with 30 aphids for 24 h. The aqueous suspension of ground neem seeds caused 50 per cent reduction in the number of aphids consumed.

Stark and Wennergren (1995) opined that toxicity of pesticides to bioagents might not be straightforward but the susceptibility of various life stages should be estimated for notwothy concusions. Banken and Stark (1997) studied the stage and age influence on the susceptibility of *C. septumpunctata* after direct exposure to neem product, 'Neemix'. Where first instars were treated by direct application of 0, 40, 100, 200, 400, 600, and 1000 ppm and fourth instars were treated with 400, 600, 800 and 1000 ppm azadirachtin, the active ingredient in Neemix. The LC_{50} for first and fourth instars were estimated as 1120 ppm and 520 ppm azadirachtin, respectively. These values were much higher than the recommended rates for control of aphids (3 weekly applications of 20 ppm), suggesting that Neemix might be used in IPM programmes because application rates that control aphids should not result in appreciable mortality of predators. Fourth-instar larvae of C. *septempunctata* were innately more sensitive to the growth disrupting effects of acute exposure to Neemix than 1st instars. It is possible for early instars to sustain the effects of Neemix as long as the pesticide is detoxified before the onset of pupation. These results suggest that it is extremely important to examine more than one life stage of a species to estimate the total effect of pesticides. Banken and Stark (1998) also studied the exposure and the risk of neem products against *C. septumpunctata* using direct sprays, residues on leaves, and pesticide-contaminated prey. The pesticide alone and the predator caused significant decrease in aphid population. However, no significant ($P<0.05$) interaction between the predator and the pesticide was detected, indicating that the chemical and biological control agents were not working synergistically. Furthermore, exposure to the pesticide in microcosms significantly reduced or completely eliminated oviposition in adult *C. septumpunctata*, and all of the larvae exposed to 100 or 600 ppm died within 10 days of treatment. Although survivorship of adult ladybird beetles was unaffected, exposure to Neemix resulted in a severe reduction in fecundity or complete sterility depending on the concentration.

Mani *et al.* (1997) studied the effect of 5 per cent neem seed kernel extracts on the predator, *Cryptolaemus montrouzieri* and no detrimental effect was observed on the progeny production of *C. montrouzieri*. Dhaliwal *et al.* (1998) tested Achook and Nimbecidine for the control of insect pests on cabbage. The neem formulations were evaluated at 1, 2 and 4 kg/ha and compared to 0.5 kg a.i./ha of endosulfan used as treated control. Among these, endosulfan was the most effective against all the insect pests, followed by Achook and Nimbecidine. The feeding efficiency of the *C. septempunctata* on *L. erysimi*, treated with neem-based insecticides was higher than for aphids treated with endosulfan. Studies on *L. erysimi* control by Neemol and nicotine sulfate applied alone or in combination with chemical insecticides dimethoate and methyl-O-demeton in mustard (Vekaria and Patel, 2000) have also revealed that both plant products were less toxic to the predators, *Diaraetiella rapae* and *C. septempunctata* than the chemical insecticides. Chakraborti and Chatterjee (1999) also found that all formulations of neem were safe to the ladybird predators even at the highest concentrations (9 ml a.i./l).

Comparing the toxicity of different insecticides to the adult of *C. sexmaculata*, Prasad and Logiswaran (1998) report neem oil as the safest insecticide, based on LT_{50} values. It was concluded that the less toxic phosalone, monocrotophos or neem oil could be integrated with the release of *C. sexmaculata* in the field. Azadirachtin and dichlorvos also induced lowest toxicity to the predator *Cryptolaemus montrouzieri* (Sundari, 1998) and Neemix and Multineem had least affect against predatory coccinellids (Mishra and Mishra, 1998). Singh and Singh (1998) tested different neem based formulations and synthetic insecticides on aphidophagous coccinellids on *Brassica juncea.* Achook (WSP), RD-9 Repelin, NeemAzal-T/S, Neemgold, Neemta 2100 and Nimbecidine at 0.03 per cent were quite safe to coccinellids than the synthetic insecticides such as endosulfan 35EC, fenvalerate 20 EC, dimethoate 30EC and Chess 25EC. The order of safety was maximum in Achook followed by RD-9 Repelin, NeemAzal-T/S, Neemgold, Neemta 2100, Nimbecidine 0.03 per cent, endosulfan, Chess 25, fenvalerate and dimethoate during the first experimental trial (1994-95) and Neemgold followed by Achook, Annona EC, Neemta 2100, Achook EC, NeemAzal-T/S, endosulfan, Nimbecidine, Chess 25, fenvalerate, and dimethoate during second experimental year (1995-96). However, Imtiaz *et al.* (1998) found two neem extracts (RB-a and RB-b at 6, 7, 8, 9 and 10%) and an extract of bakayan (*Melia* sp.) berries (1, 2, 3, 4 and 5%) toxic to the coccinellid, *Coccinella* sp. and reported that 10 per cent RB-a and RB-b induced the highest mortality (85.7 and 82.5, respectively). Neem oil was, however, quite safe for natural enemies *Aphytis melinus* and *Chilocorus nigrita* predating *Aonidiella aurantii* (Krishnamoorthy and Rajagopal, 1998)

Simmonds *et al.* (2000) investigated the effect of crude neem seed extract, a formulation of azadirachtin (Azatin), a pyrethrum extract and one of the two naphthoquinones isolated from *Calceolaria andinabenth* on the foraging behaviour of the *C. montrouzieri* larvae and adults. All the botanicals influenced the foraging behaviour of *C. montrouzieri*, at one or more concentrations. Larval and adult foraging behaviour was influenced most by neem that also affected larval behaviour; the predators contacted fewer treated leaves and spent less time on treated than on untreated leaves. Larvae also consumed fewer mealybugs treated with naphthoquinones.

Ma *et al.* (2000) assessed the toxicity of several biorational pesticides and chemicals to *H. armigera* and *H. punctigera* and also on the major predators in cotton ecosystem. Moderate rate-dependent control was obtained in plots treated with neem seed extract - azadirachtin (AZA) at rates of 30, 60 and 90 g/ha. Plots treated with Talstar EC (bifenthrin) applications achieved the best results, followed by treatment with alternation of chemicals (methomyl, bifenthrin, thiodicarb and endosulfan) and biorational insecticides (neem oil, azadirachtin and Btk). Predators, including coccinellids, chrysopids, Araneae and hemipterans were insensitive to Aza, toosendanin (Tsdn) and Bt applications. In contrast, chemicals were very toxic to predators. The toxicity of azadirachtin to predaceous insects attacking bollworm, *H. armigera* by exposing *Menochilus signatus* and lacewings, *Harmonia conformis* to neem oil (50 and 200 ppm) and endosulfan (50 and 200 ppm) through prey, which had consumed one or the other of these compounds showed that endosulfan decreased predation rates by *H. conformis* at 50 ppm. However, azadirachtin, when ingested with prey, did not affect predation rates between 50 to 200 ppm concentrations (Oi *et al.*, 2001). Neither of these pesticides caused direct mortality to adult beetles or lacewing larvae at the tested concentrations. Azadirachtin at both concentrations delayed pupation of *M. signatus* and extended duration of the larval stage, which increased the number of prey consumed by the predator causing serious morality of the pupae. However, Pupal lacewings were all killed by 200 ppm azadirachtin treatment and 50 per cent at 50 ppm azadirachtin treatment, distinctly reducing the population of the next generation.

Michaud (2001) exposed two ladybird beetles, *Cycloneda sanguinea* and *Harmonia oxyridi* in the laboratory to eight fungicide formulations commonly used in citrus production in Florida, USA. Both benomyl and the combination of copper and petroleum oil proved toxic to larvae of *C. sanguinea* that were exposed to concentrations corresponding to recommended field rates, either as leaf residues or in topical spray applications. Larvae of *C. sanguinea* also suffered lethal effects when exposed to neem oil as a leaf residue, but not after topical application. No compounds appeared repellent to adult beetles of either species.

4.1.6 Syrphids

Field trial conducted using neem emulsifiable concentrate for the control of sorghum aphid, *Melanaphis sacchari* did not show any adverse effect on syrphid larvae and adults of coccinellids (Srivastava and Parmar, 1985). Third instar larvae of the hover fly, *Episyrphus balteatus* were mostly killed when treated with 100 ppm of an enriched seed kernel extract MTB/H_2O-VR-K synergized with sesame oil combined in a ratio of 1:4 (Schauer, 1985). The larvae/pupae of syrphid flies seem to be more sensitive to neem products than those of other predators. Eisenlohr *et al.* (1992) reported that the number of syrphid larvae was not reduced in the field after spraying of NeemAzal-F on peach trees infested by *Myzus persicae*, but the survival of adults derived from larvae collected in the field on treated trees and held afterwards in the laboratory was quite low. Lowery and Isman (1995) observed that adult emergence of *Eupeodes fuscipennis* was reduced by NO (0.5%, 1%, 2%) to 35, 24 and 0 per cent, respectively in comparison with controls.

4.1.7 Cecidomyiids

Lowery and Isman (1995) reported that the number of larvae of predaceous cecidomyiids was reduced in the field after application of NSE and NO (1%) as compared to controls.

4.1.8 Lacewings

Joshi *et al.* (1982) noted that 2 per cent neem seed kernel suspension, when sprayed on tobacco plants, conserved the *Chrysopa scelestes*, an egg and larval predator of *S.*

litura. The adults of the lacewing, *Brinckochrysa scelestes* (= *Chrysopa scelestes*) were repelled from egg laying on cotton plants after they were sprayed with various commercial neem products of Indian origin and aqueous NSKE (Yadav and Patel, 1992). First instar larvae of the predator emerged normally from treated eggs. Polyphagous predator, *Chrysoperla carnea* treated in laboratory and semi-field trials with AZT-VR-K (1000 ppm) and with a mixture of this product with NO (250-30000 ppm) induced no toxicity on eggs or adults; the fecundity of the latter was also not significantly affected (Kaethner, 1990,1991). The number of eggs (fecundity) laid by adult females developed from treated larvae was normal. The mortality of larvae fed with neem-treated aphids did not differ from that of controls. On the other hand, 79 per cent mortality of larvae occurred after topical treatment in the laboratory. In contrast, spraying of potato plants together with larvae of *C. carnea* in screenhouses did not result in any toxic or morphological effects. Vogt (1993) did not find any significant influence of NeemAzal-F on the larvae of the lacewing in field trials. In laboratory experiments of Hermann *et al.* (1995) high mortality of larvae and pupae of *C. carnea* occurred if larvae were kept on NeemAzal-T/S (0.3% and 0.6%) contaminated glass plates (IOBC/WPRS standardized tests), but practically no mortality was found in semi-field trials.
Vogt *et al.* (1997) also studied the effectiveness of NeemAzal-T/S at 0.3 per cent against *Dysaphis plantaginea* on apple and on its side-effects on *C. carnea*. A single application of NeemAzal-T/S in April gave very good control of *D. plantaginea* for about 5-6 weeks. After this period *D. plantaginea* builtup new colonies and *Aphis pomi*, too, increased in abundance. Yield losses caused by *D. plantaginea* were significantly lower in the neem-treated plot than in the untreated control plot. The side-effect test revealed that in the field NeemAzal-T/S was harmless to larvae of *C. carnea*. Neem seed extract was also found safe to *C. carnea* in comparison to nine insecticidal products (Sarode and Sonalka, 1999a) where chlorpyrifos, deltamethrin and cypermethrin were found highly toxic to *Chrysoperla*. There was no mortality of *C. carnea* due to neem-based pesticides like NSE 5 per cent, Neemark, Achook, and Nimbecidine each at 0.003 per cent and neem oil at 1 per cent (Deole *et al.*, 2000). On the contrary, Srinivasan and Babu (2000) evaluated NSKE and commercial neem products viz., NeemAzal-T/S, NeemAzal-F, Nimecidine, Neemgold, TNAU neem product 0.03 per cent EC, TNAU neem product NO 60 EC and Indeem against eggs, grubs and adults of the *C. carnea*. The products caused 14.66 to 25.33 per cent egg mortality compared to 8.00 per cent in untreated controls and 6.66 to 16.66 per cent grub mortality compared to 3.33 per cent in controls. The longevity of treated adults ranged from 18.66 to 20.66 days in treatements, while it was 23.66 days in control. Fecundity was also affected slightly by all neem products (599.66 to 741.66) as against 874.66 eggs in controls.

4.2 Predatory Mites

Different solvent (pentane, acetone, ethanol, and methanol) extracts of neem seed kernel were considerably more toxic to carmine spider mite, *Tetranychus cinnabarinus* than to its predator, *Phytoseiulus persimilis* (Mansour *et al.*, 1987). Also, Mansour *et al.* (1993) studied the effect of three commercial neem-based insecticides viz., Margosan-O, Azatin and Repelin on *T. cinnabarinus* and the predaceous mite, *Typhlodromus athiasae*. None of these products had any detrimental effect on the spider, *C. mildei*. Margosan-O and Azatin were not toxic either to *T. cinnabarinus* or to its predator, *T. athiasae*. Repelin induced high toxicity to phytophagous and predaceous mites. In laboratory trials of Chiu (1985), NO (0.5%) was found effective against the citrus spider mite, *Panonychus citri*, but harmless to predaceous mites of the genus *Amblyseius*.

Sanguanpong (1992) reported that NO and NSKE caused no harm to *P. persimilis* if applied at concentrations up to 1 per cent under laboratory and greenhouse trials. The eggs of the spider mite *Tetranychus urticae* were more sensitive than those of *P. persimilis*. However, the development of predaceous mites emerged from neem-treated eggs and the fecundity of females derived from such eggs was adversely affected. After application of NO at 0.8 per cent and above no adult *P. persimilis* developed, whereas pentane extract and the AZA-enriched product AZT-VR-K gave 63 and 51 per cent adults, respectively. Surviving adults of the predator, only after application of concentrations up to 0.6 per cent NO, 0.2 per cent pentane extract and 0.4 per cent AZT extract, laid eggs. Oil from seeds of *Azadirachta excelsa* (marrango tree) was considerably more effective against *T. urticae* and *P. persimilis* than NO. Commercial products of neem seed extracts (NeemAzal-S and Margosan-O) evaluated against the predatory mites *Amblyseius barkeri* and *Typhlodromus richteri* in the laboratory were found safe to *A. barkeri* as no mortality occurred at 0.05 and 0.2 per cent. However, the treatments had concentration dependent adverse effect on the rate of oviposition and food consumption of *A. barkeri* (Dimetry *et al.*, 1994). Although Margosan-O had no significant adverse effects on *T. richteri* up to 0.2 per cent treatment, NeemAzal-S caused about 30 per cent mortality. Neither product affected the sex ratio nor the progeny of *A. barkeri*. On the other hand, Kim *et al.* (2000) tested the effect of NeemAzal-T/S on fecundity, egg mortality and host preference of two spotted spider mite, *T. urticae* and its predator *A. womersleyi* in the laboratory. Mortalities of *T. urticae* and *A. womersleyi* adults were 97.7 and 20.0 per cent at 100 ppm treatment 72h after application, respectively. These results indicated that NeemAzal-T/S was highly toxic to *T. urticae*, and was less toxic to *A. womersleyi*. NeemAzal-F also reduced the food consumption rate at 0.2 and 0.05 per cent treatments for all the predatory mites, *Amblyseius barkeri, A. swirskii* and *A. zaheri* and was highly toxic to *A. swirskii*. In contrast, the two tested concentrations were safe for *A. barkeri* and *A. zaheri* as was neem cake extract (5%), NSKE (5%) and neem oil (3%) to *Amblyseius* sp. on cotton (Chinnaiah, 1999). Mansour *et al.* (1997) studied the effects of Neemguard on predaceous mite *P. persimilis* and the predatory spider *Cheiracanthium mildei*. It was observed that Neemguard was highly toxic to phytophagous mite *T. cinnabarinus* but had no effect on *P. persimilis* and *C. mildei*.

Papaioannou *et al.* (2000) studied the effects of a NSKE (Neemark) and Bioryl(R) vegetable oils against phytophagous and predatory mites using bean leaves treated with different concentrations. Neemark (3 and 5%) was moderately toxic to *T. urticae*, and highly toxic to *P. persimilis*.

4.3 Predatory Spiders

Saxena *et al.* (1984) reported that the wolf spider, *Lycosa* (=*Pardosa*) *pseudoannulata*, an important predator of leafhoppers in rice fields in Asia, was not harmed by neem oil (NO) and alcoholic or aqueous NSKE. In fact, NO (3%) and aqueous NSKE (5%) were quite safe for the spiders, though endosulfan induced 100 per cent mortality of the predators (Fernandez *et al.*, 1992). NSKE, NO or NCE (10%) treated rice plots had better recolonization of spider *L. pseudoannulata* than in monocrotophos (0.07%) treated plots after seven days of treatment (Raguraman, 1987; Raguraman and Rajasekaran, 1996). The same neem products also spared the predatory mirid bug, *C. lividipennis* (Mohan, 1989). The population of *L. pseudoannulata* and *C. lividipennis* were reported to be unaffected by different neem seed kernel extracts in paddy crop (Saxena, 1987; Shukla *et al.*, 1988; Saxena, 1989; Jayaraj *et al.*, 1993; Mariappan *et al.*, 1993). Similar observation on rice crop was made by Nirmala and Balasubramanian (1999) who studied the effects of insecticides and neem based formulations on the predatory spiders of rice-ecosystem. It was observed that feeding efficiency of *L. pseudoannulata* was higher than *T. javana* in all the treatments except in NSKE against green leafhopper, *Nephotettix virescens* as prey, whereas rise in body weight was obtained in both predator species when they were treated with neem products indicating the safety of neem to spiders. Babu *et al.* (1998) also reported that a combination of seedling root dip in 1 per cent neem oil emulsion for 12h + soil application of neem cake at 500 kg/ha + 1 per cent neem oil spray emulsion at weekly intervals gave an effective level of control of green leafhopper (*Nephotettix virescens*) infesting rice (var. Swarna). A combination of neem oil+urea at a ratio of 1:10 when applied three times at the basal, tillering and panicle initiation stages gave a superior level of control of brown planthopper (*Nilaparvata lugens*). The treatments, urea+nimin [neem seed extract] and a seedling root dip with 1 per cent neem oil emulsion+neem cake at 500 kg/ha+1 per cent neem oil spray emulsion at weekly intervals was equally effective against *N. lugens*. All neem products had little effect on predators, *C. lividipennis* and *L. pseudoannulata* (Sontakke, 1993; Babu *et al.*, 1998). NSKE sprays at 5, 10 and 20 per cent were also substantially safe for spiders and ants in cowpea ecosystems (Sithanantham *et al.*, 1997).

In laboratory trials, Mansour *et al.* (1986,1987) studied the toxicity of NSKE from different solvents on the spider, *Cheiracnthium mildei* and found that NSKE 2 per cent did not affect the spiders. But at 4 per cent concentration the sequence of toxicity of the extracts was pentane > acetone > ethanol > methanol and water; the latter two solvent extracts were non-toxic. Mansour *et al.* (1993) reported that the commercial products namely Margosan-O, Azatin and RD9 Repelin showed no toxicity to the spider.

Wu (1986) and Serra (1992) observed that the neem products were not at all toxic to spider predators. Nanda Kumar and Saradamma (1996) observed the activity of natural enemies in cucurbit fields, where neem-based pesticides were applied for the control of *Henosepilachna vigintioctopunctata*. Natural enemies observed in considerable numbers were *Tetrastichus* sp., *Chrysocoris johnsoni*,

Tetragnatha sp., *Oxyopes* sp. and orb-web spiders, and neem product did not inflict any harm to them.

Lynx spider, *Oxyopes javanus* was less sensitive to NO (50% EC) than *L. pseudoannulata* (LC_{50} values = 9.73 and 1.18%, respectively) (Karim *et al.*, 1992), thereby confirming that NO was the safest pesticide for spiders (Wu, 1986). In cornfields (Breithaupt, 1995) and cabbage fields (Saucke, 1995) in Papua New Guinea no significant effect was observed against *Oxyopes papuanus* from aqueous NSKEs (2%) or NeemAzal-S treatments. Serra (1992) did not observe adverse effects from NSKE 4 per cent applied on unidentified spiders in tomato fields in the Caribbean.

Nanda *et al.* (1996) tested the bioefficacy of neem derivatives against the predatory spiders, wolf spiders (*L. pseudoannulata*), jumping spider (*Phidippus* sp), lynx spider (*Oxyopes* sp.), dwarf spider (*Callitrichia formosana*), orb spider (*Argiope* sp.), damselflies (*Agriocnemis* sp.) and mirid bug (*C. lividipennis*). It was observed that the neem kernel extract and oil were relatively safer than the insecticides to *L. pseudoannulata*, *Phidippus* sp. and *C. lividipennis* in field conditions.

Markandeya and Divakar (1999) evaluated the effect of a commercial neem formulation (Margosan 1500 ppm) in the laboratory against two parasitoids and two predators. The formulation was tested at the field recommended dose of 10 ml/l. The neem formulation Margosan 1500 ppm was safe to all the four bioagents studied viz., *T. chilonis, B. brevicornis, L. pseudoannulata* and *C. sexmaculata*. Spider population in rice ecosystem was the lowest in carbofuran treatment and highest in neem cake treatments. The mean predator population of *Ophionea indica, Paederus fuscipes, Lycosa* sp. and coccinellid beetles was significantly higher in plots with *Azolla* at 5 t/ha, with or without neem cake at 1.5 t/ha, in field trials conducted in southern Tamil Nadu, India under lowland rice irrigated conditions (Baitha *et al.*, 2000).

5. CONCLUSIONS

Neem products are now widely acclaimed as broad-spectrum pesticides. Schmutterer and Singh (1995) listed 417 insect pest species as sensitive to neem. In the present era of biocontrol, safety concerns predominate the agro-ecosystem besides pest control. Since neem products are now on large-scale use, their safety to natural enemies has also become a debatable issue.

In the case of microbial agents, NPV and Bt are the most successful commercial products. Neem products either pure, crude or commercial so far did not show any adverse effects when combined with NPV or Bt. Though combining neem products with antifeedant property and microbials with stomach poison activity is disputed, the vast volume of research work carried out reveals that the antifeedant principles of neem do not influence in any way the activity of the microbials inside the insect gut. In fact, the growth disrupting principles of neem were found to add to the activity inside the insect system along with microbial principles leading to quicker mortality to give a cumulative effect.

In the case of parasitoids, certain guiding principles are suggested in accordance with multi-array activities of neem products in insects. Parasitoids are also susceptible, when they come in direct contact with neem products. In such circumstances blanket application of neem products without understanding the behaviour of the parasitoid may adversely affect the beneficial capacity of the parasitoid. For example, the inundative release of the egg parasitoid *T. chilonis*, should be resorted 3-4 days before / after neem products application. The external larval parasitoids are no exception to the ill effects if they are in direct contact with neem products. To avoid this, for inundative releases, application of neem products may be followed by the release of the parasitoids and spraying may be avoided if the parasitoids are in larval stages in the field. Hence presampling is suggested to know the stage of the parasitoid, be it internal or external, for timing the application of neem products.

In the case of predatory insects, mites and spiders, certain degree of selectivity is nevertheless appararent, as adult insects show, no or relatively low sensitivity as in the case of earwigs, crickets, true bugs, beetles, lacewings and wasps. This can be explained by the fact that growth-disrupting compounds affect the first line juvenile instars of insects. The fecundity of neem-treated adult, predaceous parasitic insects and the fertility of their eggs are also not or only slightly affected by neem, in contrast to some phytophagous species. In some cases the predation efficiency may be reduced Nymphal/larval instars of beneficial insects are sensitive to neem products. When topically treated, reduction in food ingestion, delayed growth, difficulties in moulting, teretological and morphogenetic defects, reduced activity and increased mortality are normally observed in the laboratory. But, far less drastic or even no effects are observed under semi-field or field conditions. This is partly due to the fast breakdown of the active principles under field conditions.

From the foregoing text, it is no exaggeration that neem and bioagents are nature's twin gifts to mankind for their utility in the IPM of agricultural pests, without endangering the agro-ecosystem. In fact, conservation biological control has most commonly used to enhance the activity of native organisms (Landis *et al.*, 2000). NPV and Bt are highly compatible with neem products. In the case of parasitoids/predators, presampling and timing of application are necessary in order to avoid the ill effects of neem products, if any, on them. It is obvious that new mellinnium will look forward to "integrated biological control" that will include natural enemies vis-à-vis other biopesticides synchronizing with ecological and behavioural aspects of pests.

REFERENCES

Aguda, R.M., Rombach, M.C. and Shepard, B.M. (1986), Effect of neem oil on germination and sporulation of the entomophagous fungus *Metarhizium anisopliae*, *International Rice Research Newsletter*, **11**, 34 -35.

Arun Baitha, Hameed, S.F. and Singh, R. (2000), Effectiveness and economics of various treatments and their impact on spider population in rice ecosystem,

Annals of Plant Protection Scienc, **8**, 13-16.

Awad, T.I., Onder, F. and Kismali, S. (1998), Focus on natural pesticides from the neem tree, *Azadirachta indica* A. Juss (Meliaceae), *Turkiye Entomoloji Dergisi*, **22**, 225-240.

Babu, B.R. and Krishnayya, P.V. (1998), Bioinsecticides to mitigate the load of quinalphos against cauliflower caterpillars. *Pesticide Research Journal*, **10**, 231-234.

Babu, G.R., Rao, G.M. and Rao, P.A. (1998), Efficacy of neem oil and neem cake for the control of green leafhoppers, brown plant hoppers and their effect on predators of brown plant hoppers, *Shashpa*, **5**, 91-94.

Bajan, C., Kmitowa, K. and Popowska, N.E. (1998), Reaction of various ecotypes of entomopathogenic fungus *Beauveria bassiana* to the botanical preparation neem and pyrethroid Fastak, *Archive of Phytopathology and Plant Protection*, **31**, 369-375.

Balikai, R.A., Giraddi, R.S., Yelshetty, S. and Sattigi, H.N. (1997), Control of *Helicoverpa armigera* (Hubner) on redgram with neem seed kernel extract, Bellary jali leaf extract, NPV and insecticides, *Karnataka Journal of Agricultural Science*, **10**, 999-1002.

Bandara, K.A.N.P. and Kudagamage, C. (1996), Potential of neem for controlling Leaf-eating caterpillar complex on cabbage, in *Neem and Environment 1*, eds. R.P. Singh, M.S.Chari, A.K. Raheja and W. Kraus, Oxford & IBH Publishing Co., New Delhi, pp. 547-558.

Banken, J.A.O. and Stark, J.D. (1997), Stage and age influence on the susceptibility of *Coccinella septempunctata* (Coleoptera: Coccinellidae) after direct exposure to Neemix, a neem insecticide, *Journal of Economic Entomology*, **90**, 1102-1105.

Banken, J.A.O. and Stark, J.D. (1998), Multiple routes of pesticide exposure and the risk of pesticides to biological controls: a study of neem and the seven spotted lady beetle (Coleoptera: Coccinellidae), *Journal of Economic Entomology*, **91**, 1-6.

Basedow, T., Lucknhard, J. and Peters, A. (1997), Results of two experiments on the control of the Colorado potato beetle *Leptinotarsa decemlineata* Say by *Bacillus thuringiensis tenebrionis* and NeemAzal-T/S, *Proceedings Entomology Conference*, Bayreuth, Germany, **11,** 1-6.

Baskaran, R.K.M., Venugopal, M.S. and Mahadevan, N.R. (1999), Effect of certain botanicals on biological activity of nuclear polyhedrosis virus of tobacco caterpillar (*Spodoptera litura*), *Indian Journal of Agricultural Sciences,* **69,** 224-226.

Beckage, N.E., Metcalf, J.S., Nielsen, B.D. and Nesbitt, J.D. (1988), Disruptive effects of azadirachtin on development of *Cotesia congregata* in host tobacco hornworm larvae, *Archives of Insect Biochemistry Physiology*, **9,** 47-65.

Beitzen H.I. and Hofmann, R. (1992), Versuchezur wirkung des azadireachtin haltigen, formulierten neemsamenextraktes, AZT-VR-NR aug *Hylobius obietis* L., *Lymantria monacha* L. and *Drino inconspicua* Meig., *Zeitschrift fur Pflanzenkrankheiten und Pflnzenschutz,* **99,** 337-348.

Bhanukiran, Y., Rao, P.A. and Rao, K.T. (1997), Effect of certain chemicals and their combinations on natural enemies of *Spodoptera litura* (F.) in groundnut, *Pesticide Research Journal*, **9**, 238-240.

Borah, M. and Basit, A. (1996), Effect of certain insecticides on the emergence of *Trichogramma japonicum* Ashmead, *Journal Agricultural Science Society North East India*, **9**, 224-225.

Breithaupt, J. (1995), Untersuchungen zur bekampfung des Asiatischen Maiszonslers *Ostrinia furnacalis* Guenee (Lepidoptera: Pyralidae) unter Vermeidung des Einsatzes synthetischer Insektizide im Markhamtal und im sudlichen Hochland in Papua Neuguinea, Doctoral thesis, University of Giessen, Germany.

Breithaupt, J., Schmutterer, H. and Singh, P.P. (1999), Aqueous neem seed kernel extracts for control of the Asian corn borer, *Ostrinia furnacalis* in Papua New Guinea, in *Azadirachta indica* A. Juss, eds. R.P. Singh and R.C. Saxena, Oxford & IBH Pub. Co. Pvt. Ltd., New Delhi, pp 191-198.

Cano, V. and Gladstone, S.M. (1994), Efecto de insecticida botanica, NIM-20, sobre el parasitismo por *Trichogramma pretiosum* an huevos del *Helicoverpa zea* en el cultivo de melon, *Manecho Integrado de Plagas* **33**, 23-25.

Chakraborti, S. and Chatterjee, M.L. (1999), Bioefficacy of neem based pesticides against safflower aphid, *Dactynotus carthami* HRL and their impact on coccinellid predators, *Journal of Interacademicia*, **3**, 186-191.

Chandra Mohan, N. and Nanjan, K. (1990), Effect of neem oil against diamond back moth in cabbage, *Neem Newsletter*, **7**, 35-36.

Chandrababu, A., Gautam, R.D. and Garg, A.K. (1997), Residual toxicity of some insecticides against the predatory beetle, *Brumoides suturalis* (Fab.), *Annals of Plant Protection Science*, **5**, 131-136.

Chari, M.S., Sreedhar, U. and Rao, R.S.N. (1997), Differential response of *Apanteles africanus* Cameron (Braconidae) and *Telenomus remus* Nixon (Scelionidae) to various pesticides on tobacco, *Journal of Insect Science*, **10**, 181-183.

Chaudhari, S. and Singh, R.P. (1997), Bio-efficacy of nuclear polyhedrosis virus (NPV) in combination with a neem formulations against tobacco caterpillar, *Spodoptera litura* (Fab.), *Neem Newsletter,* **14**, 12-14.

Chelliah, S. and Rajendran, R. (1984) Toxicity of insecticides to the predatory mirid bug, *Cyrtorhinus lividipennis* Reuter, *International Rice Research Newsletter*, **9**, 15-16.

Chinniah, C. (1999), Evaluation of certain neem derivatives for their safety to predatory mites *Amblyseius* sp. (Acari: Phytoseiidae), *Insect Environment*, **5**, 38.

Chiu, S.F. (1985), Recent findings on Meliaceae and other promising botanical insecticides in China, *Zeitschrift fur Pflanzenkrankheiten und Pflnzenschutz*, **92**, 310-319.

Cook, S.P., Webb, R.E. and Thorpe, K.W. (1996), Potential enhancement of the gypsy moth (Lepidoptera: Lymantriidae) nuclear polyhedrosis virus with the triterpene azadirachtin, *Environmental Entomology,* **25**, 1209-1214.

Das, N.D., Sankar, G.R.M. and Biswas, K.K. (2000), Field evaluation of botanical and biopesticides against pod borer, *Helicoverpa armigera* on pigeonpea, *Annals of Plant Protection Science*, **8**, 233-234.

Dash, A.N., Senapati, B. and Tripathy, M. K. (1994), Effect of neem derivatives alone and in combination with synthetic insecticides on gall midge and its parasitization, *Environmental Ecology*, **12**, 581-583.

Deepak, K.D., Choudhary, A.K. (1998) Toxicity of some insecticides to *Tetrastichus pyrillae* Crawford (Hymenoptera: Eulophidae) an egg parasite of sugarcane leafhopper, *Pyrilla perpusilla* (Walker), *Annals of Plant Protection Science*, **6**, 214-216.

Deole, S.A., Bodhade, S.N., Mahajan, L.B., Deotale, V.Y. and Sharnagat, B.K. (2000), Residual toxicity of some pesticides used in cotton pest management against a chrysopid, (*Chrysoperla carnea*), *Journal of Soils and Crops*, **10**, 279-291.

Devi, P.S.V. and Prasad, Y.G. (1996), Compatibility of oils and antifeedants of plant origin with the entomopathogenic fungus *Nomuraea releyi*, *Journal of Invertebrate Pathology*, **68**, 91-93.

Dhaliwal, G.S., Gill, R.S. and Dilawari, V.K. (1998), Management of insect pest complex of cabbage with neem based insecticides, in *Ecological Agriculture and Sustainable Development,* Vol. 2, eds. G.S. Dhaliwal, R. Arora, N.S. Randhawa, and A.K. Dhawan, Indian Ecological Society and Centre for Research in Rural & Industrial Development, Chandigarh, pp 306-314.

Dimetry, N.Z., Amer, S.A.A. and Momen, F.M. (1994) Laboratory trials of two neem seed extracts on the predatory mites, *Amblyseius barkeri* (Hughes) and *Typhlodromus richteri* Karg, *Bollettino di Zoologia Agraria e di Bachicoltura,* **26,** 127-137.

Dobelin, H. (1997), Aphid control of cut anemones with Neem-Azal T/S taking into account its effect on beneficials, in *Proc. 5th Workshop on Practice Oriented Results on Use and Production of Neem Ingredients and Pheromones*, eds. H. Kleeberg, and C.P.W. Zebitz, Wetzlar, Germany, pp 115-121.

Drescher, K. and Madel, G. (1995), Auswirkungen eines Niem Extraktes auf den Pradator *Orius majusculus* Reuter (Heteroptera: Anthocoridae*), Mitt. Dtsch. Ges. Allg. Angew. Entomol.* **9**, 128.

Facknath, S. (1999), Application of neem extract and intercropping for the control of cabbage pests in Mauritius, in *Azadirachta indica* A. Juss, eds. R.P. Singh and R.C. Saxena, Oxford & IBH Pub. Co., Pvt. Ltd., New Delhi, pp 165-176.

Eisenlohr, K., Domange, A. L., Lenfant, C. and Sauphanor, B. (1992), Effects of NeemAzal-F on aphids and beneficial insects in peach orchards in France, in *Proc.1st Worksh. Practice Oriented Results on Use* and *Production of Neem-Ingredients*, ed. H. Kleeberg, Druck & Graphic, Giessen, Germany, pp 27-40.

Feldhege, M. and Schmutterer, H. (1993), Investigations on side-effects of Margosan-O on *Encarsia formosa* Gah. (Hymenoptera: Aphelinidae), parasitoid of the greenhouse whitefly, *Trialeurodes vaporariorum* Westw. (Homoptera.: Aley-rodidae), *Journal of Applied Entomology*, **115**, 37-42.

Fernandez, N. J., Palanginan, E. L., Soon, L. L. and Botrella, D. (1992), Impact of neem on nontarget organisms, *Proceedings Final Workshop on Botanical Pest Control,* IRRI, Los Banos, Philippines, pp 117-121.

Forster, P. (1991), EinfluB Von Pflanzenschutzmitteln auf-larven and Adulte Von *Platymus dorsalis* (Pont) (Coleoptera: Carabidae) Und auf Adulte Von *Tachyporus hyphorum* (L.) (Coleoptera: Staphylinidae) in Labor-Halbfreiland Und Freilandversuchen, *Doctoral Thesis*, University of Giessen, Germany.

Garg , D.K. and Baranwal, V.K. (1998), An integrated approach for pest management in basmati rice in Haryana, in *Ecological Agriculture and Sustainable Development*, Vol. 2, eds. G.S. Dhaliwal, R. Arora, N.S. Randhawa and A.K. Dhawan, Indian Ecological Society and Centre for Research in Rural & Industrial Development, Chandigarh, pp 216-221.

Ghelani, Y.H., Vyas, H.N. and Jhala, R.C. (2000), Toxicological impact of certain insecticides to *Rhinocoris fuscipes* Fabricius (Heteroptera: Reduviidae)-a potential predator. *Journal of Applied Zoological Research*, **11**, 54-57.

Gill, S.A. and Raupp, M.J. (1994), Using entomopathogenic nematodes and conventional and biorational pesticides for controlling bagworm, *Journal of Arboriculture*, **20**, 318-322.

Gonzalez, D.M.E., Valbuena, P.B.F., Rivera, M.A., Bustillo, P.A.E. and Chaves, B. (1996), Viability of the fungus *Metarhizium anisopliae* in mixture with agrochemicals, *Revista Colombiana de Entomologia,* **22**, 31-36.

Grimm, C. and Guharay, F. (1998), Control of leaf-footed bug *Leptoglossus zonatus* and shield-backed bug *Pachycoris klugii* with entomopathogenic fungi, *Biocontrol Science Technology,* **8**, 365 – 376.

Gupta, G.P. (1996), Use of neem in management of pest complex in cotton, in *Neem and Environment,* Vol. 1, eds. R.P. Singh, M.S. Chari, A.K. Raheja and W. Kraus, Oxford & IBH Pub. Co. Pvt. Ltd., New Delhi, India, pp 469-484.

Gupta, G.P. and Sharma, K. (1998) Neem based pest management strategy in cotton System, *Neem Newsletter*, **15,** 6.

Gupta, G.P., Katiyar, K. N. and Kirti Sharma (1999), Neem in the management strategies of insect pests of cotton, in *Azadirachta indica* A. Juss, eds. R.P. Singh and R. C. Saxena, Oxford & IBH Pub. Co. Pvt. Ltd., New Delhi, India, pp 177-190.

Hazarika, L.K. and Puzari, K.C. (1997), Field efficacy of white muscardine fungus (*Beauveria bassiana*) on rice hispa (*Dicladispa armigera*), *Indian Journal of Agricultural Sciences,* **67**, 463-465.

Hellpap, C. (1985), Populationsokologie und biologisch biotechnische Bekampfung von *Spodoptera* spp. in Nicaragua, Doctoral thesis, Univ. of Frankfurt and Giessen, Germany.

Hellpap, C. and Zebitz, C.P.W. (1986), Kombinierte Anwendung von Niem-Samenextrakten mit *Bacillus thuringiensis*-Produkten bei der Bekampfung von *Spodoptera frugiperda* und *Aedes togoi*, *Journal of Applied Entomology*, **101,** 515-525.

Hermann, P., Zebitz, C.P.W. and Kienzle, J. (1995), Wirkung verschiedener NeemAzal-Formulierungen auf larven der Florfliege *Chrysoperla carnea* Steph.

im Labor und Halbfreiland, in 7 Int. Erfahrungsaust uber fur schungserg, Zum Okologischen Obstbau, LVWO Weinsberg, pp 114-118.

Hoelmer, K.A., Osborne, L.S. and Yokumu, R.K. (1990), Effects of neem extracts on beneficial insects in greenhouse culture, in *Proceedings Neem Workshop On Neem's Potential in Pest Management Programs*, eds. J.C. Locke and R.H. Lawson, USDA, Beltsville, Madison, pp 100-105.

Holmes, M.S., Hassan, E. and Singh, P.P. (1999), The contact and systemic action of neem seed extract against green peach aphid *Myzus persicae* Sulzer (Hemiptera: Aphididae), in *Azadirachta indica* A Juss., eds. R.P. Singh and R.C. Saxena, Oxford & IBH Pub. Co. Pvt. Ltd., New Delhi, pp 93-101.

Hough, G.J. and Keil, C. B. (1991), Prospects for integrated control of the Colorado potato beetle (Coleoptera: Chrysomelidae) using *Perillus bioculatus* (Heteroptera: Pentatomidae) and various pesticides. *Journal of Economic Entomology*, **84**, 1645-1651.

Hussaini, S.S., Kavitha, J.S. and Abid Hussain, M. (2001a), Tolerance of some indigenous entomopathogenic nematode isolates to pesticides and their effect on multiplication. *Current Nematology*, **12**, 29-34.

Hussaini, S.S., Singh, S.P. and Shakeela, V. (2001b), Compatibility of entomopathogenic nematodes (Steinernema, Heterorhabditidae: Rhabditida) with selected pesticides and their influence on some biological traits. *Entomon*, **26,** 37-44.

Imtiaz, A., Naqvi, S.N.H., Rahila, T., Azmi, M.A., Yasmeen, H. and Asia, A. (1998), Efficacy of neem fractions (RB-a and RB-b) in comparison with *Melia* fraction (BB) against Coccinella sp. *Journal of Experimental Zoology*, **1**, 85-89

Jakob, G. and Dickler, E. (1996), Auswirkungen von Niempraparaten auf den Apfelschalenwickler *Adoxophyes orana* F.v. R. (Lepidoptera : Tortricidae) und die ihn parasitierende Schlupfwespe *Colpoclypeus florus* Walk. (Hymenoptera : Eulophidae), in *Proceedings 4th Workshop Practice Oriented Results on Use and Production of Neem-Ingredients and Pheromones,* Druck & Graphic, Giessen, Germany, pp 93-.

Jani, J.J., Vyas, R.V. and Yadav, D.N. (1992), Compatibility of some botanical pesticides with muscardine fungi, *Neem Newsletter*, **9,** 17.

Jayaraj, S., Bharathi, M. and Sundara Babu, P.C. (1993), Integrated pest management, in *Neem Research and Development*, eds. N.S. Randhawa and B.S. Parmar, Society of Pesticide Science, New Delhi, India, pp 154-167.

Jhansi, K. and Sundara Babu, P.V. (1987), Evaluation of effects of parasite, fungus and neem oil individually and in combination against spotted pod borer, *Maruca testulalis* (Geyer), *Andhra Agriculture Journal*, **34**, 340-342.

Joshi, B.G., Ramaprasad, G. and Sitaramaiah, S. (1982), Effect of neem seed kernel suspensions on *Telenomus remus*, an egg parasite of *Spodoptera litura*, *Phytoparasitica*, **10**, 61-63.

Joshi, B.G., Sriprakash, K. and Murthy, P.S.N. (1993), Relative efficacy of botanicals and biocides alone and in combination against tobacco caterpillar, *Spodoptera litura* F, in *Botanical Pesticides in Integrated Pest Management*, eds. M.S. Chari and G. Ramprasad, pp 154-158.

Kaethner, M. (1990) Wirkung von Niemsamenproduckten auf die Reproduktionsfahigkeit und Fitne von *Leptinotarsa decemlineata* Say, *Melolontha hippocastani* F. und *Melolontha melolontha* L. sowie Nebenwsirkungen auf die aphidophagen Nutzlinge *Coccinella septempunctata* L. und *Chrysoperla carnea* (Stephens), Doctoral Thesis, Univ. of Giessen, Germany.

Kaethner,M. (1991) Keine Nebenwirkungen von Niemprodukten auf die aphidophagen Pradatoren *Chrysoperla carnea* (Steph.) und *Coccinella septempunctata* L, Anz Schadlingsk Pflanzensch Umweltsch, **64**, 97-99.

Kareem, A.A., Saxena, R.C. and Malayba, M.T. (1988), Effect of sequential neem (*Azadirachta indica*) treatment on green leafhopper (GLH), rice tungro virus (RTV) infection, predatory mirid and spiders in rice, *International Rice Research Newsletter*, **13,** 37.

Karim, R.A.N.M., Chowdhury, M.M.A. and Haque, M.N.M. (1992) Current research on neem in rice in Bangladesh, *Proceedings Final Workshop Botanical Pest Control,* IRRI, Los Banos, Philippines, pp 30-34.

Kim, D.I., Paik, C.H., Park, J.D., Kim, S.S. and Kim, S.G. (2000), Relative toxicity of NeemAzal-T/S to the predaceous mite, *Amblysieus womersleiyi* (Acari: Phytoseiidae) and the two spotted spider mite, *Tetranychus urticae* (Acari: Tetranychidae). *Korean Journal of Applied Entomology,* **39**, 53-58.

Klemm, U. and Schmutterer, H. (1993) Wirkungen von Niempraparaten auf die Kohlmotte *Plutella xylostella* L. and ihre naturlichen Feinde aus der Gattung *Trichogramma. Zeitschrift fur Pflanzenkrankheiten und Pflnzenschutz*, **100**, 113-128.

Koul, O., Dhaliwal, G.S., Marwaha, S.S. and Arora, J.K. (2002a), *Biopesticides and Pest Management*, Vol. 1, Campus Books International, New Delhi.

Koul, O., Dhaliwal, G.S., Marwaha, S.S. and Arora, J.K. (2002b*), Biopesticides and*

Pest Management, Vol. 2, Campus Books International, New Delhi.

Krisch, K. (1986), Studies on the efficacy of neem extracts in controlling major insect pests in tobacco and cabbage, in *Natural Pesticides from the Neem Tree and Other Tropical Plants,* eds. H. Schmutterer and K.R.S. Ascher, GTZ, Eschborn, Germany, pp 495-515.

Krishnaiah, N.V. and Kalode, M.B. (1992), Environmental impact studies on neem, *Proceedings Final Workshop Botanical Pest Control,* IRRI, Los Banos, Philippines, pp 50-53.

Krishnamoorthy, A. and Rajagopal, D. (1998), Effect of insecticides on the California red scale, *Aonidiella aurantii* (Maskell) and its natural enemies, *Pest Management in Horticulture Ecosystems,* **4**, 83-88

Kroschel, J. and Koch, W. (1996), Studies on the use of chemicals, botanicals and *Bacillus thuringiensis* in the management of the potato tuber moth in potato stores. *Crop Protection*, **15**, 197-203.

Kumawat, K.C. and Jheeba, S.S. (1999), Ecofriendly management of gram pod borer, *Helicoverpa armigera*, *Annals of Plant Protection Science,* **7**, 212-214.

Lafleur, G. (1994), The effects of carbofuran, isofenphos and neem on the principal pests of rice and parasites of the rice cecidomyiid in Burkina Faso, Sahel *PV Information*, **68**, 9-14.

Lakshmi, V.J., Katti, G., Krishnaiah, N.V. and Lingaiah, T. (1998a), Laboratory evaluation of commercial neem formulations vis-a-vis insecticides against egg parasitoid, *Trichogramma japonicum* Ashmead. (Hymenoptera: Trichogrammatidae), *Journal of Biological Control*, **11**, 29-32.

Lakshmi, V.J., Katti, G. and Krishnaiah, N.V. (1998b), Safety of neem formulations and insecticides to *Microvelia douglasi atrolineata* Bergroth (Heteroptera : Veliidae), a predator of plant hoppers in rice ecosystem, *Journal of Biological Control,* **11**, 33-36.

Lakshmi, V.J., Katti, G., Krishaiah, N.V. and Kumar, K.M. (1998c), Safely of neem formulations to *Cyrtorhinus lividipennis* Reuter, a predator of brown planthoppers *Nilaparvata lugens* (Stal) in rice crop, *Journal of Biological Control*, **12**, 119-122.

Lamb, R. and Saxena, R.C. (1988), Effects of neem seed derivatives on rice leaf folders (Lepidoptera : Pyralidae) and their natural enemies, *Proceedings Final Workshop IARI- ADB-EWC Project on Botanical Pest Control of Rice Based Cropping Systems,* IRRI, Los Banos, Philippines, 47.

Landis, D., Wratten, S.D. and Gurr, G.M. (2000) Habitat management for natural enemies, *Annual Review of Entomology*, **45**, 175-201.

Latha, K., Balasubramanian, G., Sundarababu, P.C. and Gopalan, M. (1994), Field evaluation of insecticides alone and in combination against leaffolders and their effect on natural enemies in rice, *Pest Management and Economic Zoology*, **2**, 105-109.

Li, K.H., Xu, X., Li, Y.F., Meng, Q.Z. and Zhou, L.C. (1986), Determination of toxicity of 29 chemicals to *Trichogramma japonicum* at various developmental stages, *Natural Enemies of Insects,* **8**, 187-194.

Loke, W.H., Heng, C.K., Basirun, N. and Mardi, A.R. (1992), Non-target effects of neem (*Azadirachta indica* A. Juss) on *Apanteles plutellae* Kurdj., cabbage, sawi and padi, in *Proceedings 3rd International Conference Plant Protection Tropics,* **2,** 108-110.

Lowery, D.T. and Isman, M.B. (1995), Toxicity of neem to natural enemies of aphids. *Phytoparasitica,* **23**, 297-306.

Lowery, D.T. and Isman, M.B. (1996), Effects of extracts from neem on aphids (Homoptera: Aphididae) and their natural enemies, in *Neem and Environment* **1**, eds. R.P.Singh, M.S.Chari, A.K.Raheja and W.Kraus, Oxford & IBH Pub. Co. Pvt. Ltd., New Delhi, India, pp 253-264.

Lyons, B., Helson, B. and Bourchier, R. (1996), Effects of azadirachtin-based insecticides on the egg parasitoid, *Trichogramma minutum* Riley (Hymenoptera : Trichogrammatidae), *Neem Newsletter*, **13**, 50-51.

Ma, D.L., Gordh, G. and Zalucki, M.P. (2000), Toxicity of biorational insecticides to *Helicoverpa* spp. (Lepidoptera: Noctuidae) and predators in cotton field. *International Journal of Pest Management*, **46**, 237-240.

Malathi, S. and Sriramulu, M. (2000), Laboratory efficacy of biotic insecticides against lepidopterous pests fed on treated cabbage leaves, *Shashpa*, **7**, 63-66

Mani, M. (1996), Safety of neem and other plant products to parasitoids and predators of citrus insect pests, *Journal of Insect Science,* **9**, 89-90.

Mani, M. and Krishnamoorthy, A. (1996), Response of the encyrtid parasitoid, *Tetracnemoidea indica* of the oriental mealybug *Planococcus lilacinus* to different pesticides, *Indian Journal of Plant Protection,* **24**, 80-85.

Mani, M., Lakshmi, V.J. and Krishnamoorthy, A. (1997), Side effects of some pesticides on the adult longevity, progeny production and prey comsumption of *Cryptolaemus montrouzieri* Mulsant (Coccinellidae: Coleoptera), *Indian*

Journal of Plant Protection, **25**, 48-51.

Mansour, F.A., Ascher, K.R.S. and Abo, M.F. (1993), Effects of Margosan-O^{TM}, and RD9 Repelin" on spiders and on predacious and phytophagous mites, *Phytoparasitica,* **21**, 205-211.

Mansour, F.A., Ascher, K.R.S. and Abo, M.F. (1997), Effects of Neemguard on phytophagous and predaceous mites and on spiders, *Phytoparasitica*, **25**, 333-336.

Mansour, F.A., Ascher, K.R.S. and Omari, N. (1986), Toxicity of neem (*Azadirachta indica*) seed kernel extracts prepared with different solvents, on the spider *Cheiracanthium mildei*, *Phytoparasitica*, **14**, 73-76.

Mansour, F.A., Ascher, K.R.S. and Omari, N. (1987) Effect of neem seed kernel extracts from different solvents on the predaceous mite *Phytoseiulus persimilis* and the phytophagous mite *Tetranychus cinnabarinus* as well as on the predatory spider *Cheiracanthium mildei*, in *Natural Pesticides from the Neem Tree and other Tropical Plants,* eds. H.Schmutterer, and K.R.S. Ascher, GTZ, Eschborn, Germany, pp 577-587.

Mariappan, V. and Samuel, D.K. (1993), Effect of non-edible seed oils in aphid transmitted chilli mosaic virus in Tamil Nadu, in *Neem and Environment 2*, eds. R.P. Singh, M.S. Chari, A.K. Raheja and W. Kraus, Oxford & IBH Pub. Co. Ltd., New Delhi, India, pp 787-792.

Markandeya, V. and Divakar, B.J. (1999), Effects of neem formulation on four Bioagents, *Plant Protection Bulletin*, **51**, 28-29.

Matter, M.M., Marei, S.S., Moawad, S.M. and Gengaihi, S. (1993), The reaction of *Aphis gossypii* and its predator, *Coccinella undecimpunctata* to some plant extracts, *Bulletin of Faculty of Agriculture, University of Cairo*, **44**, 417-432.

McCloskey, C., Arnason, J.T., Donskov, N., Chenier, R., Kaminski, J. and Philogene, B.J.R. (1993), Third tropic level effects of azadirachtin, *Canadian Entomologist*, **125,** 163-165.

Men, U.B. (1999), Braconid parasitoids from neem semilooper, *Hypomecis* sp. in Vidarbha, Maharashtra. *Insect Environment*, **5**, 27.

Men, U.B. and Thakare, H.S. (1998), Efficacy of biocides alone and in combination with insecticides against sunflower semilooper, *Thysanoplusia orichalcea* F. *PKV Research Journal*, **22**: 1, 57-59.

Michaud, J.P. (2001), Responses of two ladybeetles to eight fungicides used in Florida citrus: implications for biological control, *Journal of Insect Science*, **1**,

1-7.

Michelakis, S.E. and Vacante, V. (1997), The citrus leafminer status in Greece, *Bulletin OILB-SROP*, **20**, 81 – 82.

Mineo, N., Mineo, G. and Sinacori, A. (2000), Side effects of sprays with azadirachtin mixed with Biolid E(R) against the natural parasitoids of *Phyllocnistis citrella* Stainton (Lepidoptera: Gracillaridae), *Bollettino di Zoologia Agraria e di Bachicoltura,* **32**, 157-162.

Mishra, N.C. and Mishra, S.N. (1998), Impact of biopesticides on the useful insect pests and defenders of okra, *Neem Newsletter,* **15**, 18.

Moar, W.J. and Trumble, J.T. (1987), Toxicity, joint action and mean time of mortality of Dipel QX, avermectin BI, neem and *B. thuringiensis* against beet army worm (Lepidoptera: Noctuidae), *Journal of Economic Entomology*, **80**, 588-592.

Mohan, K. (1989), Studies on the effect of neem products and vegetable oils against major pests of rice and safety to natural enemies, M.Sc.(Ag.) Thesis, Tamil Nadu Agricultural University, Coimbatore, India.

Mohapatra, R.N., Patnaik, N.C., Mishra, B.K. and Patnaik, H.P. (1991), Comparative toxicity of some insecticides and neem oil against the brown plant hopper, *Nilaparvata lugens* (Stal) and its predators, *Orissa Journal of Agricultural Research,* **4**, 239-240.

Momen, F.M., Reda, A.S. and Amer, S.A.A. (1997), Effect of NeemAzal-F on *Tetranychus urticae* and three predaceous mites of the family Phytoseiidae, *Acta Phytopathologica et Entomological Hungarica*, **32**, 355-362.

Moser, G. (1994), (Nutzlingskomplex um *Aphis gossypii* Versuchmd Niems menkernwasser extrakten, an Beispeel von okra.) in der Dominikanischen Republik, Diploma Thesis, University of Giessen, Germany.

Murlikrishna, T., Lal, O.P., Singh, V.S., Dhingra, S. and Srivastava, Y.N.S. (1998), Effectiveness of *Bacillus thuringiensis* var. *kurstaki* and neem alone as well as in combination with other insecticides against three major insect pests of egg plant, Abstract *ICPPMSA*, Kanpur, India, 268-269.

Murugan, K. and Jeyabalan, D. (1998), Neem enhances the activity of microbial pesticides, *Insect Environment*, **4**, 3-4.

Murugan, K., Sivaramakrishnan, S., Kumar, N.S., Jeyabalan, D and Nathan, S.S. (1999), Potentiating effects of neem on nucleopolyhedrovirus treatment of *Spodoptera litura* Fab, *Insect Science Applications*, **19**, 229-235.

Murugan, K., Balu, R. and Pandey, A. (1998), Impact of certain plant products and *Bacillus thuringiensis* Berliner subsp. k*urstaki* on the growth and feeding physiology of *Helicoverpa armigera* Hubner (Insecta: Lepidoptera: Noctuidae), *Journal of Scientific and Industrial Research*, **57**, 757-765.

Nagesh, M. and Verma, S. (1997), Bioefficacy of certain insecticides against diamondback moth *(Plutella xylostella)* on cabbage, *Indian Journal of Entomology,* **59**, 411-414.

Nanda, U.K., Parija, B., Pradhan, N.C., Nanda, B. and Dash, D.D. (1996), Bioefficacy of neem derivatives against the insect pest complex of rice, in *Neem and Environment* **1**, eds. R.P. Singh, M.S. Chari, A.K. Raheja and W. Kraus, Oxford & IBH Pub. Co. Ltd., New Delhi, India, pp 517-528.

Nandakumar, C. and Saradamma, K. (1996), Botanical pesticides are benign to the natural enemies of *Epilachna* beetle, *Insect Environment*, **2**, 27.

Nasseh, O.M., Freres, T., Wilps, J., Kirkilionis, E. and Krall, S. (1992), Field cage trials on the effects of enriched neem oil, insect growth regulators and the pathogens *Beauveria bassiana* and *Nosema locustae* on desert locusts in the Republic of Niger. *Proceedings of Bilogical Control of Locusts* and *Grasshoppers,* International Institute of Tropical Agriculture, Cotonou, Republic of Benin, pp 311-320.

Nirmala, R. and Balasubramaniam, G. (1999), Residual toxicity effects of insecticides and neem based formulations on the predatory spiders of rice-ecosystem. *Neem Newsletter*, **16**, 37-38.

Oi, B.Y., Gordon, G. and Gimme, W. (2001), Effects of neem-fed prey on the predaceous insects *Harmonia conformis* (Boisduval) (Coleoptera: Coccinellidae) and *Mallada signatus* (Schneider) (Neuroptera: Chrysopidae), *Biological Control*, **22**, 185-190.

Olivella, P.E. and Vogt, H. (1997), Seasonal occurrence, abundance, parasitism and leaf damage of leafminer moths in apple orchards in South West Germany, Mitteilungen-der-Deutschen-Gesellschaft-fur-Allgemeine-und-*Angewandte-Entomologie,* **11**, 611-617.

Osman, M.Z. and Bradley, J. (1993), Effects of neem seed extracts on *Pholeastor (Apanteles) glomeratus* I. (Hymenoptera: Braconidae), a parasitoid of *Pieris brassicae* L. (Lepidoptera: Pieridae*), Journal of Applied Entomology*, **115,** 259-265.

Oswald, S. (1989), Untersuchungen zur integrierten Bekamofung der Kokoswanze *pseudothreaptus wayi* Brown (Heteroptera: Coriedae) auf Sansibar, Doctoral Thesis, Univ., of Giessen, Germany.

Papaioannou, S.P., Markoyiannaki, P.D. and Zoaki, M.D. (2000), Side effects of Neemark (*Azadirachta indica* A. Juss) and two new vegetable oils formulations on *Tetranychus urticae* Koch and its predator *Phytoseiulus persimilis* Athias Henriot, *Bollettino di Zoologia Agraria e di Bachicoltura*, **32**, 25-33.

Patel, M.M. and Yadav, D.N. (1993), Toxicity of some botanical and chemical insecticide to *Menochilus sexmaculatusi* (Coccinellidae) an important aphidophage and its parasite, *Tetrastichus coccinellae* (Eulopidae). World Neem Conference, Bangalore, India, Abstract p 24.

Patil, P.P. and Sarode, S.V. (1996), Impact of bio-rational insecticides combined with chemical insecticides on quality and yield of cotton, *Indian Journal of Entomology*, ***58,*** 175-181.

Perera, D.R., Armstrong, G., Naylor, R.E.L. and Senanayake, N. (2000), Response of *Chrysodeixis eriosoma* (Doubleday), *Plutella xylostella* L. and the parasitoid, *Cotesia plutellae* (Kurdjumov) to feeding deterrents. *Tropical Agricultural Research,* **12**, 186-198.

Pezowicz, E., Ignatowicz, S. and Kamionek, M. (1997), Comparative effect of water extracts of seeds of the Indian neem tree, *Azadirachta indica* A. Juss, a plant derived insecticide, Neemix TM and a synthetic insecticides, Bancol 50 WP, on the efficacy of the entomongenous nematode, *Steinernema feltiae* (Filipjev), *Annals of Warsaw Agricultural University*, **18**, 31-39.

Prasad, D., Chand, P. and Haque, M.F. (1987), Parasitization of *Heliothis armigera* (Hub.) in chickpea field as affected by pesticidal seed treatment. *Research and Development Reporter,* **4**, 236-238.

Prasad, G.S. and Logiswaran, G. (1998) Toxicity of different insecticides to the adult of *Cheilomenes sexmaculata* F. (Coccinellidae: Coleoptera). *Insect Environment*, **3**, 117-118.

Rabindra, R.J., Rajasekaran, B. and Jayaraj, S. (1997) Combined action of nuclear polyhedrosis virus and neem bitter against *Spodoptera litura* (Fabricius) larvae. *Journal of Biological Control,* **11**, 5-9.

Raguraman, S. (1987), Studies on the efficacy of neem products against rice insect pests, M.Sc. (Ag.) Thesis. Tamil Nadu Agricultural University, Coimbatore, India.

Raguraman, S. (1993), Behavioral and physiological effects of neem (*Azadirachta indica* A. Juss) seed kernel extracts on egg parasitoid, Trichogramma chilonis Ishii and larval parasitoid, *Bracon hebetor* Say, Doctoral Thesis, Indian Agricultural Research Institute, New Delhi, India.

Raguraman, S. and Rajasekaran, B. (1996), Effect of neem products on insect pests of rice and the predatory spider, *Madras Agriculture Journal*, **83**, 510-515.

Raguraman, S. and Singh, R.P. (1998), Behavioural and physiological effects of neem (*Azadirachta indica*) seed kernel extracts on larval parasitoid, *Bracon hebetor*, *Journal of Chemical Ecology*, **24**, 1241-1250.

Raguraman, S. and Singh, R.P. (1999), Biological effects of neem (*Azadirachta indica*) seed oil on an egg parasitoid, *Trichogramma chilonis, Jounal of Economic Entomogyl,* **92**, 1274-1280.

Reddy, G.V.P and Manjunatha, M. (2000), Laboratory and field studies on the integrated pest management of *Helicoverpa armigera* (Hubner) in cotton, based on pheromone trap catch threshold level, *Journal of Applied Entomology*, **124**, 213-221.

Reddy, G.V.P. and Guerrero, A. (2000), Pheromone based integrated pest management to control the diamondback moth, *Plutella xylostella* in cabbage fields, *Pest Management Science*, **56**, 882-888.

Rajasekhar, P., Rao, N.V. and Venkataiah, M. (1996), Utility of insecticide additives for managing *Helicoverpa armigera* on cotton, *Pesticide Research Journal*, **8**, 119-123.

Ranaivo, F., Welling, M., Zimmermann, G. and Schmutterer, H. (1996), Fitness reduction by the African migratory locust, *Locusta migratoria*, after application of low concentrations of *Metarhizium flavoviride* blastospores and neem oil, *Bulletin-OILB-SROP*, **19**, 236-239.

Rao, M.S., Reddy, R.P. and Chawla, G. (1996), Potential use of botanicals for the development of formulation of the bio-control fungus *Verticillium lecanii* (Zimm.) Viegas, *Zeitschrift fur Pflanzenkrankheiten und Pflnzenschutz,* **103**, 436-439.

Rao, P.R.M. and Rao, R.S.P. (1979) Note on the use of *Bacillus thuringiensis* and extracts of *Eclipta alba* (Linn.) Hassak and *Azadirachta indica* A. Juss. for the control of rice brown planthopper, *Indian Journal of Agricultural Science,* **49**, 905 – 906.

Reddy, G.V.P and Manjunatha, M. (2000), Laboratory and field studies on the integrated pest management of *Helicoverpa armigera* (Hubner) in cotton, based on pheromone trap catch threshold level, *Journal of Applied Entomology*, **124**, 213-221.

Rodriguez, A.L.D., Lagunes Tejeda, A., Riestra Diaz, D., Rodriguez Maciel, C., Velazquez Mendoza, J., Becerril Romand, E. Rodriguez Colorado, S. and

Pacheco Velasco, E. (1997), Compatibility of *Beauveria bassiana* and aqueous extracts of neem (*Azadirachta indica*) to control coffee borer (*Hypothenemus h ampei*), *Manejo Integrado de Plagas*, **44,** 14-19.

Roger, C., Vincent, C. and Coderre, D. (1995), Mortality and predation efficiency of *Coleomegilla maculata lengi* Timb. (Coleoptera: Coccinellidae) following application of Neem extracts (*Azadirachta indica* A. Juss., Meliaceae), *Journal of Applied Entomology*, **199**, 439-443.

Rohan, R. (2000), The management of major insect pests *Bactrocera cucurbitae* and *Aulacophora* spp. in cucurbits under three intensive systems, *The BCPC Conference: Pests and Diseases*, **3**, 981-985.

Rovesti, L. and Deseo, K.V. (1991), Effect of neem kernel extract on Steinernematid and Heterorhabditid nematodes, *Nematologia*, **37**, 493-496.

Saikia, P. and Parameswaran, S. (2001), Contact toxicity of chemical and bio-pesticides against *Cnaphalocrocis medinalis* Guenee and *Trichogramma chilonis* Ishii, *Jornal of Applied Zooogical Research,* **12**, 86-87.

Sahayaraj, K. and Paulraj, M.G. (1999), Toxicity of some plant extracts against life stages of a reduviid predator, *Rhynocoris marginatus*, *Indian Journal of Entomology*, **61**, 342-344.

Saleem, S.A. and Matter, M.M. (1991), Relative effects of neem seed oil and Deenate on the cotton leaf worm, *Spodoptera littoralis* Boisd. and the most prevalent predators in cotton fields in Menoufyia Governorate, *Bulletin of Faculty of Agricultural University, Cairo*, **42**, 941 – 952.

Sanguanpong, U. (1992), Zur Wirkung olhaltiger Niem und Marrangosamenprodukte auf die Gemeine Spinnmilbe *Tetranychus urticae* Koch sowie Nebenwirkung auf ihren naturlichen Gegenspieler *Phytoseiulus persimilis* Athias-Henriot, Doctoral thesis, Univ. of Giessen, Germany.

Sarode, S.V. and Sonalkar, V.U. (1999a), Ovicidal effect of some insecticides against *Chrysoperla carnea* (Stephens), *Pesticide Research Journal*, **11**, 97-98.

Sarode, S.V. and Sonalkar, V.U. (1999b), Influence of different insecticides on parasitization of *Corcyra cephalonica* by *Trichogramma chilonis* Ishii. *Pesticide Research Journal*, **11**, 99-101.

Sarode, S.V., Jumade, Y.S., Deotale, R.O. and Thakare, H.S. (1996), Bioefficacy of neem seed kernel extract (NSKE) and *Heliothis* nuclear polyhedrosis virus (HNPV) against *Helicoverpa* (*Heliothis*) *armigera* (Hub) on cotton, *PKV Research Journal*, **20**, 96-97.

Sarode, S.V., Jumade, Y.S. and Borkar, S.L. (1997), Efficacy of *Heliothis* neclear polyhedrosis virus and neem seed kernel extract combinations against *Helicoverpa armigera* (Hub.) on pigeonpea, *PKV Research Journal,* **21**, 227-229.

Sasikala, K., Rao, P.A. and Krishnayya, P.V. (1999), Comparative efficacy of eco-friendly methods involving egg parasitoid, *Trichogramma japonicum*, mechanical control and safe chemicals against *Leucinodes orbonalis* Guenee infesting brinjal, *Journal of Entomological Research*, **23**, 369-372.

Satpathy, S. and Samarjit Rai. (2000), Preliminary field evaluation of NPV and insecticides against tomato fruit borer, *Helicoverpa armigera* (Hubner), *Insect Environment,* **6**, 21-22.

Saucke, H. (1995), *Biological and integrated control of diamondback moth Plutella xylostella Lin.(Lepidoptera: Yponomeutidae) and other major pests in brassica crops*, SPC/GTZ Biological Control Project: Final Report 1992-1995, Conedobu, Papua New Guinea.

Sauphanor, B., Lenfant, C. and Sureau, F. (1995), Effects d'un extrait de neem (*Azadirachta indica* A. Juss.) sur le development de *Forficula auricularia* L. (Dermoptera), *Journal of Applied Entomology*, **119,** 215-219.

Saxena, R. C. (1987), Neem seed derivatives for management of rice insect pests-a review of recent studies, in *Natural Pesticides from Neem Tree and Other Tropical Plants,* eds. H. Schmutterer, and K.R.S. Ascher, GTZ, Eschborn, Germany, pp 81-93.

Saxena, R.C. (1989), Insecticides from neem tree, in *Insecticides of Plant Origin*, eds. J.T. Arnason, B.J.R. Philogene and P. Morand, ACS Symposium series 387, American Chemical Society, Washington D.C., pp 110-135.

Saxena, R.C., Epino, P.B., Cheng, W.Z. and Puma, B.C. (1984), Neem, chinaberry and custard apple: antifeedant and insecticidal effects of seed oils on leafhopper and planthopper pests of rice, in *Natural Pesticides from Neem Tree and Other Tropical Plants*, eds. H. Schmutterer and K.R.S. Ascher, GTZ, Eschborn, Germany, pp, 403-412.

Saxena, R.C., Waldbauer, G.P., Liguido, N.J. and Puma, B.C. (1981), Effects of neem seed oil on the rice leaf folder *Cnaphalocrosis medinalis*, in *Natural Pesticides from the Neem Tree* (*Azadirachta indica* A. Juss), eds. H. Schmutterer, K.R.S. Ascher and H. Rembold, GTZ, Eschborn, Germany, pp 198-203.

Schauer, M. (1985), Die Wirkung von Nieminhaltsstoffen auf Blattlause und die Rubenblattwanze, Doctoral Thesis, Univ. of Giessen, Germany.

Schmidt, G. and Pesel, E. (1987), Studies on the sterilizing effect of neem extracts in ants, in *Natural Pesticides from Neem Tree and Other Tropical Plants*, eds. H. Schmutterer and K.R.S. Ascher, GTZ, Eschborn, Germany, pp 361-373.

Schmutterer, H. (1981), Ten years of neem research in the Federal Republic of Germany, in *Natural Pesticides from the Neem Tree* (*Azadirachta indica* A. Juss), eds. H. Schmutterer, K.R.S. Ascher, and H. Rembold, GTZ, Eschborn, Germany, pp 21-51.

Schmutterer, H. (1990), Properties and potential of natural pesticides from the neem tree *Azadirachta indica*. *Annual Review of Entomology*, **35**, 271-297.

Schmutterer, H. (1992), Influence of azadirachtin, of an azadirachtin-free fraction, of an alcoholic neem seed kernel extract and of formulated extracts on pupation, adult emergence and adults of the braconid, *Apanteles glomeratus*, *Journal of Applied Entomology,* **113**, 79-87.

Schmutterer, H. (1995), *The Neem Tree: Source of Unique Natural Products for Integrated Pest Management, Medicine, Industry and Other Purposes,* VCH Publishers, Weinheim, Germany.

Schmutterer, H. (1996), Side effects of neem products on insect pathogens, predators and parasitoids, *Neem Newsletter*, **13**, 54-55.

Schmutterer, H. (1997), Side-effects of neem (*Azadirachta indica*) products on insect pathogens and natural enemies of spider mites and insects, *Journal of Applied Entomology*, **121**, 121-128.

Schmutterer, H. (1999), Side effects of neem products on insect pathogens and natural enemies of spider mites and insects, in *Azadirachta indica* A. Juss, eds. R.P. Singh and R.C. Saxena, Oxford & IBH Pub. Co. Pvt. Ltd., New Delhi, India, pp. 147-162.

Schmutterer, H. and Singh, R.P. (1995), List of insect pests susceptible to neem products, in *The Neem Tree Azadirachta indica A. Juss. and other Meliaceous Plants : Sources of Unique Natural Products for Integrated Pest Management, Medicine, Industry and Other Purposes*, ed. H. Schmutterer, VCH Weinheim, Germany, 325-326.

Schmutterer, H., Saxena R.C. and Von der Hyke, J. (1983), Morphogenetic effects of some partially purified fractions and methanolic extracts of neem seed on *Mythimna separata* (Walker) and *Cnaphalocrocis medinalis* (Guenee). *Zeitschrift fur Angwandte Entomolgie*, **95**, 230-237.

Schneider, A. and Madel, G. (1992) Fekunditat und Vitalitat adulter Schlupfwespen nach Exposition auf Niem (*Azadirachta indica*) behandelten Flachen, *Mitt.*

Dtsch. Ges. Allg. Angew. Entomol., **8**, 273-278.

Schrod, J., Basedow, T.H. and Langenbruch, G.A. (1996), Untersuchungen zur Populationsentwicklung und biologischen Bekampfung des Kartoffelkafers (*Leptinotarsa decemlineata* Say) an zwei Standorten in Sudhessen (BRD), *Journal of Applied Entomology*, **20**, 619-626.

Serra, A.C. (1992), Untersuchungen zum Einsatz von Niem samenextrakten in Rahmen integretter Ansaltze zur Bekampfung von Tomatenschadlingen in der Dominikanischen Repubik, Doctoral Thesis, Univ. of Giessen, Germany.

Shapiro, M., Robertson, J. L. and Webb, R. E. (1994), Effect of neem seed extract upon the gypsy moth (Lepidoptera: Lymantriidae) and its polyhedrosis virus. *Journal of Economic Entomology*, **87**, 356-360.

Sharma, H.C., Leushner, K., Shankaran, AV.B., Gunasekhar, D., Marthandamoorthi, M., Bhaskaraiah, K., Subramaniyam, M. and Sulthana, N. (1984), Insect antifeedants and growth inhibitors from *Azadirachta indica* and *Plumbago zeylanica*, in *Natural Pesticides from Neem Tree and other Tropical Plants*, eds. H. Schmutterer and K.R.S. Ascher, GTZ, Eschborn, Germany, pp 291-320.

Sharma, H.C., Sankaram, A.V.B., Nwanze, K.F. and Singh, P.P. (1999), Utilization of natural pesticides derived from neem and custard apple in integrated pest management, in *Azadirachta indica* A. Juss., Oxford & IBH Pub. Co. Pvt. Ltd., New Delhi, India, pp 199-213.

Shilke, S.S., Jadhav, L.D. and Salunkhe, G.N. (1990), Toxicity of some vegetable oils/extracts to *Chelonus blackburnii* Cam., an egg-larval parasite of *Phthorimaea operculella* (Zell.), *Indian Journal of Entomology*, **52**, 709-711.

Shukla, B.C., Kaushik, U.K. and Gupta, R. (1998), Efficacy of neem products for useful management of chickpea pod borer, *Neem Newsletter*, **15**,19.

Simmonds, M.S.J., Manlove, J.D., Blaney, W.M. and Khambay, B.P.S. (2000), Effect of botanical insecticides on the foraging and feeding behaviour of the coccinellid predator *Cryptolaemus montrouzieri*, *Phytoparasitica*, **28**, 99-107.

Singh, R. P. and Saxena, R. C. (1999), *Azadirachta indica* A.Juss, Oxford & IBH Pulishing Co. Pvt. Ltd., New Delhi.

Sithanantham, S., Otieno, Z.N. and Nyarko, K.A. (1997), Preliminary observations on effects on neem sprays on some non target arthropods in cowpea ecosystem, in *Quatrieme Conference Internationale sur les Ravageurs en Agriculture*, Le Corum, Montpellier, France, pp 643-648.

Sonalkar, V.U., Deshmukh, S.D., Satpute, U.S. and Ingle, S.T. (1998), Efficacy of nuclear polyhedorsis virus in combination with adjuvants against *Helicoverpa armigera* (Hub.), *Journal of Soils and Crops*, **8**, 67-69.

Sontakke, B.K. (1993), Field efficacy of insecticide alone and in combination with neem oil against insect pests and their predators in rice. *Indian Journal of Entomology,* **55**, 260-266.

Sreenivasa, A.G. and Patil, B.V. (1998), Role of commercial neem products in cotton insect pest management, *Neem Newsletter,* **15**,17.

Srinivasa Babu, K., Murthy, M.S.N. and Ramesh Babu, T. (1996), Effect of Botanicals on certain parasitoids, in *Neem and Environment,* **2**, eds. R.P. Singh, M.S. Chari, A.K. Raheja and W. Kraus, Oxford & IBH Co. Pvt. Ltd., New Delhi, pp 1117-1126.

Srivastava, K.P. and Parmar, B.S. (1985), Evaluation of neem oil emulcifiable concentrate against sorghum aphids, *Neem Newsletter,* **2**, 7.

Srivastava, M., Paul, AV.N., Rengasamy, S., Kumar, J. and Parmar, B.S. (1997), Effects of neem (*Azadirachta indica* A. Juss) seed kernel extracts on the larval parasitoid *Bracon brevicornis* Wesm. (Hymenoptera: Braconidae), *Journal of Applied Entomology*, **121**, 51-57.

Srinivasan, G. and Babu, P.C.S. (2000), Effect of neem products on predatory lacewing, *Chrysoperla carnea* Stephens (Chrysopidae: Neuroptera), *Pesticide Research Journal*, **12**, 123-126.

Stansly, P.A. and Liu, T.X. (1997), Selectivity of insecticides to *Encarsia pergandiella* (Hymenoptera: Aphelinidae), an endoparasitoid of *Bemisia argentifolii* (Hemiptera: Aleyrodidae), *Bulletin of Entomological Research,* **87**, 525-531.

Stark, J.D. (1997), Entomopathogenic nematodes (Rhabditida: Steinernematidae): Toxicity of neem. *Neem Newsletter*, **14**, 18.

Stark, J.D. and Wennergren, U. (1995), Can population effects of pesticides be predicted from demographic toxicological studies? *Journal of Economic Entomology*, **88**, 1089-1096.

Stark, J.D., Vargas, R.I. and Wong, T.Y. (1990), Effects of neem seed extracts on tephritid fruiflies (Diptera: Tephritidae) and their parasitoids in Hawaii, in *Neem Potential in Pest Management Programme*, eds. J.C. Lock and R.H. Lawson, USDA, ARS 86, pp.106-111.

Stark, J.D., Wong, T.Y., Vargas, R.I. and Thalman, R.K. (1992), Survival, longevity and reproduction of tephritid fruitfly parasitoids (Hymenoptera:

Braconidae) reared from fruit flies exposed to azadirachtin, *Journal of Economic Entomology*, **84**, 1125-1129.

Stathers, T.E., Moore, D. and Prior, C. (1993), The effect of different temperatures on the viability of *Metarhizium flavoviride* conidia stored in vegetable and mineral oils, *Journal of Invertebrate Pathology*, **62**, 111-115.

Sundari, M.S.N. (1998), Residual toxicity of insecticides against *Cryptolaemus montrouzieri* Mulsant (Coleoptera: Coccinellidae), *Insect Environment*, **4**, 45

Tedeschi, R., Alma, A. and Tavella, L. (2001), Side-effects of three neem (*Azadirachta indica* A. Juss) products on the predator *Macrolophus caliginosus* Wagner (Heteroptera: Miridae), *Journal of Applied Entomology,* **125**, 397-402.

Teggelli, G.R., Biradar, A.P., Vastrad, S.M. and Rajashaker, D.W. (1998), Impact of pesticides on an endoparasitoid, *Campoletis chlorideae* Uchida, *Advances in Agricultural Research India*, **10**, 25-29.

Tewari, G.L. and Moorthy, K.P.N. (1985), Plant extracts as antifeedant against *Henosepilachna viginctioctopunctata* (F.) and their effects on the parasites, *Indian Journal of Agricultural Science*, **55,** 120-124.

Thakare, A.V., Ameta, O.P., Tripathi, N.N., Vyas, A. and Tauro, P. (1992), Relative efficacy of indigenous plant extract against gram pod borer, *Heliothis armigera* (Hub.), in *Proc. First Nat. Symp. Allelopathy in Agroecosystems (Agriculture & Forestry*), ed. S.S. Narwal, CCS Haryana Agricultural University, Hisar, India, pp. 162- 167.

Thakur, J.N. and Pawar, A.D. (2000), Comparative toxicity of different insecticides against *Trichogramma chilonis* Ishii, *Journal of Biological Control*, **14**, 51-53.

Tillmann, M. (1992), Kombinierte Anwendung der Praparate *Bacillus thuringiensis* var. *tenebrionis* und Neem Azal-S gegen Larven von *Leptiontarsa decemlineata* Say (Coleoptera, Chrysomelidae) and *Epilachna varivestis* Muls. (Coleoptera: Coccinellidae), Diploma thesis, Univ. of Giessen, Germany.

Tomar, R.K.S. (1998), Efficacy and economics of biopesticides and insecticide combinations against okra shoot and fruit borer, *Indian Journal of Entomology*, **60**, 25-28.

Trisyono, A. and Whalon, M.E. (1999), Toxicity of neem applied alone and in combinations with *Bacillus thuringiensis* to Colorado potato beetle (Coleoptera: Chrysomelidae), *Journal of Economic Entomology*, **92**, 1281-1288 .

Vekaria, M.V. and Patel, G.M. (2000), Bioeffficacy of botanicals and certain chemical insecticides and their combinations against the mustard aphid,

Lipaphis erysimi, Indian Journal of Entomology, **62**, 150-158.

Velayudhan, R., Senrayan, R. and Rajadurai, S. (1988), Parasitoid-host interactions with reference to *Anastatus ramakrishnae* (Mani) (Hymenoptera: Eupelmidae) in relation to pentatomid and coreid hosts. *Proc. Indian National Science Academy.- Biological Sciences*, **54**, 145-153.

Villanueva, J.A. and Hoy, M.A. (1998), Toxicity of pesticides to the citrus leafminer and its parasitoid *Ageniaspis citricola* evaluated to assess their suitability for an IPM program in citrus nurseries, *Biocontrol,* **43**, 357-388.

Vogt, H. (1993), Einsalz von Niempraparaten gagen *Adoxophyes arana* F.v. R und Untersuchungen zu Nebenwirkungen, in Intern. Erfahrungsaust. uber For schungserg. Zum Okologischen Obstbau, LVWO Weinsberg, pp 51-55.

Vogt, H., Handel, U., Vinuela, E. and Kleeberg, H. (1997), Field investigations on the efficacy of NeemAzal-T/S against *Dysaphis plantaginea* (Passerini) (Homoptera: Aphididae) and its effects on larvae of *Chrysoperla carnea* Stephens (Neuroptera: Chrysopidae), in *Proc. 5th Workshop on Practice Oriented Results on Use and Production of Neem-ingredients and Pheromones*, ed. C.P.W. Zebitz, Wetzlar, Germany, pp 105-114.

Vyas, R.V., Jani, J.J. and Yadav, D.N. (1992), Effect of some natural pesticides on entomogenous muscardine fungi, *Indian Journal of Experimental Biology,* **30**, 435-436.

Wanjari, R.R., More, G.D., Supare, N.R., Turkar, K.S. and Agarkar, V.K. (1998), Management of *Helicoverpa armigera* (Hub.) on chickpea with some herbal, c hemical and bio-pesticides, *Journal of Soils and Crops*, **8**, 34-37.

Wu, H.L. (1986), Determination of toxicity of some insecticides to *Lycosa pseudoannulata* and *Apanteles cypris*, *Natural Enemies of Insects,* **8,** 230-231.

Wilps, H., Nasseh, O., Krall, S., Kabo, O. and Nasseh, O.M. (1992), Control of *Schistocerca gregaria* adults in the area of new preparations and methods, *Sahel PV Information,* **44,** 7-18.

Yadav, D.N. and Patel, A.R. (1992), Effect of some botanical insecticides on oviposition of *Chrysopa scelestes* and their ovicidal action, in *Proceedings of Symposium on Botanical Pesticides in IPM*, eds. M.S. Chari and G. Ramaprasad, CTRI, Rajahmundry, India, pp 166-169.

Chapter 9

NEEM FOR PLANT PATHOGENIC FUNGAL CONTROL: THE OUTLOOK IN THE NEW MILLENNIUM

GOVINDARAGHAVAN SURESH, GEETHA GOPALAKRISHNAN AND SELLADURAI MASILAMANI
Centre for Natural Products, SPIC Science Foundation, New 64, Mount Road, Chennai 600 032, India

1. INTRODUCTION

Rachael Carson's "Silent Spring" strongly recommended the benign biological control as a viable and useful alternative to environmentally toxic synthetic chemicals. Yet, the future is predicted to belong to a middle path, i.e., Integrated Pest Management (van Emden and Peakall, 1996). Use of non-toxic and environmentally safe natural products forms a part of benign biological control strategies in crop protection management.

Angiosperms have adopted mechanical, phenological and an array of chemical defenses to ward off microbial attack. Based on the disease resistance factors in higher plants several attempts have been made to classify the chemical defenses of plants, primarily as pre-infectional or constitutive and post-infectional or induced chemical constituents (Harborne, 1982). In search of chemical leads of environmental friendly antimicrobials, diverse classes of secondary chemistries have been studied in detail (Harborne, 1982). Among these, "if a census were taken one would probably find that the number of different terpenes present in plants is greater than that of any other group of natural products " (Goodwin, 1967). Enormous structural diversity of terpenoids evidenced in angiosperms, if viewed as evolution in progress, it will not be surprising to understand the chemoecological functions for the plant terpenoids especially in the context of constitutive defense strategy against phytopathogens.

To date, one of the most extensively studied plant species for its prodigality in terpenoid diversity is the Indian neem tree, *Azadirachta indica*. *Azadirachta indica* A. Juss, an arboreal species native to the Indian sub-continent, has found use

O. Koul and S. Wahab (eds.), Neem: Today and in the New Millennium, 183–207.

in agriculture and medicine for centuries in India. Renaissance of neem research was initiated with the chemical investigations (Siddiqui, 1942) and insect control properties (Pradhan *et al.*, 1962) of neem, which resulted in the subsequent isolation of the most potent insect antifeedant and growth regulatory triterpenoid, azadirachtin (AZA) from the neem seed kernels (Butterworth and Morgan, 1968). Subsequently, over 120 triterpenoids have been isolated from different parts of the neem tree and more than 400 insects susceptible to neem have been identified (Kraus, 1995; Schmutterer and Singh, 1995).

Recognition of fungicidal properties of neem is known traditionally in India and the information is based on application of green and mature neem plant residues and compost (Singh and Pandey, 1966; Dath, 1982; Prakash *et al.*, 1985), neem seed oil and seed extract (Jain and Agarwal, 1978; Singh *et al.*, 1980; Singh *et al.*, 1984; Muthusamy *et al.*, 1988; Nasem and Lanjewar, 1990), leaf powder/extracts (Sinha and Saxena, 1987; Krishna *et al.*, 1986; Ghewande, 1989) and oil cake (Singh, 1968; Singh and Vyas, 1984) for the control of phylloplane and rhizosphere fungi in cropping systems. Results on the *in vitro* antifungal assays and field evaluation of neem extractives have been shown to be highly variable (Parveen and Alam, 1993; Locke, 1995) which are in part due to lack of standardization in extraction protocols and innate variability of the seed material. Extensive studies on the chemistry and biology of the neem tree has yielded information that in part explains such variability in bioactivity against phytopathogens. Such information includes

- diversity of chemical constituents in different parts of neem,
- variability in concentration and quality of the chemical constituents
  - due to agroclimatic conditions, seasonal variations and genetic variability
  - use of diverse extraction solvents and methods
  - storage conditions,
- differential susceptibility of different species of phytopathogens, and
- diversity of bioassay design and target variability.

2. DIVERSITY OF CHEMICAL CONSTITUENTS IN DIFFERENT PARTS OF NEEM

Neem seeds are a rich source of protolimonoids, intact, D-ring modified and C-ring modified apoeuphol tetranortriterpenoids of which the latter is characteristic of the Sub-family Melieae to which neem belongs. Number of limonoids has been isolated till date (Kraus, 2002). Other than triterpenoids, neem seed oil also contains 20-45 per cent oil consisting of a number of long chain fatty acids (Devkumar and Mukherjee, 1983; Rukmini, 1987) alkyl sulfides, disulfides, cyclic tri and tetra sulfides (Balandrin *et al.*, 1988; Riar *et al.*, 1990). A number of intact and C-ring modified limonoids have been isolated from the leaves and the bark yielded a number of phenolic diterpenoids (Schmutterer, 1995; Devkumar and Sukh Dev, 1993).

Concentration of oil in seeds varied from 38–54 per cent in seeds collected from various parts of the world (Ermel, 1995). Research on the variability in concentration of chemical constituents in seed / leaves / bark are fragmentary. But, comprehensive information on the azadirachtin content in seeds collected from different parts of the world has been provided (Ermel, 1995) showing a variation of AZA content from 2.05 mg to as high as 6.10 mg per g seed kernel.

Based on preparative HPLC analyses, the authors have found that the content of nimonol, an intact apoeuphol limonoid, to be comparatively meager in foliage during November – December, while the fresh foliage yielded as much as 2 g / 5 kg of leaves. The former did contain higher concentrations of compounds such as nimbolide and epoxynimonol while in the fresh foliage comparatively lesser concentrations of these limonoids were recorded (unpublished).

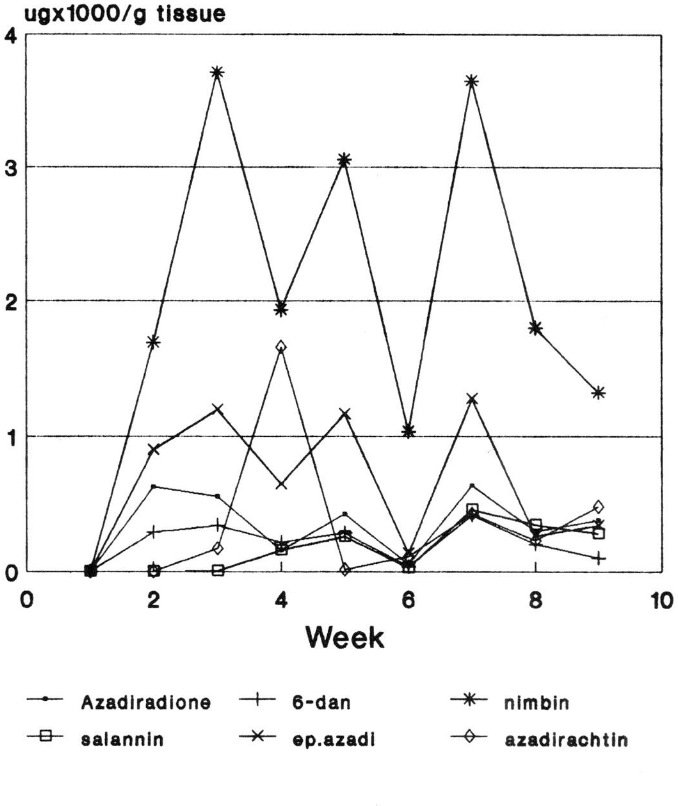

Figure 1. Neem seed limonoids in seeds during fruit development (Samples from a single tree)

During an exercise to study the appearance of major tritepenoids in neem seeds during seed development, the authors found that appearance of nimbin and azadiradione preceded azadirachtin and salannin (Fig. 1). It was also found that there is a wide variation in concentration of the four limonoids in seeds of the same age collected from different bunches of the same tree (unpublished). Addition to the innate variability in chemical constituents and their concentrations in various plant parts, the procedures of extraction results in increased differences in the content and concentration. Use of organic solvents such as n-hexane, chloroform, acetone, ethyl acetate, methanol and ethanol are normally utilized for the extraction of active triterpenoids (Feuerhake, 1985), while the farmers traditionally use aqueous extracts. Fresh neem seed kernel powder is packed in muslin cloth bags and left in containers with water for 24 hours and the aqueous extracts then directly sprayed on plants for control of herbivorous insects. Constituents of such aqueous extracts also show considerable amounts of tetranortriterpenoids (Table 1).

3. DIFFERENTIAL SUSCEPTIBILITY OF PHYTOPATHOGENIC FUNGI TO NEEM EXTRACTIVES

A perusal of literature clearly indicates differential susceptibility of diverse species of phytpathogenic fungi to neem extractives. A number of soil-borne, foliar and post harvest pathogenic fungi have been controlled by soil amendment or foliar spray of extractives of neem cake, neem leaves, neem seed kernel and oil, both under laboratory and field evaluations. Fungi thus controlled would include *Rhizoctonia solani* (Singh, 1968; Kannaiyan and Prasad, 1981), *Fusarium oxysporum, Helminthosporium nodulosum, Alternaria tenuis* (Khan *et al*, 1974) *Sclerotium rolfsii, S.sclerotiorum* (Singh *et al.*, 1980), *Botrytis cinerea, Penicillium expansum* and *Glomerella cingulata* (Moline and Locke, 1993), *Penicillium italicum, Alternaria alternata, Aspergillus niger* (Ali *et al.*, 1992). Downy mildew of grape due to *Plasmopara viticola* was reported susceptible to several neem derived extracts (Achimu and Schlosser, 1992) and a number of rust pathogens have also been reported to be controlled using neem seed oil application (Locke, 1990; Parveen and Alam, 1993).

In contrast, soil amendment of neem cake resulted in the increase of the total saprophytic fungal population (Khan *et al.*, 1974). Growth of *Pythium aphanidermatum* was stimulated in natural soils amended with oil cake and it was suggested that alcohol soluble portions of oil has growth stimulatory properties (Singh and Pandey, 1967). Neem kernel suspension as spray did not control damping-off of tobacco (Nagarajan and Reddy, 1980). Clarified neem oil did not reduce rot disease incidence in sweet potato caused by *Rhizopus* and may have increased rot severity (Locke, 1995). The differential susceptibility of sixteen phytopathogenic fungi to 90 per cent methanol extractive of neem seed oil has been demonstrated by Govindachari *et al* (1998), (Table 2).

Table 1. Concentration of triterpenoids in the aqueous extracts of neem kernels (A) and in the water extracted kernels (B)[a] (*Source:* Govindachari *et al.*, 1999a)

Compound	% in A	In mg	Concentration[b] Ppm	% in B	In mg
Azadirachtin A	1.03	43	14	1.29	165
Azadirachtin B	0.91	38	12	1.6	204
Azadirachtin D	0.69	28	9	1.12	143
Azadirachtin H	0.20	8	2	0.25	32
Azadirachtin I	0.21	8	2	0.45	60
Desacetylnimbin	0.09	4	1.3	2.88	368
Azadiradione	1.65	69	23	0.36	46
Nimbin	0.84	35	12	2.9	371
Salannin	1.65	69	23	5.97	764

[a] A – residue (4.2 g) obtained by the CH_2Cl_2 extraction f the aqueous extract of neem seed kernel powder (2 kg/3 lit); B – residue (12.8 g) obtained by extraction of marc with MeOH.
[b] Concentration in aqueous extract.

4. IDENTIFICATION OF ANTIFUNGAL COMPOUNDS IN NEEM EXTRACTIVES: THE NECESSITY

Locke (1995) while summarizing the work on the use of neem for fungal control, emphasized the understanding of chemical compositional diversity influenced by biotype differences, geographic distributional patterns, edaphic and climatic conditions and the resultant changes in efficacy of neem extractives. These factors will be decisive in order to extract and separate antifungals and to formulate them to maximize the benefit. Subsequent approaches, then, should be to identify the 'chemical entities' that confer fungitoxicity and to understand their mechanism(s) of action. The clues on the chemical entities conferring fungistatic/fungitoxic activity have been demonstrated by the use of clarified neem oil as an effective fungistat against a number of phytopathogens and chemical characterization of neem oil/neem seed kernel extracts. The advancement in chromatographic techniques and structure elucidation procedures has helped, to a large extent, in the identification of phytochemical constituents of neem seed oil/kernel extractives. Till recently, there was a lack of concerted effort to take up antifungal activity guided isolation of active compounds from the extractives of neem seeds, leaves and other parts of the tree. Before 1990s, the only reference to antifungal activity of neem constituents relate to nimbidin (supposedly a mixture of a number of triterpenoids from seed oil) (Khan *et al,* 1974) against *Rhizoctonia solani, Alternaria tenuis, Fusarium oxysporum, Helminthosporium nodulosum* and *Curvularia tuberculata.* Subsequently, Steinhauer (1996) attempted, for the first time, to identify antifungal constituents from the methanolic extract of neem seed kernels subsequent to defatting with petroleum-benzene. Water-soluble portion of the methanolic extract and dichloromethane fraction of DCM : H_2O partition of the methanol extract were initially tested against a number of phytopathogens. The active dichloromethane phase was further fractionated through low-pressure column chromatography and most active fraction was identified. Further preparative HPLC resulted in a pure compound that was found to inhibit the growth of *Drechslera teres* and *Alternaria porri* completely. Elucidation of the structure of the active compound was not completed in this study. It was shown that the activity of the isolated compound was much stronger than the crude extract and it was concluded that other compounds in the crude extract must have counteracted the inhibitory activity of the isolated compound. Using the groundnut rust disease (*Puccinia arachidis*) as the bioassay system, two intact apoeuphol limonoids, nimonol and isomeldenin with antifungal activity were isolated through extraction, solvent fractionation and preparative HPLC from the uncrushed green neem leaves (Fig. 2), (Suresh *et al.*, 1997).

Groundnut leaves treated with the n-hexane wash of the fresh uncrushed green leaves, had reduced rust disease incidence as evidenced by reduced number of rust pustules (Fig. 3). By the 13th day almost 50 per cent reduction in the number of

Table 2. Antifungal activity of 90% MeOH extract (at 1000 ppm) of neem oil against phytopathogenic fungi

Fungus	Radial growth (in mm)		%
	Control	1000 ppm	Inhibition
Bipolaris indica Rai,Wadhwani &Tewari	56.4 ±8.4 [a]	48.5 ± 7.1	14.0
Botryodiplodia theobromae Pat.	72.3 ± 6.0 [e]	64.9 ± 9.5	10.2
Colletotrichum lindemuthianum (Sacc.& Magnus)	51.6 ±10.4 [b]	41.2 ±8.4	20.7
Colletotrichum dematium (Pers. : Fr.) Grove	48.3 ± 9.0 [b]	36.5 ±6.8	24.4
Curvularia lunata (Wakker) Boedijn	61.2 ± 9.4 [c]	45.3 ±7.2	26.0
Fusarium equisetii (Corda) Sacc.	55.4 ± 10.6 [d]	44.5 ±7.1	19.6
Fusarium solani (Mart.) Sacc.	41.6 ± 11.8 [d]	32.4 ±1.1	22.1
Kolerago noxia Donkae	69.2 ± 7.5 [e]	51.7 ±6.4	25.3
Nectria galligena Bresad	53.0 ± 10.0 [f]	34.06 ±6.0	35.8
Pestalotiopsis mangiferae (Henn.) Stey.	49.6 ± 6.5 [c]	26.6 ±2.9	46.5
Phoma betae Frank.	56.5 ± 9.6 [c]	41.1 ±6.0	27.2
Pyricularia oryzae Cavara	35.4 ± 6.4 [g]	29.3 ±4.6	17.2
Pythium aphanidermatum (Edson) Fitzp.	27.4 ± 5.9 [a]	7.7 ±2.8	71.8

% Inhibition values and the fungal radial growth in diameter (mm) ±SE derived from Neumann-Keul means of growth (mm i.d) of the fungi at 1000 ppm vs control at maximum growth. [a] after 238 h, [b] after 252 h, [c] after 214 h, [d] after 142 h [e] after 94 h, [f] after 166 h, [g] after 190 h of growth. (*Source:* Govindachari *et al.*, 1998).

Figure 2. Fractionation of antifungal triterpenoids from fresh, green leaves of A. *indica*

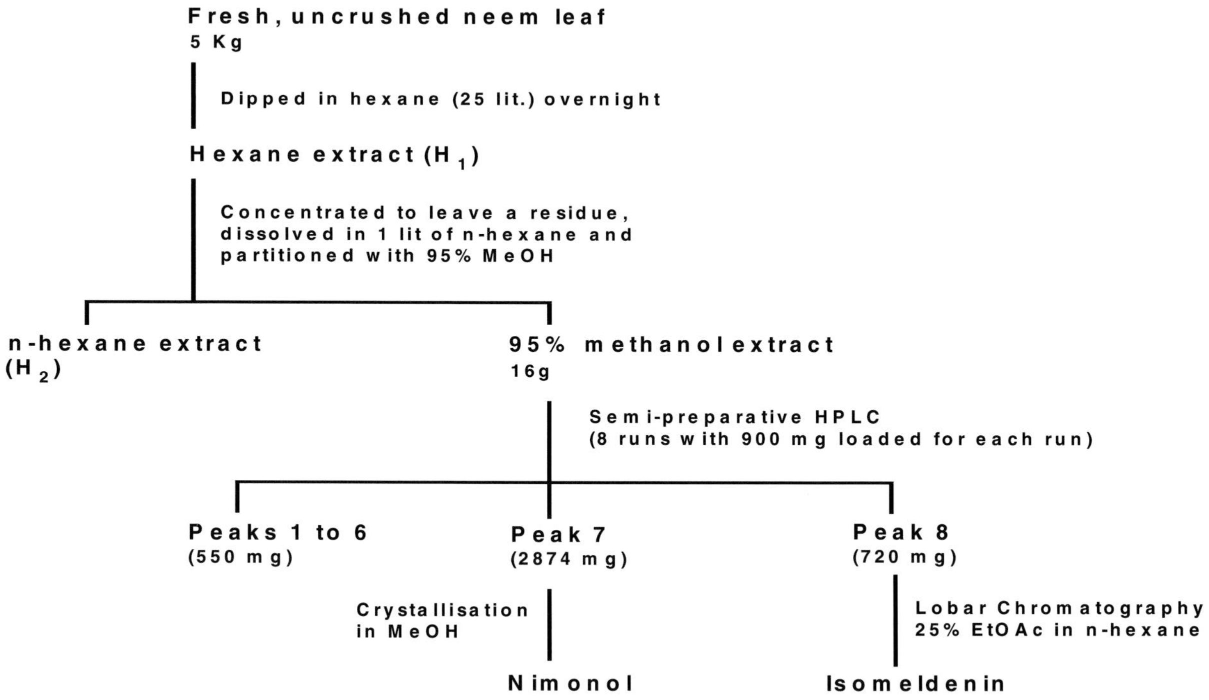

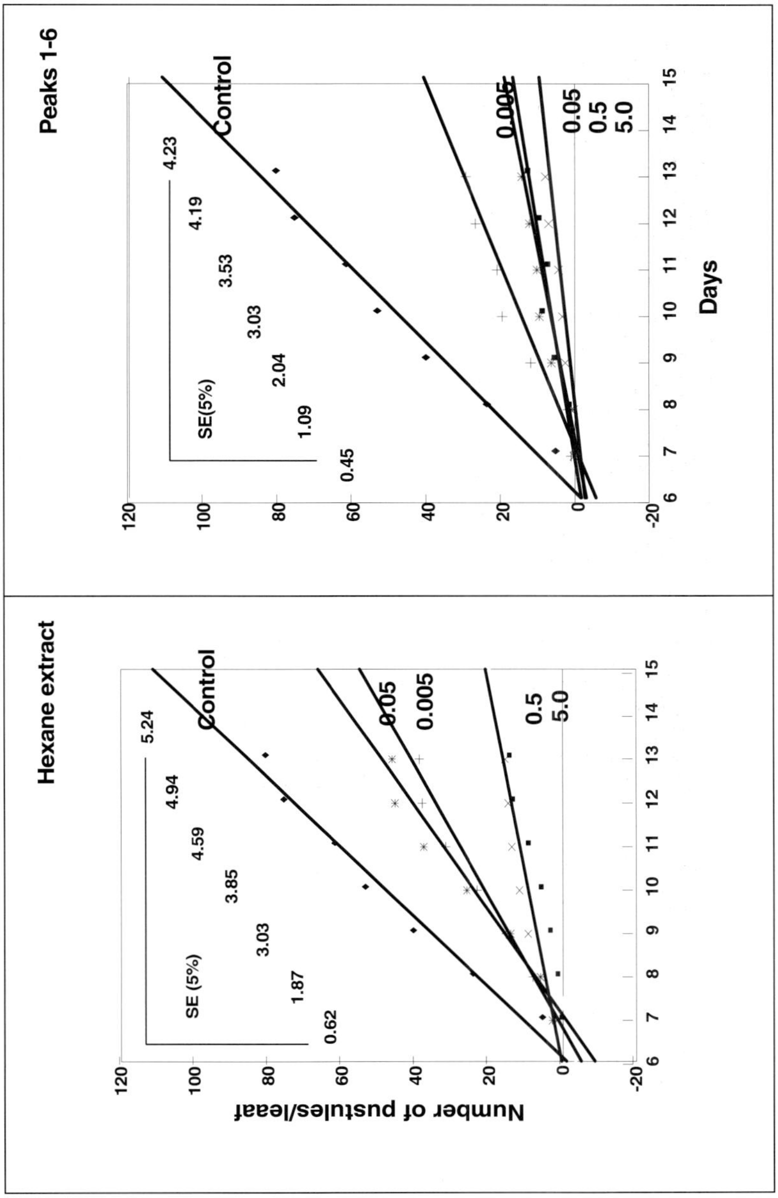
Hexane extract
Peaks 1-6
Number of pustules/leaaf
Days
SE (5%)
5.24
4.94
4.59
3.85
3.03
1.87
0.62
SE(5%)
4.23
4.19
3.53
3.03
2.04
1.09
0.45
Control
0.05
0.005
0.5
5.0

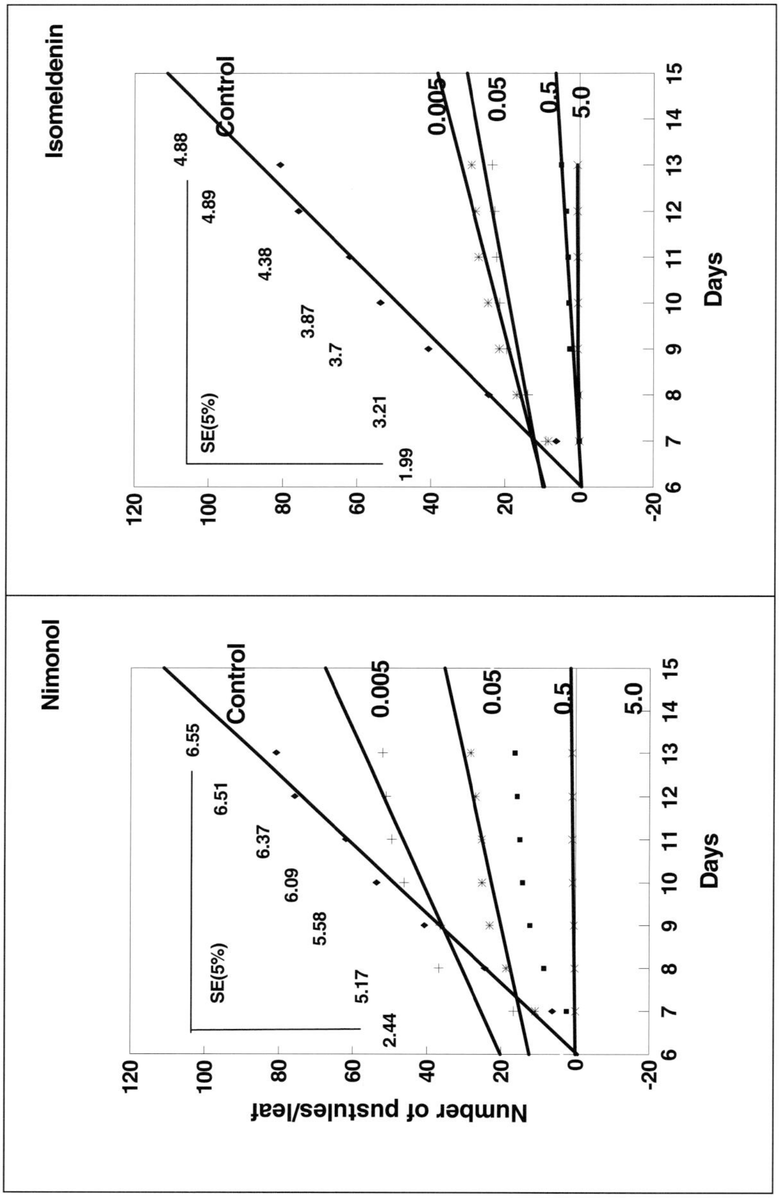

Figure 3. Effect of neem leaf extractives on the incidence of rust pustules on groundnut leaves.

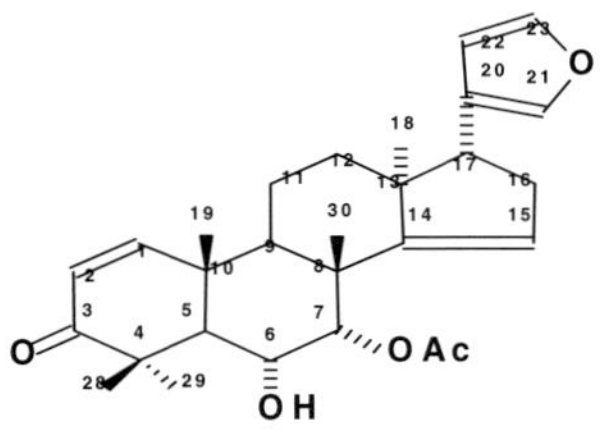

Nimonol

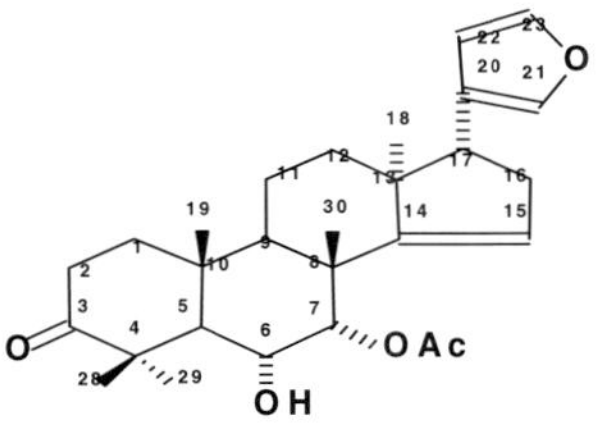

Isomeldenin

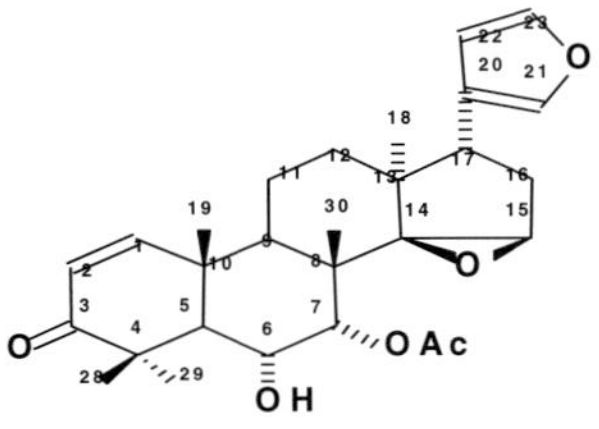

14,15-β-epoxynimonol

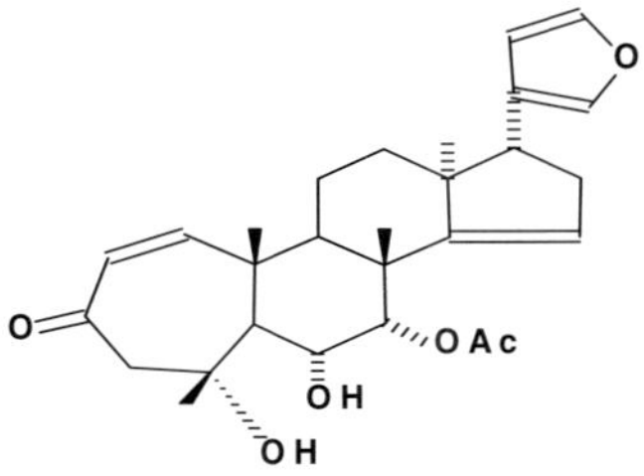

Homoazadiradione

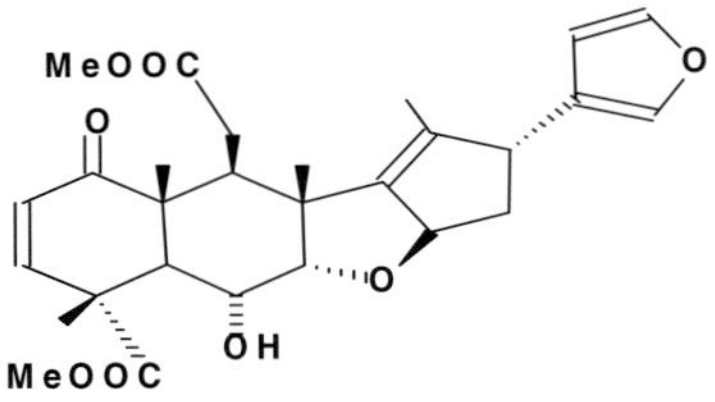

6-deacetylnimbin

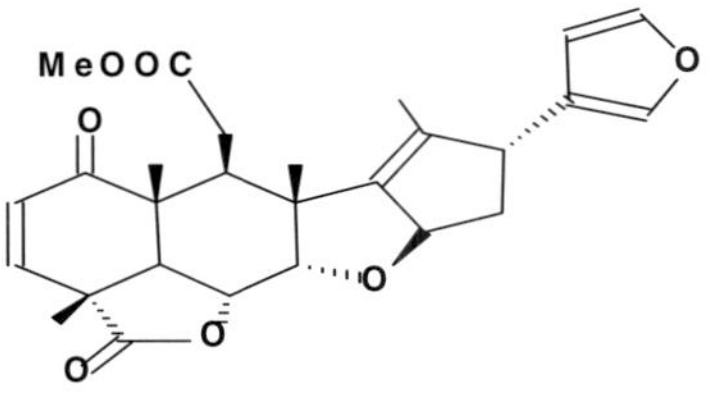

Nimbolide

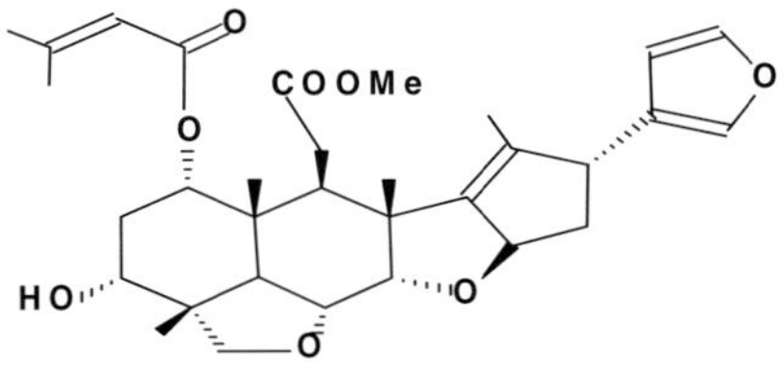

2',3'-Dihydrosalannol

O

Figure 4. Limonoids isolated from fresh green leaves of *A.indica* pustules was noticed even at a concentration of 0.005 μg/cm^2 leaf area. This clearly indicated that n-hexane wash contained compounds with antifungal activity. Partitioning of the n-hexane residue between n-hexane and 95 per cent methanol resulted in the concentration of the antifungal compounds in the methanol layer and enrichment of the fatty and waxy material in the n-hexane fraction. The methanol fraction showed a marked increase in antifungal activity, which was comparable to the original n-hexane (H1) extract (Fig. 3). Fractionation of the polar fraction by semi-preparative HPLC resulted in at least eight distinct peaks, of which peak 7 (rt. 39.7 min.) and peak 8 (rt. 52.9 min.) were the major ones (Fig. 2). Individual analysis of peaks 1 to 6 by HPLC revealed that it was a mixture of azadirone, nimbolide, 14,15-epoxynimonol and homoazadirone alongside unidentified compounds (Fig.4). The mixture of peaks 1 to 6, as eluted from HPLC, was assayed for antifungal activity. At 0.005 μg/cm^2 leaflet, the mixture of peaks 1-6 brought about 65 per cent reduction in pustule number on day 13. At higher concentrations of 0.05, o.5 and 5.0 ug/cm2, between 80-90 per cent reduction in pustules was recorded. Peak-7 and peak-8, which were the major constituents of the methanol extract, brought about drastic reduction in pustule number. With both peak-7 and peak-8, even at 0.005 ug/cm2, over 80 per cent reduction in pustules was noticed. Peak-8 was the most effective among the three HPLC fractions, followed by peak-7 and peaks 1-6. Peak-7, which contained negligible impurities of peaks 1-6, was distinctly more active than pure nimonol. Similarly peak-8, which contained impurities of peaks 1-6 and peak-7, was more active than isomeldenin. It would appear that though these limonoids are active in reducing the number of pustules, in mixtures decreased the disease intensity even better (Suresh *et al.*, 1997).

The n-hexane layer (H1) of the neem leaf extract after partition on Si-gel chromatography resulted in fractions (F-1, F-2, F-3), which were analyzed by GC-MS (Table 3) and tested for inhibition of conidial germination of *Fusarium oxysporum* and *Colletotrichum lindemuthianum*. H-1 and F-1 were active only at 1000-2000 μg/cm^2. Fractions F-2 and F-3 completely inhibited the conidial germination of the two fungi at 400 μg/cm^2 (Table 4). The antifungal activity of both these fractions can be correlated to the high concentration of 10-undecyn-1-ol (Govindachari *et al.*, 1999b).

Neem leaf volatiles collected through steam distillation were shown to contain a number of sulfur compounds such as linear alkyl and cyclic di, tri and tetrasulfides. The neem volatiles, thus collected, were found to be effective against *Trichophyton mentagrophytes* (Pant *et al.*, 1986). In an attempt to decipher the neem leaf volatiles, a much milder method of collection over Poropak-Q was attempted by the authors and the results (unpublished) show the presence of a number of sulfur volatiles emanated by the fresh neem leaves (Table 5). It is of interest to note that a number of such sulfur volatiles from the genus *Allium* have been shown to be potent antimicrobials.

Although neem oil has been used for control of phytopathogenic fungi, concentrations needed for complete field control were shown to be as high as 2 to 10 per cent. High concentrations of neem oil are known to induce phytotoxicity (Locke, 1995). Cold expeller neem oil at 1000 ppm either brought about no inhibition against the test fungi (Govindachari *et al.*, 1998). While this conforms to earlier literature, it may, in part, explain the reason for the use of higher concentrations of neem oil for field control. In an effort to identify the component(s) that actually impart antifungal activities, solvent partitioning of the neem oil between n-hexane and 90 per cent MeOH was resorted to (Fig. 5) (Govindachari *et al.*, 1998). The 90 per cent MeOH extract, inhibited growth of the test fungi to varying degrees. Analysis of the 90 per cent MeOH extract by analytical HPLC revealed the presence of major triterpenoids. Fungal inhibitory activity may hence be attributed to the triterpenoidal fraction. Preparative HPLC resolved the 90 per cent MeOH extract into ten peaks and analysis of all the ten peaks using analytical HPLC revealed that peaks 1 and 2 contained mainly azadirachtins A, B, D, H and I. Peak 1 and 2 did not show any appreciable inhibitory activity. It was surmised that azadirachtins do not possess any antifungal activity. Peaks 3 and 4 yielded small amounts of material, with little or no activity against *F. oxysporum* and *A. tenuis*, but with considerable activity against *D. oryzae.* Peak 5 was identified as 6-deacetylnimbin of 96 per cent purity (by analytical HPLC) and showed appreciable inhibition against *D. oryzae* (59.9%), *A. tenuis* (30.6%) and *F. oxysporum* (49.2%) at 1000 ppm. Pure 6-deacetylnimbin (purified by HPLC), retained antifungal activity against *F. oxysporum* and *D. oryzae*, but showed drastic reduction in activity against *A. tenuis*. Peak 6 (azadiradione as the major component), peak 7 (nimbin as the major constituent), peak 8 (salannin as the major constituent) showed excellent inhibitory activities against *D. oryzae* and moderate activity against *A. tenuis, F. oxysporum* at 1000 ppm. When purified, azadiradione and salannin showed drastic reduction in activity against all the test fungi. Nimbin in pure form had reduced activity against *F. oxysporum* and *A. tenuis*, but retained the activity against *D. oryzae*. Peak 9 showed excellent inhibitory activity against *D. oryzae* and was moderately active against *F. oxysporum* and *A. tenuis* at 1000 ppm. Peak 10 (epoxyazadiradione as the major constituent) was most effective against *D. oryzae,* and least effective against both *A. tenuis* and *F. oxysporum.* Epoxyazadiradione in pure form exhibited no inhibitory activity against all the test fungi. It is possible that in pure form the major triterpenoids from oil, have very low or no antifungal activity, while in combination they show excellent activity against all the three test fungi, suggesting additive / synergistic effects.

In order to evaluate the finding that the major terpenoids act additively/ synergistically, five pure terpenoids were mixed in the following proportion based on our studies of the concentrations of various triterpenoids as they occur naturally in neem oil: epoxyazadiradione (1 part): salannin (5 parts): nimbin (4 parts): azadiradione (2 parts): 6-deacetylnimbin (4 parts). Assays of the antifungal activities of the mixture against the three test fungi revealed again that maximum inhibition was observed with *D. oryzae* at 1000 ppm (Fig. 6). It is also not surprising that differences existed in inhibition percentages among the test fungi. Hence, concentrations needed to effectively inhibit each fungal species have to be worked out independently. Both the natural triterpenoidal mixture from neem oil, as well as

Table 3. GC-MS analysis of neem leaf hexane extract (H-1) and its chromatographic fractions (F-1, F-2, F-3)

Compound	H-1 (%)	F-1 (%)	F-2 (%)	F-3 (%)
Unknown		0.6		
Isocaryophyllene	0.8	2.66	-	-
Ocimene		0.8		
2,5-Dimethyl-3-methylene-1,5 heptadiene		0.5		
Unknown		0.5	0.5	
Germacrene	9.6	25.78		
Copaene	0.6			
2,3,4-Trimethylheptane	0.3	1.05		
Unknown		1.0		
Unknown	0.6	2.0		
Unknown		0.5		
2,6,10,14-Tetramethyl heptadecane	0.4	1.4		
Palmitic acid	6.47		2.68	18.4
1-Dodecanol		0.5		
2,6-Dimethyl heptadecane	0.87	2.5		
Eicosatrienoic acid, methyl ester				0.5
Unknown		1.3		0.8
10-Undecyn-1-ol	34.5		90	65.1
Linolenic acid, methyl ester		1.3		
2,6,11-Trimethyl dodecane	0.75	1.7		
Nonacosane	0.4	1.1		
Unknown		0.6		
2,6-Dimethyl heptadecane		1.3		

Table 4. Antifungal activity of neem leaf hexane extract (H-1) and its fractions (F-1, F-2, F-3)

Fraction	Conidial germination (%) at Concentration (μg / cm^2 glass slide surface)					
Fusarium oxysproum						
	Control	2000	1000	400	200	100
Mancozeb	86.3 ± 21.7	0	0	0	0	0
H-1	91.8 ± 5.5	25.1 ± 20.3	49.9 ± 30.9	75.1 ± 29.1	76.8 ± 35.1	75.2 ± 12.2
F-1	87.0 ± 10.3	12.4 ± 8.6	74.1 ± 11.1	56.7 ± 26.2	66.5 ± 28.2	66.2 ± 6.9
F-2	80.6 ± 6.9	0	0	0	49.0 ± 18.0	61.4 ± 19.0
F-3	50.1 ± 24.8	0	0	0	10.6 ±12.7	27.2 ± 20.7
Colletotrichum lindemuthianum						
Mancozeb	89.1 ± 3.7	0	0	0	0	0
H-1	40.6 ± 3.6	7.9 ± 9.3	29.0 ± 38.8	60.0 ± 33.4	72.3 ± 32.4	75.6 ± 10.0
F-1	87.5 ± 4.2	22.2 ± 12.0	62.4 ± 28.8	92.7 ± 7.4	93.8 ± 7.7	78.6 ± 4.8
F-2	64.6 ±1 5.5	0	0	0	0	82.9 ± 6.1
F-3	59.8 ±1 9.6	0	0	0	57.9 ± 26.7	48.4 ± 18.4

Control : n-hexane; Values are mean ± SD ; n = 25. (*Source:* Govindachari *et al.*,1999b)

Table 5. Neem leaf volatiles collected on poropak Q and analysed by GC-MS

Pk.No	Retention time (min)	Compound
1	3.65	n-hexenol
2	5.15	Methyl 2-propenyl disulfide
3	5.39	Methyl 1-propenyl disulfide
4	7.92	2,6 nonadienol
6	8.74	Eucalyptol/7-decenone
7	11.15	Methyl n-pentenyl disulfide
8	11.53	1,2 dithiacyclopentenone
9	11.65	1,6 dimethoxy hexane
10	11.85	Unidentified
11	12.88	2,4-hexadecanoic acid
12	15.71	Unidentified
13	17.66	1,2,4-trithiolane
14	17.86	3,5-diethyl1,2,4-trithiolane
15	18.07	Trithiolane
16	18.37	Allyl monosulfide
17	19.88	Copaene
18	20.98	Isocaryophyllene
19	21.84	a-caryophyllene
20		Unidentified
21		Unidentified
22		Unidentified
23		Unidentified
24	24.14	Germacrene

(Source : Suresh, unpublished)

Figure 5. Fractionation of antifungal triterpenoids from seed oil of *A. indica*

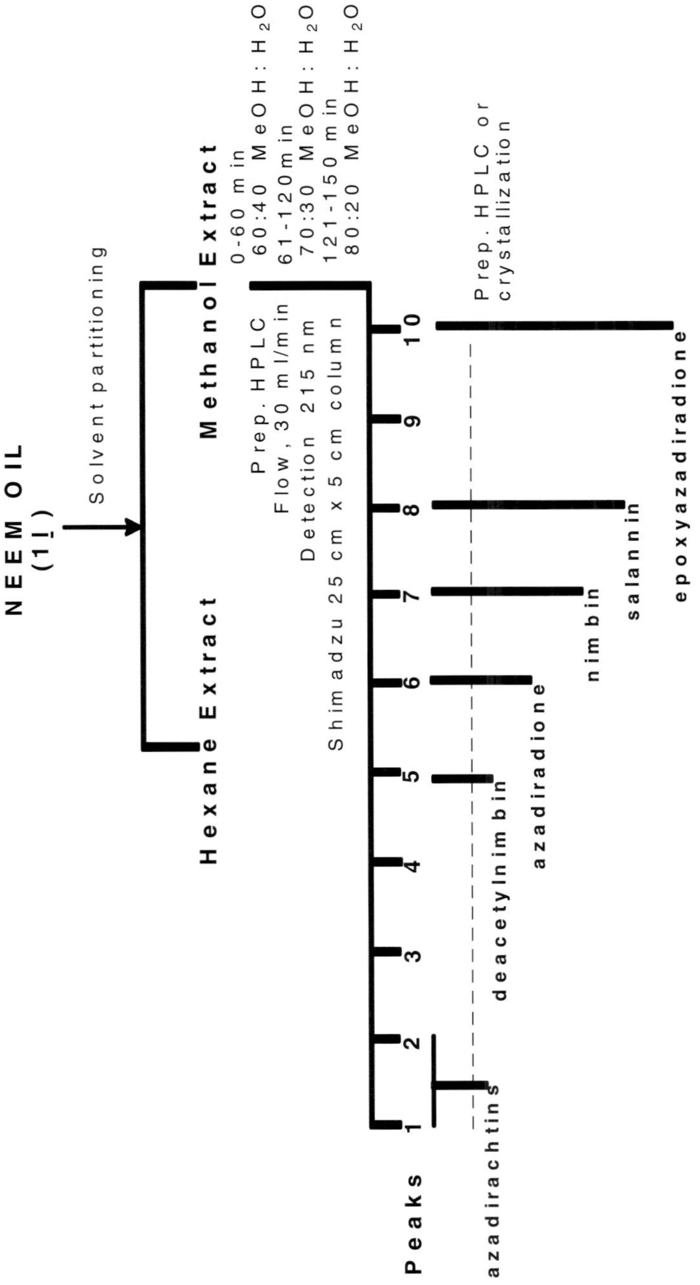

a mixture made up from pure salannin, nimbin, azadiradione, deacetyl nimbin and epoxyazadiradione were similar in their antifungal activity against the test fungi.

5. CONSTITUTIVE ANTIFUNGALS OF NEEM: APPROACHES FOR THE NEW MILLENIUM

5.1 Interactions Among Neem Constitutive Chemistry: Need for Data Generation

The effectiveness of triterpenoidal mixtures as fungistatic against a number of phytopathogens has been amply demonstrated based on our work with chemically defined leaf and seed oil extractives (Govindachari *et al*, 1998; Suresh *et al.*, 1997). It is interesting to note that the approach of bioassay guided isolation mainly concentrated on the major triterpenoids and nothing is known on the relative

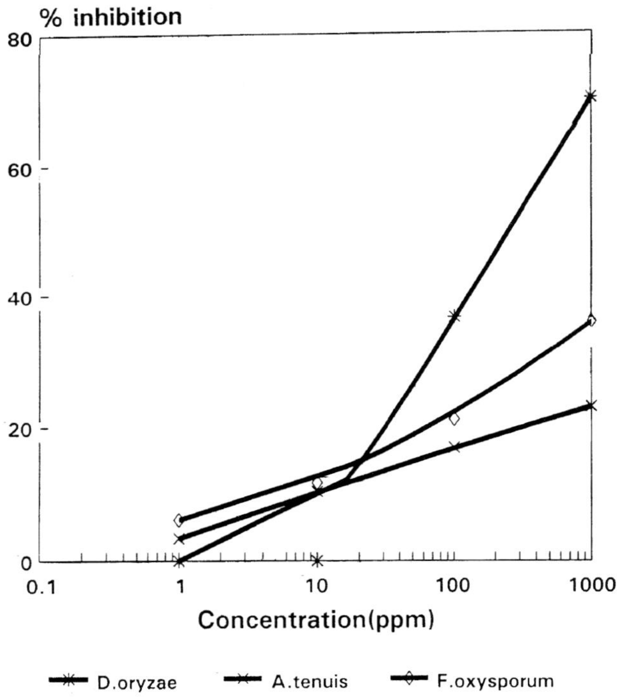

Figure 6. Antifungal activity of mixture of limonoids isolated from neem seed oil (*Source:* Govindachari *et al.*, 1998)

contributions of minor triterpenoids, which in total exceed 100 compounds from different parts of the neem tree. It is also relevant to note that the differing activity profiles due to structural diversity of tetranortriterpenoids of different neem extract mixtures become more complex due to the variability in concentrations of individual terpenes contributed by genetic, biochemical, ontogenetic and ecological conditions. Such varying constitutive compositional profiles of mixtures are of considerable ecological significance. These may in part explain the enormous variability in effectiveness of neem extractives against phytopathogenic fungi recorded in literature. It is then imperative; that any new approach to control phytopathogenic fungi using neem should address structural and compositional variability of extractives and interactions since there is a high probability of interactions among the compounds in the neem extracts that influence bioefficacy. Specifically, compound interactions, i.e., synergism, additive or antagonistic interactions need to be studied in order to develop effective formulations that contain useful compounds. The primary limitation in studying compound interactions is due to the difficulty in detecting, analyzing and displaying such interactions. A simple method of analysis has been suggested for identifying compound interactions among plant defense compounds based on isobolographic analysis (Nelson and Kursar, 1999). In order to understand interactions among compounds in mixtures, bioassay guided isolation, identification of compounds need to be established.

5.2 Mode of Action of Neem Antifungals and Identification of Target Site(s)

In any search for biologically active natural molecule, subsequent to establishing the bioefficacy and identification of active compounds, it is necessary to understand the mode of action of the active ingredients and to elucidate the target sites in order to maximize the bioefficacy through enrichment or by modifications of active components. Understanding mode of action and target sites also give leads for synthesis of novel molecules.

Clues on the mode of action of neem extractives have come through the studies on aflatoxin production by aflatoxigenic fungi, i.e., *Aspergillus flavus* and *A. parasiticus.* Blended neem leaf extracts have been shown to reduce growth and aflatoxin production by *A. flavus* in vitro (Bhatnagar and McCormick, 1988). Aflatoxin production in *A.flavus* infected cotton bolls was significantly reduced when treated with neem leaf extractives (Zeringue and Bhatnagar, 1993). It is interesting to note that neem leaf extracts did not affect fungal growth, but essentially blocked (>95%) aflatoxin synthesis suggesting the effect of neem leaf extractives on the biochemical pathway leading to aflatoxin biosynthesis (Zeringue and Bhatnagar, 1993). Volatiles containing mainly, 3-methyl-2-buten-1-ol, did not inhibit either the growth or aflatoxin biosynthesis. Boiling and autoclaving the extracts resulted in considerable reduction in activity and Zeringue and Bhatnagar (1993) concluded that the aflatoxin biosynthesis inhibitor might be a volatile compound. Neem leaf contains a number of limonoids and some of which are shown to be antifungal as well (Suresh *et al.*, 1997). Neem leaf limonoids are also susceptible to high temperature or light regimes. Secondary metabolism of resting mycelial mats are ideally targeted to monitor the effect on biosynthesis (Bhatnagar and McCormick, 1988) and the effect of neem leaf extractives showed that the enzymes involved in the conversion of precursors of aflatoxin production were not inhibited, while the aflatoxin titres were drastically reduced in treated mycelia. Similar work should be initiated with other phytopathogenic fungi as well to understand the molecular mechanisms involved in the antifungal activity of neem extractives.

5.3 Targeting Plant – Pathogen Interactions: Novel Approach for Disease Control Using Neem Constitutive Chemistry

Lead identification from natural or synthetic sources always targeted growth and reproductive phases for controlling phytopathogenic fungi. This in turn resulted in search of biochemical targets that affect growth and reproduction of disease causing pathogens. While it is surmised that this approach will continue and become efficient through development of precise biochemical and molecular screens, there is a renewed emphasis on developing lead molecules that interfere with the successful plant – pathogen interactions. This approach will target specific stages of pathogen development in plant tissues with relatively less broad spectrum toxicity that is associated with fungicides that target fungal growth (Bailey, 1995).

The two approaches of targeting plant – pathogen interactions (Bailey, 1995) are manipulation of natural plant defenses and targeting specific stages of pathogenesis. For, eg. tricyclazole, inhibits melanization of appressoria that are necessary for successful infection (Viviani *et al.*, 1993).

NeemAzal (5% azadirachtin), a commercial formulation based on neem kernel extractive significantly reduced the number of germlings producing multiple germ tubes of *Erysiphe pisi* on excised pea leaves. Over 85 per cent of conidia formed multiple germ tubes in control in contrast to neem formulation treatment resulted in 3.6 per cent conidia formed multiple germ tubes. NeemAzal did not inhibit conidial germination or appressorium formation on excised leaves; but reduced the number of secondary branches of hyphae in a dose-dependent manner. Additionally, the leaves showed higher number of hypersensitive cells subsequent treatment with NeemAzal (Singh and Prithiviraj, 1997).

Phenylalanine Ammonia Lyase, an enzyme that converts phenylalanine to trans cinnamic acid, is the key enzyme in the shikimic acid pathway and an increase in the activity in PAL along with peroxidases and polyphenol oxidases lead to heightened synthesis of phenolics – a measure of increased resistance to infection by host plants. Neem seed kernel extractives significantly increased activity of these three enzymes in rice plants and chillies resulting in the reduction of infection by *Pyricularia oryzae* and chilli mosaic virus (Manickam and Mariappan, 1995 and references cited therein). PAL has been implicated as the key enzyme in for pisatin production in *Pisum* and it is shown to be inducible (Singh and Prithviraj, 1997). PAL activity increased in pea plants that were treated with NeemAzal (100 ppm) as a pre-inoculation treatment.

The clues on the effect of compound interactions in mixtures, mode of action and target sites as well as the ability to induce SAR activity in host-pathogen systems will form the basis for novel approaches towards maximizing the potential of neem for crop disease control in the new millennium.

REFERENCES

Achimu, P. and Schlosser, E. (1992), Effect of neem extract (*Azadirachta indica* A. Juss.) against downy mildew (*Plasmopara viticola)* of grapevine, in *Practice Oriented Results on Use and Production of Neem-Ingredients*, ed. H. Kleeberg, Druck & Graphic, Giessen, Germany, p. 99-107.

Ali, T.E.S., Nasir, M.A. and Shakir A.S. (1992), In vitro evaluation of certain neem products as mould inhibitors against post-harvest fruit rotting fungi of tomato, *Pakistan Journal of Phytopathology,* **4**, 58-61.

Bailey, J.A. (1995), Plant pathogen interactions : a target for fungicide development, in *Antifungal agents. Discovery and Mode of action*, eds. G.K. Dixon, L.G. Copping and D.W. Hollowman, Bios Scientific Publishers Ltd., UK, pp. 233-244.

Balandrin, M.F., Lee, S.M. and Klocke, J.A. (1988) Bilogically active volatile organosulfur compounds from seeds of the neem tree, *Azadirachta indica* (Meliaceae), *Journal of Agricultural and Food Chemistry* **36**, 1048-1054.

Bhatnagar, D. and McCormick, S.P. (1988), The inhibitory effect of Neem (*Azadirachta indica*) leaf extracts on aflatoxin synthesis in *Aspergillus parasiticus*, *Journal of American Oil Chemical Society,* **65,** 1166-1168.

Butterworth, J.H. and Morgan, E.D. (1968), Isolation of a substance that suppresses feeding of locusts, *Chemical Communication*, 910-911.

Dath, A.P. (1982), Effect of soil amendment with some green manures on the survival of sclerotia of *Corticum sasakii*, *Indian Phytopathology* **35**, 523-525.

Devakumar C. and Mukerjee, S.K. (1983), Chemistry of neem bitter principles. *IARI Research Bulletin,* **40**, 1-17.

Devakumar C. and Sukh Dev (1993), Chemistry, in *Neem Research and Development*, ed. N.S. Randhawa and B.S. Parmar, Society of Pesticide Science, India. pp. 63-96.

Ermel, K. (1995), Azadirachtin content of neem seed kernels from different regions of the world, in *The Neem Tree*, ed. H. Schmutterer, VCH, Weinheim, Germany, pp. 89-92.

Feuerhake, K.J. (1985), Effectiveness and selectivity of technical solvents for the extraction of neem seed components with insecticidal activity, in *Natural*

Pesticides from the Neem Tree and OtherTtropical Plants, eds. H. Schmutterer and K.R.S. Ascher, GTZ, Eschborn, Germany, pp. 103-114.

Ghewande, M.P. (1989), Management of foliar diseases of groundnut (*Arachis hypogaea*) using plant extracts. *Indian Journal of Agricultural Science* **59**, 133-134.

Goodwin, T.W. (1967), The Biological Significance of Terpenoids in Plants, in *Terpenoids in Plants,* ed. J.B. Pridham, Academic Press, London, pp. 1-23.

Govindachari, T.R., Suresh, G., Gopalakrishnan, G., Banumathy, B. and Masilamani, S. (1998), Identification of antifungal compounds from the seed oil of *Azadirachta indica* A. Juss., *Phytoparasitica* **26,** 109-116.

Govindachari, T.R., Gopalakrishnan, G. and Suresh, G. (1999a), Triterpenoidal constituents of an aqueous extract from neem kernels, *Fitoterapia* **70**, 558-560.

Govindachari, T.R., Suresh, G. and Masilamani, S. (1999b), Antifungal activity of *Azadirachta indica* leaf hexane extract, *Fitoterapia* **70**, 417-420.

Harborne, J.B. (1982), *Introduction to Ecological Biochemistry*, Acadmeic Press, London.

Jain P.C. and Agarwal, S.C. (1978), Note on the activity of some odoriferous organic compounds against some keratinophilic fungi, *Prodceedings of the National Academy of Sciences*, **44(B)**, 273-276.

Kannaiyan S. and Prasad, N.N. (1981), Effect of organic amendments on seedling infection of rice caused by *Rhizoctonia solani*, *Plant and Soil*, **62** , 131-133.

Khan, M.W., Alam, M.M., Khan, A.M. and Saxena S.K. (1974), Effect of water soluble fractions of oil-cakes and bitter principles of neem on some fungi and nematodes. *Acta Botanica Indica*, **2**, 120-128.

Krishna A., Prasad and Ojha, N.L. (1986), Antifungal evaluation of leaf extracts for the control of some cucurbitaceous fruit rot diseases, *Indian Phytopathology*, **39**, 153.

Kraus, W. (2002), Biologically active ingredients, in *The Neem Tree*, 2nd edition, ed. H. Schmutterer, Neem Foundation, Mumbai, India, pp. 39-111.

Locke, J.C. (1990), Activity of extracted neem seed oil against fungal plant pathogens, in *Neem's Potential in Pest Management* Programs, ed. J.C. Locke, USDA, ARS, Maryland, pp. 132-136.

Manickam, K. and Mariappan, V. (1995), Effect of application of plant products on the host plant resistance, in *Neem for the Management of Crop Diseases*, Associated Publishing Co., New Delh, pp. 107-114.

Moline, H.E. and Locke, J.C. (1993), Comparing neem seed oil with calcium chloride and fungicides for controlling postharvest apple decay. *Horticulture Science*, **28**, 719-720.

Muthusamy, M., Easwaramurthy, S., Muthusamy, S. and Mariappan, V. (1988), Evaluation of neem products agains rust disease of groundnut. *Neem Newsletter*, **5**, 48.

Nagarajan,K. and Reddy T.S.N. (1980), Preliminary evaluation of Ridomil against damping off and leaf blight diseases of tobacco in seed beds, *Tobacco.Research* **6**, 127-131.

Nasem, M. and Lanjewar, R.D. (1990), Effect of neem oil on seedling growth of two rice cultivars in association of *Aspergillus niger. Indian Phytopathology,* **43**, 304.

Nelson, A.C. and Kursar, T.A. (1999), Interactions among Plant defense compounds : a method for analysis, *Chemecology,* **9**, 81-92.

Pant, N., Garg, H.S., Madhusudanan, K.P. and Bhakuni, D.S. (1986), Sulfurous compounds from *Azadirachta indica* leaves, *Fitoterapia,* **57**, 302-304.

Parveen G and Alam, M.M. (1993), Bioactivity against Plant pathogens, in *Neem Research and Development*, ed. N.S. Randhawa and B.S. Parmar, Society of Pesticide Science, India, pp. 144-153.

Pradhan, S., Jotwani, M.G. and Rai, B.K. (1962), The neem seed deterrent to locust. *Indian farming*, **12**, 7-11.

Prakash, D., Alam M.M. and Ashraf, S. (1985), Studies on rhizosphere mycoflora of tomato in relation to soil amendments, *National Academy of Sciences Letters*, **8,** 133-134.

Riar, S.S., Devakumar, C., Ilavazhagan, G., Bardhan, J., Kain, A.K., Thomas, P., Singh R. and Singh, B. (1990), Volatile fraction of neem oil as a spermicide, *Contraception,* **42**, 479-487.

Rukmini, C. (1987), Chemical and nutritional evaluation of neem oil. *Food Chemistry*, **26**, 119-124.

Schmuterer, H. (1995), *The Neem Tree*,VCH, Weinheim, Germany.

Schmutterer, H and Singh, R.P. (1995), List of insect pests susceptible to neem products, in *The Neem Tree*, ed. H. .Schmutterer, VCH, Weinheim, Germany, pp. 326-365.

Siddiqui, S. (1942) A note on the isolation of three new bitter principles from the nim oil, *Current Science*, **11**, 278-279.

Sinha, P. and Saxena, S.K. (1987) Effect of neem leaf powder and extract on the development of fruit rot caused by *Aspergillus niger, Neem Newsletter*, **4,** 45-47.

Singh, R. and Vyas, K.M. (1984), Influence of certain oil cakes on betelvine *Phytophthora*, *Indian Phytopathology*, **37**, 359-361.

Singh, R.S., and Pandey, K.R. (1966), Effect of green and mature plant residues and compost on population of Phytium apahnidermatium in soil, *Indian Phytopathology*, **19**, 367-371.

Singh, R.S. and Pandey, K.R. (1967), Population dynamics of *Pythium aphanidermatum* in oil cake amended soils, *Canadian Journal of Microbiology*, **13**, 601-610.

Singh, R.S. (1968), Incidence of black scurf on potatoes in oilcake amended soil, *Indian Phytopathology*, **21**, 120-121.

Singh S.P., Ahmad, M., Khan, A.M. and Saxena, S.K. (1980), Effect of seed treatments with certain oilcakes or nematicides on the growth of tomato and on rhizosphere population of nematodes and fungi, *Nematology Mediteranean* **8,** 193-198.

Singh, U.P., Singh HG.B. and Chauhan, V.B. (1984), Effect of some funcidies, plant extracts and oils on inoculum density of different nodal leaves of pea infected by *Erisiphe polygoni*, *Zeitschrift fur Pflanzenk.Pflanzensch*, **91**, 20-26.

Singh, U.P. and Prithiviraj, B. (1997), Neemazal, a product of neem (*Azadirachta indica*) induces resistance in pea (*Pisum sativum*) against *Erisiphe pisi*, *Physiological and Molecular Plant Pathology*, **51**, 181-194.

Steinhauer, B. (1996), Antimycotic activity of a methanolic extract from *Azadirachta indica* A.Juss, in *Neem and Environment*, Vol.2, eds. R.P. Singh, M.S. Chari, A.K. Raheja and W. Kraus, Oxford & IBH Publishing Co. Pvt. Ltd., New Delhi, pp. 799-808.

Suresh, G., Narasimhan, N.S., Masilamani, S., Partho, P.D. and Gopalakrishnan, G. (1997), Antifungal limonoids from the uncrushed green leaves of *Azadirachta indica*, *Phytoparasitica*, **25,** 33-39.

van Emden, H. F. and Peakall D.B. (1996), *Beyond Silent Spring : Integrated Pest Management and Chemical Safety*, Chapman and Hall, London.

Viviani, F., Vors, JP., Gaudry, M and Marquet A. (1993), Doxygenation of polyphenols by ascomycetes – kinetic behaviour of the NADPH-dependent naphthol dehydrogenase and inhibition by tricyclazole and its analogues, *Bulletin SocietyChim.France*, **130,** 395-404.

Zerinque, H.J. and Bhatnagar, D. (1993), Neem and control of aflatoxin contaimination, in *Neem and Environment*, Vol.2, eds. R.P. Singh, M.S. Chari, A.K. Raheja and W. Kraus, Oxford & IBH Publishing Co. Pvt. Ltd., New Delhi, pp. 713-727.

Chapter 10

NEEM PLANT TISSUE CULTURE TODAY AND THE OUTLOOK FOR THE NEW MILLENNIUM

S. ANDREW VAN DER ESCH AND GERMINA GIAGNACOVO
ENEA, C.R. Casaccia, Rome, Italy

1. INTRODUCTION

The neem tree (Tewari, 1992; Schmutterer, 1995; Puri, 1999) has been widely studied mainly for its biopesticidal properties. Indeed, Neem products are natural insecticides and are known to affect feeding, growth, reproduction and metamorphosis (Rembold, 1989; Rembold *et al.*, 1989; Mordue and Blackwell, 1993; Rembold and Annadurai, 1993; Sayah *et al.*, 1998; Guerrini and Kriticos, 1998). Because of the broad-spectrum control of insects and the relatively low nontarget toxicity to mammals (Isman, 1997), fish (Wan *et al.*, 1996), and beneficial invertebrates (Schmutterer, 1995), neem-based products are potential candidates for use in agriculture (Isman *et al.*, 1991; Wan *et al.*, 1996; Isman, 1997) and possibly forestry (Sundaram, 1996) and medicine (Koul, 1996a). Furthermore, neem's biopesticidal properties aren't only limited to phytophagous insects but are also active on other pathogenous organisms like nematodes (Akhtar, 2000), fungi (Govindachari *et al.*, 2000) and micro-organisms (eukaryotic as well as prokaryotic) (Ahmad and Ahmad, 1995; Samy and Ignacimuthu, 1998). In fact, the latter mentioned bioactivities imply the potential of neem-derived products for both veterinary and medical applications well known in India (Ayurvedic medicine), though for such applications more scientific data will have to sustain the validity of these claims. For some applications, such as antifertility and dental cure, scientific articles exist which sustain the veracity of these claims (Wolinsky and Mania, 1996; Talwar *et al.*, 1997a, 1997b), for most other claims the evidence is only of an anecdotal nature. Veterinary applications will prove to be of very high interest as in

O. Koul and S. Wahab (eds.), Neem: Today and in the New Millennium, 209–227.

the western world an increasing trend towards biological rearing and husbandry of animals is coming into evidence (after BSE and Foot and Mouth Disease (FMDV)). Indeed from *Melia azedarach* a peptide, meliacine (MA) has been isolated which inhibits uncoating of the FMDV virus *in vitro* (Wachsman *et al.*, 1998). The fundamental hurdle to overcome, in biological rearing of animals, is the lack of substantially efficacious alternative medicines for treating both exo- as well as endo-parasites which are accepted by western regulations as "biological". The use of neem derived products for "biological agriculture" and "integrated pest management" (IPM) on crops are already accepted by European Union legislation.
At this point all different production processes of neem trees and products derived thereof are of great importance. In many countries efforts are underway for increasing the propagation of planted neem trees. Clearly a better understanding of all the factors contributing to better growing, harvesting, extraction and formulation practices -plus biological activities including mode-of-action studies (Koul, 1996a,b) will be essential for developing a viable future for neem derived products. Therefore, biotechnological approach can be very useful for reaching the goals and deeper understanding of different aspects of the neem tree.

2. PLANT CELL BIOTECHNOLOGY

The definition of biotechnology that has been adopted throughout this review is that "Biotechnology is any technological application that uses biological systems, living organisms, or derivatives thereof, to make or modify products or processes for specific use." This is the definition that has been adopted at the Convention on Biological Diversity (Burhenne-Guilmin *et al.*, 1994). Broadly interpreted, this definition includes traditional methods either utilising microorganisms for food and beverages (cheese, beer, etc.) or traditional plant and animal breeding. The interpretation used more precisely in the present content referes to the technologies such as:

- recombinant DNA (genetic engineering)
- rapid screening techniques for natural products
- sophisticated culture processes
- hybridoma technology (monoclonal antibodies)
- down stream processing (DSP)

These technologies also include the development of methods by which biological processes may be controlled such that their rate of production enables economic industrial production, or by which living material is obtained that can be utilised in industry, agriculture and forestry, as well as in gardening and breeding. Development of sophisticated culture processes includes culturing of tissue fragments, cell aggregates and individual cells, and the production of biomass by unconventional methods for industrial production of physiologically active substances. It thus includes all biological reactions carried out with living organisms, plant or animal cells or tissues, or with enzymes derived from them (Endreß, 1994).

Starting from the whole plant, biotechnological approaches can lead to very different applications. The totipotency of plant cells has already been predicted in 1902 by Haberlandt and in 1935 the first true plant tissue culture on agar was established. Since then plant tissue culture techniques have greatly evolved. The tissue culture technique most widely in use today is micropropagation, which allows the production of large numbers of "true-to-type" plants from small pieces of the stock plant in relatively short period of time. Depending on the species in question, the original tissue piece may be taken from shoot tip, leaf, lateral bud, stem or root tissue. In most cases, the original plant is not destroyed in the process. A single explant can be multiplied into several thousand plants in less than one year. Once established, actively dividing cultures are a continuous source of microcuttings, which can result in plant production under greenhouse conditions without seasonal interruption. Thus micropropagation can help in the multiplication of superior genotypes especially in the case of plants that have seeds with a relatively rapid loss of the germination ability.

The most important impact of these tissue culture techniques and micropropagation is the controlled manipulations of plant germplasm at the cellular level. These techniques can be used for selection of plants either with enhanced stress or pest resistance or for the creation of pathogen free plants and somatic hybridisation. Somatic hybridisation can help in crossing the genetic barriers for the production of new recombinants, which are not available in nature (transgenic plants).

The ability to disorganise and thus to obtain undifferentiated plant cells has led to applications similar to those widely used in microbial biotechnology. One can obtain liquid suspension cultures of plant cells, which can be grown in large quantities in bioreactors, for the production of secondary metabolites (Endreß, 1994). In the next sections we will illustrate how some of these different plant tissue culture techniques have been applied to the neem tree and what the future perspectives might be.

3. NEEM AND MICROPROPAGATION

Plants that reproduce themselves through seeds undergo considerable variations from one generation to the other. Furthermore in case of the neem tree the seeds are vital for only a short period, two to four weeks (Schmutterer and Ermel, 1995) with possibly low germination ability. Considerable progress has been made in the last 30 years in the application of tissue culture techniques to tree species (Ahuja, 1989). Clearly the plant tissue culture approach is important for trees of economic importance in view of establishing plantations of high quality (Liew and Teo, 1998) because it allows to overcome the limitations presented, when using seeds.
Micropropagation has been defined as "in vitro regeneration of plants from organs, tissues, cells or protoplasts" (Beversdorf, 1990) and "the true-to-type propagation of a selected genotype using *in vitro* culture techniques" (Debergh and Read, 1991). The *in vitro* propagation of *Azadirachta indica* can be carried out by different plant tissue culture approaches. Those successfully applied are (i) induction of the

formation of multiple shoots from axillar buds (Subramani *et al.* 1993; Mohamed-Yasseen, 1994), (ii) direct regeneration from somatic tissues (Ramesh and Padhya, 1990; Subramani *et al.*, 1993; Kearney, 1993; Eeswara *et al.* 1998), (iii) direct somatic embryogenesis (Murthy and Saxena, 1998) and (iv) indirect somatic embryogenesis (Shrikhande *et al.*, 1993; Wei Wen Su *et al.*, 1997).

Either *in vitro* induction of multiple shoots from axillar buds or direct somatic regeneration from leaf discs or stem are certainly the fastest methods to propagate a huge number of genetically identical plantlets true-to-type to the mother plant.

Different micropropagation experiments carried out with *Azadirachta indica* are summarized in Table 1. The culture medium most widely used for the *in vitro* cultivation of *Azadirachta indica* is the Murashige and Skoog medium (Murashige and Skoog, 1962). It is fundamental that growth regulators are added to the medium in order to obtain a morphogenetic response. No morphogenetic response was obtained from leaf discs cultivated on MS in the absence of growth regulators (Subramani *et al.*, 1993). The best response, from nodal segments, was obtained when the MS medium was supplemented with 6-Benzylaminopurine (BAP) (1 mgl^{-1}) and Kinetin (0,5 mgl^{-1}) with the formation of 3.37 ± 0.66 shoot buds/explant with a response of the explants equal to 80 per cent. After four weeks 3 to 5 shoot buds were obtained for every nodal segment. These shoot buds were once more sub-cultivated on the same medium and after eight weeks 20 to 25 buds were obtained.

Ramesh and Padhya (1990) obtained a major morphogenetic response from leaf discs when using the Wood and Brauns medium (Wood and Brauns, 1961) supplemented with BAP (0.9 mgl^{-1}) and Kinetin (0.86 mgl^{-1}) with a response frequency of 70 per cent after 14 days. In both these studies addition of adenine sulphate increased the shoot production. The optimal combination for increasing shoot bud formation from leaf disc explants was Kinetin (0.86 mgl^{-1}), BAP (0.9 mgl^{-1}) supplemented with 2.95 mgl^{-1} of adenine sulphate.

For *in vitro* plant tissue culture the molar ratio between the different PGR's used is very important. In neem tissue culture an increase of the Kinetin concentration up to 1.29 mgl^{-1} induces the formation of shoot buds with callus. A further increase of adenine sulphate up to 3.68 mgl^{-1} in the culture medium induces callus formation from leaf disc explants (Ramesh and Padhya, 1990). Other studies (Sanyal *et al.*, 1981; Narayan and Jaiswal, 1985; Gautam *et al.* 1993; Wewetzer and Schultz, 1994) which have attempted micropropagation from neem leaf explants have used an intermediate callus stage which may possibly have led to somaclonal variation (Allan, 1991). A standard procedure for the micropropagation of the neem tree, which avoids this problem, has been developed by Eeswara *et al.* (1998). Micropropagated shoots were initiated from leaf explants cultured on Murashige and Skoog medium containing BAP (1 mgl^{-1}), Kinetin (0.8 mgl^{-1}) and adenine sulphate (6 mgl^{-1}) in complete darkness. These shoots were further multiplied on MS medium containing BAP (0.1 mgl^{-1}), Kinetin (008 mgl^{-1}) and adenine sulphate (0.6 mgl^{-1}). Within 32 weeks, 80 shoots could be produced from a single leaf explant. Fifty-five percent of these shoots rooted on MS medium supplemented with indolebutyric acid (IAA) (1 mgl^{-1}) established successfully after transfer to soil.

Another very important plant tissue culture approach, which has been applied to the neem tree for whole plantlet regeneration is the induction of somatic embryogenesis (Su *et al.*, 1997). The first report on successful neem regeneration via somatic embryogenesis was by Shrikhande *et al.* (1993). In that study, somatic embryos were induced from immature cotyledonary tissues via indirect somatic embryogenesis, and the experiments were conducted using semi-solid media. Wei Wen Su and co-workers developed a modified protocol for the induction of somatic embryogenesis in *Azadirachta indica*, which involved the use of both agar and liquid media. The results they obtained were reproducible and the culture protocol has the potential for mass propagation of neem trees in bioreactors and for probing the mechanism of somatic embryogenesis in neem.

Murthy and Saxena (1998) describe a protocol for both direct and indirect embryogenesis. The conversion frequency of somatic embryos to plantlets was approximately 60-70 per cent. These kind of studies are very important if one wants to develop synthetic seed technology (Onishi *et al.*, 1994). Indeed two principal pathways of plant regeneration *in vitro* can be followed. The most common, and described above for neem, is micropropagation and involves the abolition of apical dominance resulting in the depression and multiplication of axillary buds. In the other, called somatic embryogenesis, cotyledonary embryos with a root-shoot axis are formed from somatic cells. Both result in the production of non-chimeric and true-to-type plants that comprise clonal populations (Vasil, 1994).

4. SECONDARY METABOLITES PRODUCTION FROM NEEM PLANT TISSUE CULTURE

The first report on the application of plant tissue culture (PTC) techniques for the production of secondary metabolites (Table 2) from neem is of 1983 (Schulz, 1984) where callus proliferation from various parts of the plant by improving culture media was obtained. Subsequently, Sarkar and Datta (1986) studied the relationship between biosynthesis of nimbin and β-sitosterol in bark and bark-originated callus of increasing age. They also studied the effect of glycine on *in vitro* biosynthesis of nimbin and β-sitosterol in tissues (Sanyal *et al.*, 1988). In this work on cotyledons, which contained nimbin, glycine, other amino acids and β-sitosterol it was observed that glycine affected the synthesis of both. Nimbin was also isolated from leaves and callus cultures by Ramesh Kumar and Padhya (1988). Indole acetic acid (IAA) and indole butyric acid (IBA) interaction showed a linear increase in nimbin content. This was the first indication to demonstrate that the expression of the tetranortriterpenoids produced by the neem tree in tissue culture could be under hormonal control.

Allan *et al.* (1994) also studied the induction of callus cultures from neem from Ghanian origin from leaf explants and the production of azadirachtin by these callus cultures (0.0007 % of the dry weight) after a six week culture period was reported. The feeding deterrent effect was also tested on the desert locust,

Table 1. In vitro propagation of *Azadirachta indica* A. Juss

Type of explant	Culture medium	Growth regulators mgl^{-1}				Shoot (No.)	Year	Authors
		cytokinin	Auxin	giberelline	others			
Leaf discs	WB	BAP 0.9, KIN 0.86				10-12	1990	Ramesh and Padhya
Leaf discs	WB	BAP 0.9 KIN 0.86			2.95 Ads	18-20	1990	Ramesh and Padhya
Nodal segments	MS	BAP 1 KIN 0.5				3.37±0.6	1993	Subramani *et al.*
Stem Nodes	MS		NAA 0.0931		TDZ 0.11	Multiple shoots	1994	Mohamed-Yasseen
Cotyledons	MS				TDZ 2.2	34,3 somatic embryos	1998	Murthy and Saxena
Cotyledonary tissue	MS	BAP 5.63				7.1 Adventitious buds	1998	Murthy and Saxena
Leaf discs	MS	BAP 1 KIN 0.8			6 Ad	9.73±1.4	1998	Eeswara *et al.*

BAP= 6-benzyladeninopurine, KIN= Kinetin , NAA= α-naphthaleneacetic acid, Ads= adenine sulphate, TDZ= Thidiazuron, MS=Murashige and Skoog medium, WB= Wood and Braun medium.

Schistocerca gregaria. Feeding suppression occurred when whole extracts of seed, leaf, callus, and suspension cultures were tested in no-choice feeding bioassays. Azadirachtin was present in the seed extracts but was not detected in any of the other extracts (Kearney *et al.*, 1994). Similar results were obtained by Zypman *et al.* (2001) where they compared the antifeedant effects of extracts both from whole plants as well as from callus cultures. Clearly biological effects are present when extracts from cultured neem tissue are used, although in the last study mentioned no analysis was done to establish, which substances were present in the cultures.

Azadirachtin production from callus tissue culture from neem has also been reported by Wewetzer (1998). Four different media were tested (MS, Nitsch's and Nitsch's, White's and McCown's) with different sucrose concentrations (15 and 30 gl^{-1}) supplemented with 0.2 mgl^{-1} of IAA and 0.1 mgl^{-1} BAP. Callus formation was initiated either from leaf or bark segments from trees of different origin (Nicaragua, Nigeria and Togo). Analysis of azadirachtin content showed that the production of azadirachtin A in callus cultures of *A. indica* depended on the age, cell line, the medium and the carbohydrate source employed. The main objective was to determine if differentiation was necessary for the expression of azadirachtin in culture and it could be shown that morphological differentiation is not a prerequisite for azadirachtin A production; the highest concentrations were detected in completely undifferentiated cells.

Another article reporting production of azadirachtin in callus cultures (Veeresham *et al.*, 1998) was surprisingly high. They established callus cultures both from leaves and flowers using MS medium supplemented with 2,4-dichlorophenoxyacetic acid (1 mgl^{-1}) and kinetin (0.5 mgl^{-1}). The presence of azadirachtin could be measured from the 12th week onwards. The highest levels reached corresponded to 2.68 per cent of dry weight for leaf callus (20 weeks old) and 2.46 per cent of dry weight for flower callus (12 weeks old). Such levels have never been reported either for seed or tissue culture material.

The first reports of secondary metabolites production in suspension cultures are of 1993 and 1994 (van der Esch *et al.*, 1993,1994a,b). These suspension cultures were initiated from callus cultures induced from seedling hypocotyls (seeds originated from Togo). A sure chemically acceptable proofs that azadirachtin A was produced in suspension cultures of neem was reported by Jarvis *et al.* (1997). Other compounds also produced by the suspension cultures were azadirachtin I, nimbin, 6-desacetylnimbin, salannin, 3-desacetylsalannin, 3-tigloylazadirachtol (azadirachtin B) and 3-acetyl-1-tigloylazadirachtinin.

Our studies on neem tissue culture in the laboratory eventually led to a European project supported by the European Commission called AZTEC (contract No. AIR2-CT94-1343). Very briefly we will describe what the fundamental aim of the project was followed by the main findings strictly limited to the plant tissue culture work.

The fundamental aim of the AZTEC project was to harness multidisciplinary and multi-European co-operation in view of demonstrating -at a pre-competitive level- the feasibility of the biotechnological approach for producing the environmentally sound insecticides azadirachtin and marrangin from PTC

Table 2. Secondary metabolites production from neem plant tissue culture

Type of explant	Culture type	Metabolites identified	Yield	Year	Authors
Young stem bark	Callus with differentiating roots	Nimbin	0.025% of DW	1981	Sanyal *et al.*
Young stem bark	Callus Grown with low kinetin concentrations	Nimbin, β-sitosterol	0.25% of DW	1986	Sanyal and Datta*
Cotyledons	Callus	Nimbin glycine and other aminoacids and β-sitosterol		1988	Sanyal *et al.*
Leaves	Callus	Nimbin		1988	Ramesch and Padhya
Leaves	Callus	Azadirachtin	0.0007% of DW	1994	Allan *et al.*
Shoots apex	Cell suspension	Azadirachtin A	0.66 μg/mg of DW (grown in the light) 0.19μg/mg of DW (grown in the dark)	1993	van der Esch *et al.*
Shoots apex	Cell suspension	Azadirachtin A	0.28 μg/mg of DW (grown in the light) 0.016μg/mg of DW (grown in the dark)	1994a 1994b	van der Esch *et al.*
Leaf	Callus	Azadirachtin	0.0007% DW	1994	Allan *et al.*
Bark	Callus	Azadirachtin Nimbin Gedunin	0.12% DW 0.21% DW 0.13% DW	1996	Bajagopal and Ramaswamy
Shoots apex	Cell suspension	Azadirachtin A Azadirachtin B Azadirachtin I Nimbin 6-desacetylnimbin Salannin 3-desacetylsalannin 3-acetyl-1-tigloylazadirachtinin	No yields determined but structures confirmed through appropriate spectral data (NMR, IR etc..)	1997	Jarvis *et al.*
Leaf and bark	Callus	Azadirachtin	0.5-64 μg/gDW	1998	Wewetzer
Leaf and flowers	Callus	Azadirachtin	2.68% of DW	1998	Veeresham *et al.*
Parts of whole plants	Callus	Azadirachtin	Not determined.	2001	Zypman
Shoots tip	Callus	Azadirachtin	0.5 μg/g DW	2000	Schaaf *et al.*

During the 3 year project (1995-1997) protocols for inducing callus formation from plantlets, either *Azadirachta indica* or *Azadirachta excelsa* were developed. In the case of *Azadirachta indica* 9 different callus cell lines were developed and characterised for growth capabilities and target compound expression. The most promising cell line was Vm as it showed the best properties (friability and presence of target compounds) in suspension cell culture development and thus it was selected for all future upstream and onstream work.

In the case of *Azadirachta excelsa* successful callus induction could be obtained but during the whole three-year period all attempts to obtain a regularly growing suspension culture failed. All the same protocols for the regeneration of whole plantlets through indirect organogenesis have been developed (Giagnacovo *et al.*, 2001).

For developing a biotechnological process, which uses plant cells derived from the *Azadirachta indica* whole plant it is necessary to create a suspension culture that allows the use of this biomass in properly designed bioreactors. As described above the cell line Vm was selected for this purpose. Suspension cultures were established successfully and the fundamental culture conditions (incubation conditions and nutritive demand) were defined which has lead to the development of a model system. Of the different culture media tested MS (Murashige and Skoog, 1962) supplemented with IBA (4 mgl^{-1}) and BAP (2mgl^{-1}) and sucrose 3 per cent was the most indicated. A good growth occurred in this model system and that target compound expression was mainly concentrated during the acceleration and exponential phase of the culture period, although in some experiments production was concentrated in the early and late plateau phase. In different experiments carried out azadirachtin production ranged from 0.01 per cent on dry weight basis to a maximum of 0.23 per cent. Using this model system it was possible to establish how some culture conditions influence azadirachtin and other target compound production during a regular culture period. It was established that the influence of carbon source was apparently fundamental. All carbon sources tested (sucrose, glucose and fructose) favoured a good biomass growth while only in the presence of sucrose significant azadirachtin production could be observed. An increase in the sucrose usually present in the model system (3% sucrose) led both to an increase of the dry weight of the biomass produced and to an increase in the azadirachtin production (sucrose at 6%). These two factors had a cumulative effect, which led to a 9-fold increase in azadirachtin production if compared to the model system, i.e. from 2 mgl^{-1} (3% sucrose) to 18 mgl^{-1} (6% sucrose).

With regard to other macronutrients present in the culture medium it has been established that, considering the nitrogen sources, the cells seem to have relied mostly on the nitrate (most abundant form of nitrogen in this medium) as a source of nitrogen throughout the growth cycle and depletion never reached more than 2/3 of the amount initially present. Thus it can be said that NO_3^- certainly isn't a rate-limiting factor. On the contrary phosphate is assimilated by the cell cultures in a very short time. It has been repeatedly observed that phosphate is completely depleted from the medium within a period of 3 to 7 days. Phosphate concentration may thus be a rate-limiting factor during the culture period.

The effect of plant growth regulators (PGR) has also been investigated. During the work for the definition of the model system it was already established that the two PGRs, Indole-3-butyric acid (IBA, an auxin) and 6-Benzylaminopurine (BAP, a kinetin) were the most indicated for obtaining a good biomass growth (respectively 4 mgl^{-1} and 2 mgl^{-1}). If the IBA concentrations were changed over a range from 0 – 6 mgl^{-1} no effect on azadirachtin expression was observed (in all cases about 0.2% of azadirachtin on dry weight basis was obtained). On the other hand if BAP concentration was altered over the same range a strong inverse correlation occured with azadirachtin expression. Indeed at 0 mgl^{-1} of BAP azadirachtin produced was 0.23 per cent on dry weight basis, while at the highest doses tested (6 mgl^{-1}) azadirachtin expression only reached 0.005 per cent.

Another approach that has been followed is an attempt to influence secondary metabolism through elicitation. This approach has been applied on many different plant systems to increase the amount of desired target compound in plant cell suspension cultures. The main compounds that are stimulated through elicitation are the phytoalexins (secondary metabolites which function as so-called post-infection defensive substances within the plant's defensive system against attack by micro-organisms). The signals triggering the formation of phytoalexins are called elicitors. In the widest sense, these are "molecules" inducing a reaction in plant cells assumed to be characteristic of its defensive responses. It is considered that an elicitor molecule combines with a plant membrane receptor and that the complex activates a series of specific genes, resulting in the synthesis of phytoalexins. Formation of phytoalexins is only one of several possible reactions. Indeed, phytoalexins are compounds that are not constitutively present but are synthesised only in response to attack. This response is very fast (within a few hours synthesis is observed). At the same time other mechanisms are being identified which have a much slower reaction time (3 days and more) and which involve constitutive secondary metabolites. In the case of azadirachtin it is the latter kind of mechanism that has to be activated if a result is to be obtained.

The studies show that elicitation had a positive effect on the amount of extractable material, confirming that elicitation has an enhancing effect on secondary metabolism as a whole. The elicitors tested were chitosan, jasmonic acid (JA) and salicylic acid (SA). Chitosan was able to enhance azadirachtin production two times, 48 to 72 hours after elicitation. Jasmonic acid had an inhibiting effect on azadirachtin production, although some other target compounds like azadirachtin I and azadirachtinin were stimulated. Salicylic acid showed the same enhancement pattern as chitosan. The latter is considered a general elicitor while JA and SA are both considered to be implied in the signal transduction pathways leading to defence gene activation. Probably different reaction pathways are activated in relation to which elicitor is applied. The important information one obtains from all the elicitation experiments is that it is possible to manipulate production rates of the target compounds within the cell cultures, both positively as well as negatively.

Another approach that had to be studied during the project was immobilisation. Immobilisation can be a very useful technique for enhancing secondary metabolite production as it allows a prolonged use of cell biomass and the creation of a continuous process. A fundamental necessity, however, is that the desired target product is excreted into the culture medium and that it be produced in maximum amount. In our system, however, the amount of azadirachtin excreted into the culture medium represented never more than 10 per cent of the total amount produced by the suspension culture. Thus the approach of cell immobilisation was of little use for this plant cell system and no more work was done in this direction.

Some final considerations have to be made on the production of the target compounds within the suspension cultures. Firstly, over the three-year period more than 30 growth experiments have been carried out generating each a minimum of 6 to 30 samples to be extracted. The presence of azadirachtin was almost always confirmed but a high variability in the degree of production was present (range 0.001 – 0.2% on dry weight basis in the model system). This variability applied also to the other target compounds monitored. In view of developing an industrial process this remains a major hurdle to be passed. Secondly, the results from the carbon source, elicitation and PGR experiments clearly indicate that the level of expression of the target compounds is amenable to manipulation by external factors. This may indicate that with further efforts in this direction it might be possible to enhance production of the target compounds to a sufficient level and in a stable enough manner so that the economical feasibility becomes viable.

It was also of great importance to develop extraction protocols, which were adapted to the very small amounts of biomass to be analysed for the presence of target compounds. This was also achieved during the AZTEC project and the results are described in Jarvis and Morgan (2000). A rapid and sensitive analysis of azadirachtin and related triterpenoids by high-performance liquid chromatography-atmospheric pressure chemical ionisation mass spectrometry from tissue culture samples has also been developed (Schaaf *et al.*, 2000).

The fermentation experiments at laboratory scale (6 litre bioreactor equipped with marine blade impeller and the necessary probes for O_2, redox, pH, etc.) were carried out with the Vm B cell line incubated in the presence of either 3 per cent or 6 per cent sucrose, as the prior shake flask experiments had shown that these conditions favoured biomass accumulation and target compound expression. The important findings of these fermentation experiments are that

- The cell line VmB is not destroyed by the change from shakeflask conditions to bioreactor conditions, indeed cell viability is about at 50 per cent at the onset of the fermentation run and increases to about 70 per cent.
- Growth occurs although the doubling times are much larger than those observed in the shake flasks system.
- Production of some of the characteristic secondary metabolites of *Azadirachta indica* occurs (nimbin, salannin) although no azadirachtin production has been observed also if the medium containing 6 per cent sucrose is used. The optimum agitation speed has been determined to be 110 rpm and the optimum air flow 0.4 vvm.

However, the major problem encountered was the continuous sterility necessary for more than 20 days. Most fermentations had to be interrupted at 7 or 10 days. This may also account for the lack of azadirachtin production, which could have occurred at a later stage in this particular production process.

The cell line Vm has also been used in a 26l bioreactor. The bioreactor was equipped with a 26l (20 litres working volume) vessel, protected by a stainless steel coating on half its height and by a glass coating on the other half. The stirring system consisted of a basal fulcrum stirrer, equipped with a rushton double impeller and a vibromixer vertical stirrer. The cell line Vm was used with the medium containing 3 per cent sucrose. In this system satisfactory growth was obtained (four times increase in biomass over a 21 day period). However target compound production was not determined. These experiments indicate that the scaling-up of the process (from 5 to 20 litres) does not pose major problems if considered just under the light of biomass production. As mentioned before production of target products still has to be assessed.

Another important approach pursued by the group at Aberdeen University of A.J. Mordue and E.J. Allan is the induction of hairy root cultures (Zounos *et al.*, 1999). They successfully transformed *Azadirachta indica* and were able to establish hairy root cultures both in 250 mL flasks as in 2 L flasks. In the 250 mL flasks maximum azadirachtin production coincided with the end of the exponential growth phase (after six weeks of culture) and was 0.035 mg/g of dry weight, which corresponds to 0.0035 per cent of dry weight. In the two liter flasks production was less, 0.0044 mg/g dry weight (0.00044% of dry weight), and peaked again at the end of the exponential growth phase. These data are interesting as the azadirachtin production is comparable to callus cultures. Clearly the differentiation had no positive effect on secondary metabolite expression and confirms Wewetzer's and our own results, which indicate that the most undifferentiated systems have higher secondary metabolite expression.

5. CONCLUSIONS

The biotechnological approach has been only developed in the last 15 years for neem. Micropropagation can be carried out successfully on different usable protocols. The most indicated is the protocol, which avoids callus formation and thus the risk of somaclonal variation (Eeswara *et al.*, 1998). To our knowledge none of the protocols described have been tested out in an industrial scale micropropagation setting. Probably minor changes will be necessary to these protocols in this kind of production but one can be fairly confident that the industrial approach will be successful. This is very important because of the recalcitrant nature of the neem seeds and also if one wants to maintain a healthy plantation. It is recommended to use seeds or plantlets originating from different provenances (increasing the overall resistance of the plantation to pathogen attack). Also, if it can be proven that elite producing trees exist this will be a very powerful tool for multiplying "true-to-type" plantlets.

The results obtained via somatic embryogenesis are of great interest as it may represent one of the major future developments for neem, i.e. the possibility to mass propagate using synthetic seed technology.

The biotechnological approach for the production of the biologically active compounds present in the neem seeds has also been demonstrated. In this case it must be stressed that an industrial application of this kind of technology is still far off. Nevertheless, suspension cultures of neem will prove to be very powerful tools for the unravelling of the biosynthetic pathway leading to the compounds of interest. The fact that production levels of azadirachtin (and related compounds) are amenable to manipulation in suspension cultures might lead to better practices in neem tree growing which may also enhance the contents of the compounds of interest in the seeds.

It is also of great interest that all the bioassays done with extracts originating from plant tissue cultures (callus, suspension cultures, embryos, hairy roots) all have proven to be active. In many cases the azadirachtin content was very low. Clearly many more bio-active substances are present -some of which are known and have been identified in the tissue cultures- of which some are still to be characterised.

The biotechnological approach has, in the last ten years, shown that it most certainly is a useful activity and in this new millennium will prove to be more and more useful as the tools that are being developed (molecular biology, biochemistry, chemistry, tissue culture techniques, etc.) are becoming ever more powerful. This will allow a comprehensive approach in order to study different aspects of the neem tree such as biodiversity, genetics, secondary metabolites production and mode of action of the bioactive compounds, which will probably help further in better acceptance of the use of its products globally and will inevitably evolve a new path towards sustainable development.

REFERENCES

Ahmad, I. and Ahmed, F. (1995), *In vitro* antimicrobial activity of leaf and bark extracts of *Azadirachta indica* A. Juss, *Indian Veterinary Medical Journal*, **19,** 204-206

Ahuja, M.R. (1989), *Woody Plant Biotechnology*, NATO ASI Series, Plenum Press, NY.

Akhtar, M. (2000), Evaluation of the nematicidal effects of a neem-based product against root-knot nematode on tomato, *International Pest Control*, **42**, 16-17.

Allan, E.J. (1991), Plant cell culture, in *Plant cell and tissue culture*, eds. A. Stafford and G. Warren, Open University Press, Milton Keynes, pp. 1-24.

Allan, E.J., Eeswara, J.P., Johnson, S., Mordue, A.J., Morgan, E.D. and Stuchbury, T. (1994), The production of azadirachtin by *in vitro* tissue culture of neem *Azadirachta indica*, *Pesticide Science*, **42**, 147-152.

Bajagopal B. and Ramaswamy (1996), Studies on *in vitro* culture and production of secondary metabolites in Neem (*Azadirachta indica* A. Juss), Abstract *International Neem Conference*, Queensland, February 4-9.

Beversdorf, W.D. (1990), Micropropagation in crop species, in *Progress in Plant Cellular and Molecular Biology*, eds. H.J.J. Nijkamp, L.H.W. Van Der Plas and J. Van Aartrijk, Kluwer Academic Publishers, Dordrecht, pp. 3-12.

Burhenne-Guilmin, G.L.F., Synge, H., Macneely, J. and Gundling, L. (1994), *A guide to the Convention on Biological Diversity*, World Conservation Union, Ghand, Switzerland.

Debergh, P.C. and Read, P.E. (1991), Micropropagation, in *Micropropagation – Technology and Application*, eds. P.C. Debergh and R.H. Zimmerman, Kluwer Academic Publishers, Dordrecht, pp. 1-14.

Eeswara, J.P., Stuchbury, T., Allan, E.J. and Mordue (Luntz), A.J. (1998) A standard procedure for the micropropagation of the neem tree (Azadirachta indica A. Juss). *Plant Cell Reports*, **17**, 215-219.

Endreß, R. (1994), *Plant Cell Biotechnology*, Springer Verlag, Berlin, Germany

Gautam, V.K., Nanda, K. and Gupta, S.C. (1993), Development of shoots and roots in anther-derived callus of *Azadirachta indica* A. Juss. - a medicinal tree, *Plant Cell Tissue Organ Culture*, **34**, 13-18.

Giagnacovo, G., Pasqua, G., Monacelli, B., van der Esch, S.A., Maccioni, O. and Vitali, F. (2001), Organogenesis and embryogenesis from callus cultures of *Azadirachta excelsa*, *Plant Biosystems*, **135**, 13-18.

Govindachari, T.R., Suresh, G., Gopalakrishnan, G., Masilamani, S. and Banumathi, B. (2000), Antifungal activity of some tetranortriterpenoids, *Fitoterapia*, **71**, 317-320

Guerrini, V.H. and Kriticos, C.M. (1998), Effects of azadirachtin on *Ctenocephlides felis* in the dog and the cat, *Veterinary Parasitology*, **74**, 289-297.

Isman, M.B. (1997), Neem Insecticides, *Pesticide Outlook*, **8**, 32-38.

Isman, M.B., Koul, O., Arnason, J.T., Stewart, J. and Salloum, G.S. (1991), Developing a neem-based insecticide for Canada, *Memoirs of Entomological Society of Canada*, **159**, 39-47.

Jarvis, A.P. and Morgan, E.D. (2000), Analysis of small samples of limonoids of neem (*Azadirachta indica*) using solid phase extraction from tissue culture, *Phytochemistry Analytical*, **11**, 184-189.

Jarvis, A.P., Johnson, S. and Morgan, E.D. (1999), Rapid separation of triterpenoids from Neem seed extracts, *Phytochemistry Analytical*, **10**, 39-43.

Jarvis, A.P., Morgan, E.D., van der Esch, S.A., Maccioni, O., Vitali, F., Ley, S.V. and Pape, A. (1997), Identification of Azadirachtin in tissue cultured cells of neem (*Azadirachta indica*), *Natural Product Letters*, **10**, 95-98.

Kearney, M.L. (1993), Plant cell and tissue culture of *Azadirachta indica* for the production of antifeedant compounds, Msc Thesis, University of Aberdeen.

Kearney, M.L., Allan, E.J., Hooker, J.E. and Mordue (Luntz), A.J. (1994), Antifeedant effects of *in vitro* culture extracts of the neem tree, *Azadirachta indica* against the desert locust (*Schistocerca gregaria* (Forskål)), *Plant Cell, Tissue and Organ Culture*, **37**, 67-71.

Koul, O. (1996a), Advances in neem research and development – present and future scenario, in *Supplement to Cultivation and Utilization of Medicinal Plants*, eds. S.S. Handa and M.K. Koul, National Institute of Science Communication, CSIR, New Delhi, pp. 583-611.

Koul, O. (1996b), Mode of azadirachtin action, in *Neem*, eds. N.S. Randhawa and B.S. Parmar, New Age International Publishers Ltd., New Delhi, pp. 160-170.

Liew, T.K. and Teo, C.K.H. (1998), Multiple shoot production *in vitro* of the tropical timber tree Sentang (*Azadirachta excelsa* Linn.), *Horticulture Science*, **33**, 1073-1075

Mohamed-Yasseen (1994), Shoot proliferation and plant formation from neem (*Azadirachta indica* A. Juss) with thidiazuron, *Horticulture Science*, **29**, 215

Mordue, A.J. and Blackwell, A. (1993), Azadirachtin: An update, *Journal of Insect Physiology*, **39**, 903-924.

Murashige, T. and Skoog, F. (1962), A revised medium for rapid growth and bioassay with tobacco tissue culture, *Physiology Plant*, **15**, 473-497

Murthy, B.N.S. and Saxena, P.K. (1998), Somatic embryogenesis and plant regeneration of neem (*Azadirachta indica* A. Juss), *Plant Cell Reports*, **17**, 469-475.

Narayan, P. and Jaiswal, V.S. (1985), Plantlet regeneration from leaflet callus of *Azadirachta indica* A. Juss, *Journal of Tree Science*, **4**, 65-68.

Onishi, N., Sakamato, Y.and Hirosawa, T. (1994), Synthetic seeds as an application of mass production of somatic embryos, *Plant Cell Tissue Organ Cult*ure, **39**, 137-145.

Puri, H.S. (1999), *Neem: The Divine Tree Azadirachta indica*, Harwood Academic Publishers, Amsterdam.

Ramesh, K. and Padhya, M.A. (1988), Isolation of nimbin from *Azadirachta indica* leaves and its callus culture, *Indian Drugs*, **25**, 526-527.

Ramesh, K. and Padhya, M.A. (1990), *In vitro* propagation of neem, *Azadirachta indica* A.Juss., from leaf discs, *Indian Journal of Experimental Biology*, **28**, 932-935.

Rembold, H. (1989), Azadirachtins, their structure and mode of action, in *Focus on Phytochemical Pesticides. Vol. I, The Neem Tree*, ed. M. Jacobsen, CRC Press, Boca Raton, FL, pp. 47-86

Rembold, H. and Annadurai, R.S. (1993), Azadirachtin inhibits proliferation of Sf 9 cells in monolayer culture, *Zeitschrift fur Naturforschung*, **48C**, 495-499.

Rembold, H., Subrahmanyam, B. and Muller, T. (1989), Corpus cardiacum- a target for azadirachtin, *Experientia*, **45**, 361-363.

Samy, R.P. and Ignacimuthu S. (1998), Antibacterial activity of different extracts of *Azadirachta indica* Juss. (Meliaceae), *Uttar Pradesh Journal of Zoology*, **18**, 71-75

Sanyal, M., Das, A., Banerjee, M. and Datta, P.C. (1981), *In vitro* hormone induced chemical and histological differentiation in stem callus of neem, *Azadirachta indica* A. Juss, *Indian Journal of Experimental Biology*, **19**, 1067-1068.

Sanyal, M.S. and Datta, P.C. (1986), Nutritional environmental and redifferentiation from cultured tissue of *Azadirachta*, *Acta Horticultures*, **188**, 99-103

Sanyal, M., Mukherji, A. and Datta, P.C. (1988), Effect of glycine on *in vitro* biosynthesis of nimbin and β-sitosterol in tissues of *Azadirachta indica*, *Current Science*, **57**, 40-41

Sarkar, M.S. and Datta, P.C. (1986), Age factor in biosynthesis of nimbin and sitosterol in the bark and callus of *Azadirachta indica*, *Indian Drugs*, **24**, 62-63

Sayah, E., Idaomar, M., Soranza, E and Karlinsky, A. (1998), Endocrine and neuroendocrine effects of Azadirachtin in adult females of the earwig *Labidura riparia*, *Tissue Cell*, **30**, 86-94.

Schaaf, O., Jarvis,A.P., van der Esch, S.A., Giagnacovo, G. and Oldham, N.J. (2000), Rapid and sensitive analysis of azadirachtin and related triterpenoids from Neem (*Azadirachta indica*) by high-performance liquid chromatography-atmospheric pressure chemical ionization mass spectrometry, *Journal of Chromatography*, **886A**, 89-97.

Schmutterer, H. (1995), *The Neem Tree: Source of Unique Natural Products for Integrated Pest Management, Medicine, Industry and Other Purposes*, VCH Verlag, Weinheim, Germany.

Schmutterer, H. and Ermel, K. (1995), The Sentang or Marrango tree: *Azadirachta excelsa* (Jack), in *The Neem Tree Azadirachta indica A. Juss. and Other Meliaceous Plants- Source of Unique Products for Pest Management and Other Purposes*, ed. H. Schmutterer, VCH Verlagsgesellschaft mbH, Weinheim, Germany, pp. 598-604.

Schulz, F.A. (1984), Tissue culture of *Azadirachta indica*, in *Natural Pesticides from the Neem Tree (Azadirachta indica A. Juss) and other tropical plants*, eds. H. Schmutterer and K.R.S. Ascher, GTZ, Eschborn, Germany, pp. 539-542.

Shrikhande, M., Thengane, S.R. and Mascarenhas, A.F. (1993), Somatic embryogenesis and plant regeneration in *Azadirachta indica* A. Juss, *In Vitro Cell Development Biology,* **29P**, 38-42.

Su, W.W., Hwang, W.I., Kim, S.Y.and Sagawa, Y. (1997), Induction of somatic embryogenesis in *Azadirachta indica*, *Plant Cell, Tissue and Organ Culture*, **50**, 91-95.

Subramani, J., Arun Raj Nair, C., Gopinathan, M.C.and Pilai, S.S. (1993), Large scale production of neem through *in vitro* methods, in *Practice Oriented Results on Use and Production of Neem-Ingredients and Pheromones*, ed. H. Kleeberg, Druck Graphic, Giessen, pp. 137-143.

Sundaram, K.M.S. (1996), Azadirachtin biopesticide: a review of studies conducted on its analytical chemistry, environmental behaviour and biological effects, *Journal of Environmental Science and Health*, **B31**, 913-948.

Talwar, G.P., Raguvanshi, P., Misra, R, Mukherjee S. and Shah S (1997a), Plant immunomodulators for termination of unwanted pregnancy and for contraception and reproductive health, *Immunology and Cell Biology*, **75**, 190-192.

Talwar, G.P., Raguvanshi, P., Misra, R, Mukherjee S. and Shah, S. (1997b),

Induced termination of pregnancy by purified extracts of *Azadirachta indica* (Neem): Mechanisms involved, *American Journal of Reproductive Immunology*, **37**, 485-491

Tewari, D.N. (1992), *Monograph on Neem (Azadirachta indica A. Juss)*, International Book Distributors, Dehra Dun, India

van der Esch, S.A., Giagnacovo, G., Maccioni, O. and Vitali F. (1994b), Plant tissue culture of *Azadirachta indica* (A. Juss) for the production of azadirachtin and related compounds. Abstracts *VIIIth International Congress of "Plant Tissue and Cell Culture"*, Firenze, p 240.

van der Esch, S.A., Giagnacovo, G., Maccioni, O. and Vitali, F. (1993), Preliminary results on the production of azadirachtin by plant tissue culture of *Azadirachta indica*, *Giornale Botanico Italiana*, **127**, 927-928.

van der Esch, S.A., Giagnacovo, G., Maccioni, O. and Vitali, F. (1994a), Plant Tissue Culture of *Azadirachta indica* (A. Juss), in *Practice Oriented Results on Use and Production of Neem-Ingredients and Pheromones*, ed. H. Kleeberg, Wetzlar, Germany, pp.125-131

Vasil, I.K. (1994), Automation of plant propagation, *Plant Cell Tissue Organ Culture*, **39**, 105-108.

Veeresham, C., Raj Kumar, M., Sowjanya, D., Kokate, C.K. and Apte, S.S. (1998), Production of azadirachtin from callus cultures of *Azdirachta indica*, *Fitoterapia*, **LXIX**, 423-424.

Wachsman, M.B., Castilla, V. and Coto, C.E. (1998), Inhibition of foot and mouth disease virus (FMDV) uncoating by a plant-derived peptide isolated from *Melia azedarach* L leaves, *Archives of Virology*, **143**, 581-590

Wan. M.T., Watts, G.G., Isman, M.B. and Strub, R. (1996), Evaluation of the acute toxicity to juvenile Pacific Northwest salmon of azadirachtin, neem extracts, and neem-based products. *Bulletin of Environmental Contamination and Toxicology*, **56**, 432-439.

Wei Wen Su, Wen-Ing Hwang, Se Young Kim and Yoneo Sagawa (1997), Induction of somatic embryogenesis in *Azadirachta indica*, *Plant Cell Tissue Organ Culture,* **50**, 91-95.

Wewetzer, A. (1998), Callus cultures of *Azadirachta indica* and their potential for the production of azadirachtin, *Phytoparasitica*, **26**, 47-52.

Wewetzer, A. and Schultz, S.A. (1994), Plantlet regeneration from tissue culture of

Azadirachta indica (Juss) *Abstracts VIIIth International Congress of "Plant Tissue and Cell Culture"*, Firenze,. p 12

Wolinsky, L.E. and Mania S, (1996), The inhibiting effect of aqueous *Azadirachta indica* (neem) extract upon bacterial properties influencing *in vitro* plaque formation, *Journal of Dental Research*, **75**, 816-822

Wood, H.N. and Braun, A.C. (1961), Studies on the regulation of certain essential biosynthetic systems in normal and crown-gall tumor cells, *Proceedings National Academy of Sciences USA*, **47**,1907.

Zounos, A.K., Mordue(Luntz), A.J. and Allan, E.J. (1999), Neem (*Azadirachta indica* A. Juss) hairy root cultures as a potential bioinsecticide production system, Abstract C-3-5, Neem 99, World Neem Conference, 19-21 May 1999, Vancouver, Canada.

Zypman, S., Applebaum, S.W. and Ziv, M. (2001), Production of desert locust feeding deterrents from *in vitro* cultured Neem, *Phytoparasitica*, **29**, 284-291.

Chapter 11

PRESENT CONCEPTS OF THE MODE OF ACTION OF AZADIRACHTIN FROM NEEM

A. J. MORDUE (LUNTZ)
Department of Zoology, University of Aberdeen
Tillydrone Avenue, Aberdeen
AB24 2TZ, UK

1. INTRODUCTION

Neem (*Azadirachta indica*) is a multi-purpose tree whose products have been used for centuries for insecticidal, antiseptic, contraceptive, antipyretic and antiparasitic purposes. It is also used as a source of wood in reforestation programs and a provider of shade. The fruit produces oil that is used in soaps and detergents while other by-products are used in fertilisers. The tree contains well over 300 plant secondary compounds, which are responsible for many of its wide-ranging properties. Most of the active principles are terpenoids and are found in neem fruit, seeds, twigs, stem and root bark. The seed kernels produce the complex tetranortriterpenoid azadirachtin (and related analogues), which is responsible for the main insecticidal and insect antifeedant properties of neem (Isman, 1997). Azadirachtin, which occurs in 4-6g/kg amounts in neem seed kernels, was first isolated by Butterworth and Morgan (1968) and its full structure was elucidated in 1987 (Bilton *et al.*, 1987; Kraus *et al.*, 1987; Turner *et al.*, 1987). The chemistry (Ley *et al.*, 1993) and biological activity of azadirachtin (Mordue (Luntz) and Blackwell, 1993; Mordue (Luntz) and Nisbet, 2000; Mordue (Luntz), 2003) have been reviewed many times. *In vitro* tissue culture techniques have been investigated for their all year round production for pesticide use (Van der Esch *et al.*, 1993; Allan *et al.*, 1999; Van der Esch *et al.*, this volume).

The aim of this chapter is to describe the mode of action of azadirachtin against insects, at the whole animal level which is well documented for a large number of insect species, and also at the cellular level where the basic lesion(s) occur. It is important to compare efficacies of azadirachtin action across species, to

O. Koul and S. Wahab (eds.), Neem: Today and in the New Millennium, 229–242.

highlight, when using azadirachtin as a neem insecticide, the important safety margins between insects and vertebrates. Examples in other animal Classes and Phyla also need to be explored that may throw light onto its novel mode of action.

2. ACTIONS OF AZADIRACHTIN AGAINST INSECTS

The antifeedant effects of neem were first described in 1952 by Heinrich Schmutterer who recorded desert locusts (*Schistocerca gregaria*, Forskal) refusing to feed on the neem tree. Since then, seven international conferences on neem, starting with the first in Germany in 1980, and a vast scientific literature report on both the antifeedant and physiological effects on neem. Such effects include primary and secondary antifeedancy, growth reduction, increased mortality, abnormal and delayed moults and sterility effects. Such wide-ranging biological effects in insects come about in two different ways: firstly by direct effects of azadirachtin on cells and tissues and secondly by indirect effects exerted via the endocrine system following direct effects in the neuroendocrine tissues themselves.

2.1 Effects on Feeding

Insects from different Orders are markedly different in their responses to azadirachtin at the feeding deterrence level. "Primary" (or gustatory) antifeedancy - the inability to ingest resulting from the perception of the antifeedant at a sensory level (Schmutterer, 1985), is important in many species of Lepidoptera and some Orthoptera. The desert locust (*Schistocerca gregaria*) is very sensitive to azadirachtin and will not feed on sugar-impregnated discs when azadirachtin is present at concentrations above 0.01ppm (Mordue (Luntz) *et al.* 1998). *Spodoptera littoralis* (African cotton leafworm), *S. frugiperda* (fall armyworm), *Heliothis virescens* (tobacco budworm) and *Helicoverpa armigera* (old world bollworm) all respond behaviourally to azadirachtin and do not feed on discs impregnated with the compound at concentrations of 0.1-10 ppm dependent upon species (ED_{50} values: *S. gregaria* $1x10^{-3}$ ppm, Lepidoptera $1x10^{-7}$ to $1x10^{-2}$ ppm. The antifeedant effect observed in these species is highly correlated with the sensory response of chemoreceptors on the insect's mouthparts (Blaney and Simmonds, 1988; Mordue (Luntz) *et al.,* 1998).

Feeding behaviour depends upon both neural input from the insects' chemical senses (taste receptors on tarsi, mouthparts and oral cavity) and central nervous integration of this 'sensory code'. Azadirachtin stimulates specific 'deterrent' cells in chemoreceptors and also blocks the firing of 'sugar' receptor cells, which normally stimulate feeding (Simmonds and Blaney, 1984; Blaney *et al.*, 1990; Simmonds *et al.*, 1990). This results in an inhibition of feeding, culminating in starvation and death of these species by feeding deterrence alone.

In most other species of phytophagous insects antifeedant together with physiological effects after some ingestion is the norm. Hemipteran insects (e.g. leafhoppers and aphids) are less susceptible to the primary antifeedant properties of azadirachtin and neem products, than Lepidoptera and, as a result, consume sufficient quantities of the compound to produce profound physiological effects. Such effects include "secondary" antifeedant effects. These constitute a reduction in food consumption subsequent to, and as a consequence of, ingestion, application or injection of the antifeedant (Schmutterer, 1985), which results from the disturbance of hormonal and/or other physiological systems. For example, aphids feeding on tobacco seedlings systemically treated with azadirachtin fed normally initially, but, after the initial feed on plants treated with 500 ppm azadirachtin, the next feed was significantly delayed and subsequent feeding activity was suppressed (Nisbet *et al.*, 1993).

2.2 Effects on Growth and Moulting (IGR effects)

The effects of azadirachtin on growth and moulting have been explained in detail elsewhere (Mordue (Luntz) and Blackwell, 1993) and consist of reduced growth, increased mortalities, abnormal moults and delayed moults in insects. These IGR effects are thought to be caused by disruptions of the complex interactions between moulting hormone (20-0H ecdysone from the prothoracic glands) and juvenile hormone (JH from the corpora allata) at the moult. The disruption can be explained by a blockage of release of morphogenetic peptides from the brain, which controls the release of the hormones from their endocrine glands. These peptide hormones, are prothoracicotropic hormone (PTTH) from the pars intercerebralis neurosecretory cell - corpus cardiacum complex, which stimulates the synthesis and release of ecdysone from the prothoracic glands; the allatotropins from the brain - cc complex which stimulates JH release; and the allatostatins, also from the brain (lateral neurosecretory cells), which inhibit JH release (Fig. l).

It is the modification of haemolymph ecdysteroid levels by these indirect effects of azadirachtin that in large part causes the well described insect growth regulatory (IGR) effects of azadirachtin. Delayed or reduced JH titres in addition cause subtle changes of cuticle structure with either larval or adult characteristics being displayed.

2.3 Effects on Reproduction

Azadirachtin can be shown to cause profound effects on reproductive processes of both male and female insects. Insects treated with azadirachtin have degenerate ovaries and a high degree of yolk resorption (Koul, 1984; Dorn *et al.*, 1987; Schlüter, 1987; Schmutterer, 1987). Azadirachtin interferes with both the synthesis of vitellogenin by the fat body and its uptake by the eggs, resulting in reduced fecundity and sterility (Rembold and Sieber, 1981; Tanzubil and McCaffrey, 1990), again due to disruption in JH levels and ovarian ecdysteroid production (Feder *et al.*, 1988). When female aphids were fed on diets containing azadirachtin at below antifeedancy levels (5-40ppm), their fecundity decreased dramatically within 48h of feeding and any nymphs that were produced were non-viable (Lowery and Isman 1994,1996; Mordue (Luntz) *et al.*, 1996; Koul, 2003). With male reproductive behaviour not as much is known regarding the involvement of the endocrine system, as spermatogenesis tends to occur as part of morphogenesis and males are often able to mate soon after emergence. However, effects of azadirachtin on female fecundity have been noted and testes development has been shown to be inhibited. The testes of male desert locusts injected with low concentrations of azadirachtin during development were significantly smaller than controls. In addition the meiotic processes responsible for the production of mature sperm in adult males were interrupted and blocked prior to metaphase (Linton *et al.,* 1997).

2.4 Effects on the Neuroendocrine System

Evidence for direct effects of azadirachtin on the insect neurosecretory system comes from classical endocrine manipulations of extirpation (or azadirachtin treatment) followed by re-implantation of endocrine organs (eg Barnby and Klocke, 1990); or from histological studies using ^{3}H dihydroazadirachtin , paraldehyde fuchsin staining, or antibodies to peptide hormones to follow the accumulation of neurosecretory materials in the brain (Subrahmanyam and Rembold, 1989; Subrahmanyam *et al.*, 1989; Sayah *et al.*, 1996,1998). These studies have shown that azadirachtin disrupts normal synthesis but especially transport and release mechanisms of the peptide hormones (PTTH, (prothoracicotropic hormone), allatostatins and allatoinhibins) that control synthesis and release of ecdysteroid moulting hormones and juvenile hormone (Fig. 1). While such studies are convincing it must be stated however that much more research work is required on the effects of azadirachtin on the neuroendocrine system of insects before a full understanding can be reached on its mode of action at this level.

2.5 Effects on Cell Lines

It has been known for some time that azadirachtin causes cell toxicity and inhibits the proliferation of insect cells grown in culture (Table 1). Effects on insect cell lines mirror the orders of magnitude of effects seen in whole insects. Cell responses show a strong dose dependency that is dependent upon the time of incubation with azadirachtin. EC_{50} for Sf9 cells range between 1.5 x 10^{-10} M azadirachtin at 96h to $5x10^{-8}$ M azadirachtin at 48h. With mammalian cell lines all studies have shown azadirachtin to be very inactive (Table 1). It is clear that there is an extremely large differential in effect of azadirachtin on insect and mammalian cells. Insect cells are much more susceptible to the effects of azadirachtin than are mammalian cells with orders of magnitude between them. EC_{50} values varying from 10^{-10} - 10^{-8} M would be classed as highly toxic (Sf9 cells, *Aedes albopictus* cells), whereas effects varying from 10^{-5} - 10^{-3} M would be classed as mildly to practically non toxic (all mammalian cell lines). Lack of mammalian toxicity of azadarachtin is clearly borne out at the whole animal level in a recent report demonstrating an 'no-observed-effect' level of 1500mg kg^{-1} day^{-1} (the highest quantity tested) when azadirachtin was administered orally to rats for 90 days (Razada *et al.,* 2001).

2.5.1 Cell division in differentiated tissues

The cytotoxic effects of azadirachtin on insect cell lines are the result of both direct cell death and a blockage of mitosis. These significant and important effects have been demonstrated in tissues and organs of insects and also in protozoa such as *Tetrahymena thermophila,* which showed inhibition of cell proliferation and RNA synthesis inhibition with azadirachtin treatment (Fritzsche and Cleffmann, 1987). In the Mexican bean beetle, *Epilachna varivestis*, massive tissue degeneration was seen after azadirachtin treatment in those tissues where active cell division was occurring i.e. the wing discs of developing larvae, the fat body, ovary and testes (Schlüter, 1985, 1987). Also in locust midgut epithelial cells, cell necrosis and a reduction in the number of regenerative cells occurs after treatment with azadirachtin (Nasiruddin and Mordue (Luntz), 1993). In testes of the desert locust, *S. gregaria*, major cytogenetic effects on spermatogenesis in insects injected in the Vth instar twenty days previously were demonstrated, by the use of testes squashes stained with acetic orcein. Meiosis was shown to be blocked at pro-metaphase I, the stage at which spindle formation normally occurs (Linton *et al.,* 1997).

Of importance in trying to understand the site of action of azadirachtin is the report that azadirachtin inhibits spermatogenesis in the malarial parasites *Plasmodium falciparum* and *Plasmodium berghei* with an EC_{50} of 3.5×10^{-6} M (Jones *et al.*, 1994). When malarial parasites enter the insect gut with the blood meal they undergo an extremely rapid, explosive series of events, which leads to the formation of gametes. The application of azadirachtin at this time inhibits the exflagellation of *Plasmodium* microgametocytes to form free-swimming male gametes. Azadirachtin blocks the separation of mitotic spindle poles and the onset of motility probably by disrupting the functioning of the microtubule organising centre (MTOC). This causes large arrays of disorganised microtubules that distort the cell shape and block the onset of sinusoidal movement of the flagellar axonemes (Billker *et al.*, 2003). This effect was only observed if azadirachtin was applied at a critical point in the development prior to the three endomitotic divisions. These observations are consistent with the observed disfunction at prometaphase I of developing locust testes (Linton *et al.*, 1997) and mitotic *Spodoptera* cell lines (Salehzadeh *et al.*, 2002), the stage where the separation and movement of centrioles by astral arrays takes place. Thus the effects are again on packaging and organizational cytoskeletal elements of the cell.

2.6 Protein Synthesis

Azadirachtin directly inhibits protein synthesis in a variety of tissues where cells are producing enzymes: e.g. midgut cells producing trypsin (Timmins and Reynolds, 1992, Koul *et al.*, 1996); midgut and fat body cells producing 20-mono-oxygenases for ecdysone catabolism (Bidmon *et al.*, 1987; Mitchell *et al.*, 1997): midgut and fat body cells producing detoxification enzymes in insecticide resistant insects (Lowery and Smirle, 2000). These effects in differentiated tissues of whole insects are seen

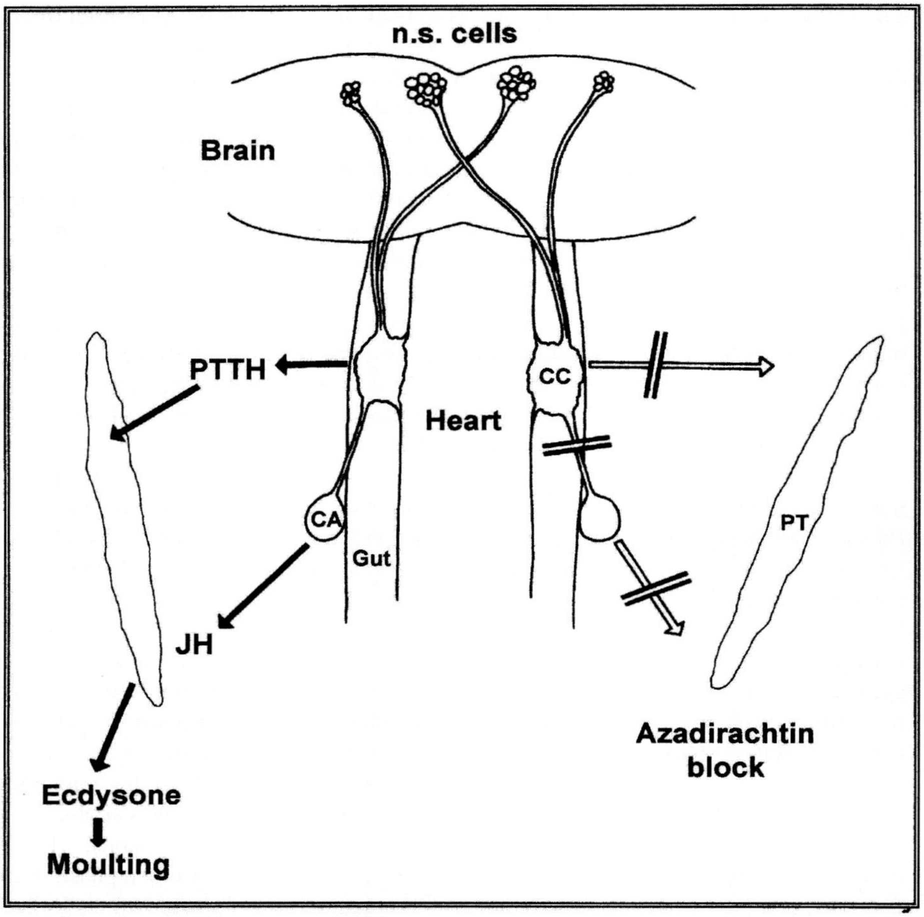

Figure 1. Diagram of a dorsal view of the locust neurosecretory system revealing the control of ecdysone and juvenile hormone (JH) release by the brain morphogenetic peptide hormones prothoracicotropic hormone (PTTH) and the allatostatins and allatotropins. Azadirachtin treatment is thought to block the transport and release of those hormones thus affecting the synthesis and release of moulting hormone, and juvenile hormone; CC, corpus cardiacum; CA, corpus allatum; PT, prothoracic gland.

Table 1. Effects of azadirachtin on cultured cell lines (molar concentrations of azadirachtin required), NCR = neuronal cell response

Cell line	Inhibition of cell growth & proliferation	Nuclear damage	Inhibition of protein synthesis	Effects on NCR	Authors
INSECT CELL LINE					
Spodoptera frugiperda (Sf9)	$1X10^{-6}$ M		$1X10^{-6}$ M		Rembold and Annadurai, 1993
	$< 1X10^{-5}$ M				Mordue and Nisbet, 2000
	$5X10^{-8} – 5X10^{-7}$ M				Reed and Majumdar, 1998
	EC_{50} 96h, $1.5X10^{-10}$ M				Salehzadeh *et al.*, 2002
	EC_{50} 48h, $5X10^{-8}$ M	$1X10^{-8}$ M			Robertson *et al.*, unpublished
Aedes albopictus (c6/36)	EC_{50} 96h, $6.3X10^{-9}$ M				Salehzadeh *et al.*, 2002
Drosophila melanogaster	EC_{50} 48h, $9.4X10^{-7}$ M	$1X10^{-7}$ M			Robertson *et al.*, unpublished
MAMMALIAN CELL LINE					
Chinese hamster ovary	$> 1X10^{-5}$ M				Rembold and Annadurai, 1993
Mouse erythroleukemia (MEL-GM86)	$> 1X10^{-5}$ M				Reed and Majumdar, 1998 Salehzadeh *et al.*, 2002
Mouse fibroblast (L929)	$1X10^{-3}$ M				Salehzadeh *et al.*, 2002
Hepatocyte cells	$1X10^{-3}$ M				Salehzadeh *et al.*, 2002
Human breast cancer (MCF7)	$1X10^{-3}$ M				Salehzadeh *et al.*, 2002
Mouse mammary acini cells	$> 1X10^{-5}$ M		$5X10^{-4}$ M		Nisbet *et al.*, unpublished
6 human gioblastoma (TP53)	$2.8X10^{-5}$ M	$2.8X10^{-5}$ M			Akuduga *et al.*, 2001
Rat dorsal root ganglion				$>1X10^{-5}$ M	Scott *et al.*, 1999

also in insect lines (Table 1) where both cell division and protein synthesis are inhibited in a dose dependant fashion. Using *S. gregaria* injected with azadirachtin, Annadurai and Rembold (1993) were able to study polypeptide profiles of the brain, haemolymph, corpora cardiaca and suboesophageal ganglion using high-resolution 2D electrophoresis. They concluded that differential effects on protein synthesis occur. Whereas overall levels of protein synthesis are reduced, some protein bands disappear after azadirachtin treatment, while others appear and some remain the same. Azadirachtin may be exerting its effects at a transcriptional level where the protein synthetic machinery is switched on for some vital function.

2.6.1 Binding studies

In order to gain an insight into the cellular site of action of azadirachtin binding studies using tritiated dihydroazadirachtin have been carried out on insect tissues and cell lines. Nisbet *et al.* (1995) showed a high level of specific binding of tritiated dihydroazadirachtin to membranes from mature locust testes homogenates. As found with Sf9 cells the binding was time dependent, saturable, of high affinity and indicated one population of binding sites. Also dissociation of the azadirachtin:membrane complex was incomplete showing very tight binding. Autoradiographic studies revealed that the binding was localised on the tail portions of developing sperm, but significantly, motility of mature sperm was not affected by azadirachtin except at very high doses (10^{-3} M) (Nisbet *et al.*, 1996). The preferential binding sites that were visualised and characterised reflect the site of action in developing sperm prior to their release as spermatozoa and the switching on of motility just prior to mating i.e. azadirachtin binds to the axoneme (cell cytoskeletal element) of sperm tails.

Binding studies have also been performed with Sf9 cells and tritiated dihydroazadirachtin (Nisbet *et al.*, 1997) to attempt to identify the specific biochemical lesion at the cellular level. Centrifugation showed that the tritiated ligand bound specifically to the nuclear fraction, rather than other cellular components (microsomes, mitochondria, cytosol). The binding, as for locust sperm, was specific, time depenedent, saturable, of high affinity (K_D 1.8×10^{-8}M) and also most likely to a single population of binding sites (Nisbet *et al.*, 1997). Azadirachtin binding characteristics are to most receptor-ligand binding interactions apart from the rate of dissociation, which is very slow (dissociation rate constant K^{-1} 0.007min^{-1}). This essentially irreversible binding reflects the mode of action of azadirachtin at the whole animal level where its effects are permanent after treatment. The binding site is proteinaceous and associated with RNA as seen by its enrichment with RNase. Unsuccessful attempts to solubilize the azadirachtin-binding complex for further characterisation suggest that its 3-dimensional integrity within membranes is essential for its activity (Mordue (Luntz) *et al.*, 1999).

3. THOUGHTS ON MODE OF ACTION

It would appear from current knowledge that azadirachtin has more than one mode of action. Firstly, azadirachtin alters or prevents the formation of new assemblages of organelles or cytoskeleton resulting in disruption of cell division, blocked transport and release of neurosecretory peptides and inhibition of formation of spermatozoa. Secondly, it inhibits protein synthesis in cells, which are metabolically active and have been switched on to produce large amounts of protein such as midgut cells for digestion, insect cell lines at high proliferation, midgut and fat body cell production of mixed function oxidases for detoxification processes.

Thus at the molecular level azadirachtin may act by altering or preventing transcription and/or translation of proteins expressed during particular stages of the cell cycle. Future work to elucidate the novel mode of action of azadirachtin must now concentrate on characterization and identification of binding sites using proteomic, microarray and differential display techniques.

REFERENCES

Akuduga, J., Gade, G. and Bohm, L. (2001), Cytotoxicity of azadirachtin A in human glioblastoma cell lines, *Life Sciences*, **68**, 1153-1160.

Allan, E.J., Stuchbury, T. and Mordue (Luntz), A.J. (1999), *Azadirachta indica* A. Juss. (Neem Tree), in vitro culture, micropropagation, and the production of azadirachtin and other secondary metabolites, in *Biotechnology in Agriculture and Forestry*, ed. Y.P.S. Bajaj, Springer-Verlag, Heidelberg, pp. 11-41.

Annadurai, R.S. and Rembold, H. (1993), Azadirachtin A modulates the tissue specific 2D polypeptide patterns of the desert locust, *Schistocerca gregaria, Naturwissenschaften*, **80,** 127-130.

Barnby, M.A. and Klocke, J.A. (1990), Effects of azadirachtin on levels of ecdysteroids and prothoracicotropic hormone-like activity in *Heliothis virescens* (Fabr.) larvae, *Journal of Insect Physiology*, **36**, 125-131.

Bidmon H.J., Käuser G., Möbus P. and Koolman J (1987), Effect of azadirachtin on blowfly larvae and pupae, in *Natural Products from the neem Tree and Other Tropical Plants, Proceedings 3rd International neem Conference*, eds. H. Schmutterer and K.R.S. Ascher, GTZ, Eschborn, Germany, pp. 232-272.

Billker, O., Shaw, M.K., Jones, I.W., Ley, S.V., Mordue (Luntz), A.J. and Sinden, R.E. (2002), Azadirachtin disrupts formation of organised microtubule arrays during microgametogenesis of *Plasmodium*, *The Journal of Eukaryotic Microbiology*, **49**, 489-497.

Bilton, J.N., Broughton, H.B., Jones P.S. Ley, S.V., Lidert, Z., Morgan, E.D., Rzepa, H.S. and Sheppard, R.N. (1987), An X-ray crystallographic, mass spectroscopic and NMR study of the limonoid insect antifeedant azadirachtin and related derivatives, *Tetrahedron*, **43**, 2805-2815.

Blaney W.M. and Simmonds M.S.J.(1988), Food selection in adults and larvae of three species of Lepidoptera: a behavioural and electrophysiological study, *Entomologia Experimentalis et Applicata*, **49,** 111-121.

Blaney, W.M., Simmonds, M.S.J., Ley, S.V., Anderson, J.C. and Toogood, P.L. (1990), Antifeedant effects of azadirachtin and structurally related compounds on lepidopterous larvae, *Entomologia Experimenta;is et Applicata*,. **55,** 149-160.

Butterworth, J.H. and Morgan, E.D. (1968), Isolation of a substance that suppresses feeding in locusts, *Journal of Chemical Society, Chemical Communications*, 23-24

Dorn A., Rademacher J.M. and Sehn, E. (1987), Effects of azadirachtin on reproductive organs and fertility in the large milkweed bug, *Oncopeltus fasciatus*, in *Natural Products from the neem Tree and Other Tropical Plants, Proceedings 3rd International neem Conference*, eds.H. Schmutterer and K.R.S. Ascher, GTZ, Eschborn, Germany, pp. 273-288.

Feder D., Valle D., Rembold H. and Garcia S.E. (1988), Azadirachtin-induced sterilization in mature females of *Rhodnius prolixus*, *Zeitschrift fur Naturforschung*, **43C**, 908-913.

Fritzsch U. and Cleffmann G. (1987), The insecticide azadirachtin reduces predominantly cellular RNA in *Tetrahymena*, *Naturwissenschaften*, **74**, 191.

Isman, M.B. (1997), Neem insecticides, *Pesticide Outlook*, 32-38.

Jones, I.W., Denholm, A.A., Ley, S.V., Lovell, H., Wood, A. and Sinden, R.E. (1994), Sexual development of malaria parasites is inhibited *in-vitro* by the neem extract azadirachtin, and its semisynthetic analogs, *FEMS Microbiology Letters*, **120**, 267-273.

Kraus, W., Bokel, M., Bruhn, A., Cramer, R., Klaiber, I., Klenck, A., Nagl, G., Pöhnl, H., Sadio, H. and Vogler, B. (1987), Structure determination by NMR of azadirachtin and related compounds from *Azadirachta indica* (Meliaceae), *Tetrahedron*, **43**, 2817-2830.

Koul, O. (2003) Variable efficacy of neem-based formulations and azadirachtin to aphids and their natural enemies, in *Biopesticides and Pest Management*, eds. O. Koul, G.S. Dhaliwal, S.S. Marwaha and J.K. Arora, Campus Books International, New Delhi, pp. 64-74.

Koul O. (1984), Azadirachtin I. Interaction with the development of red cotton bugs, *Entomologia Experimentalis et Applicata*, **36**, 85-88.

Koul, O., Shankar, J.S. and Kapil, R.S. (1996), The effect of neem allelochemicals on nutritional physiology of larval *Spodoptera litura., Entomologia Experimentalis et Applicata*, **79**, 43-50.

Ley, S.V., Denholm, A.A. and Wood, A. (1993), The chemistry of azadirachtin, *Natural Products Reports*, 109-157.

Linton, Y.M., Nisbet, A.J. and Mordue (Luntz), A.J. (1997), The effects of azadirachtin on the testes of the desert locust, *Schistocerca gregaria*, *Journal of Insect Physiology*, **43**, 1077-1084.

Lowery, D.T. and Isman, M.B. (1994), Insect growth regulating effects of neem extract and azadirachtin on aphids, *Entomologia Experimentalis et Applicata*, **72**, 77-84.

Lowery, D.T. and Isman, M.B. (1996), Inhibition of aphid (Homoptera: Aphididae) reproduction by neem seed oil and azadirachtin, *Journal of Economic Entomology*, **89**, 602-607.

Lowery, D.T. and Smirle, M.J. (2000), Toxicity of insecticides to obliquebanded leafroller, *Choristoneura rosaceana,* larvae and adults exposed previously to neem seed oil, *Entomology Experimentalis et Applicata*, **95**, 201-207.

Mitchell, M.J., Smith, S.L., Johnson, S. and Morgan, E.D. (1997), Effects of the neem tree compounds azadirachtin, salannin, nimbin, and 6-desacetylnimbin on ecdysone 20-monooxygenase activity, *Archives of Insect Biochemistry and Physiology*, **35**, 199-209.

Mordue (Luntz), A.J. (2002), The cellular actions of azadirachtin, in *The Neem Tree*, 2nd edition, ed. H. Schmutterer, The Neem Foundation, Mumbai, pp 266-274.

Mordue (Luntz), A.J. and Blackwell, A. (1993), Azadirachtin: an update, *Journal of Insect Physiology,* **39**, 903-924.

Mordue (Luntz), A.J. and Nisbet, A.J. (2000), Azadirachtin from the neem tree *Azadirachta indica*: its actions against insects, *Annals of the Society of Entomology of Brasil*, **29**, 615-632.

Mordue (Luntz), A.J., Nisbet, A.J., Nasiruddin, M. and Walker, E. (1996), Differential thresholds of azadirachtin for feeding deterrence and toxicity in locusts and an aphid, *Entomologia Experimentalis et Applicata*, **80**, 69-72.

Mordue (Luntz), A.J., Simmonds, M.S.J., Ley, S.V., Blaney, W.M., Mordue, W., Nasiruddin, M. and Nisbet, A.J. (1998), Actions of azadirachtin, a plant allelochemical, against insects, *Pesticide Science*, **54**, 277-284.

Mordue(Luntz), A.J., Nisbet A.J., Jennens L., Ley S.V. and Mordue W. (1999), Tritiated dihydroazadirachtin binding to *Schistocerca gregaria* testes and *Spodoptera* Sf9 cells suggests a similar cellular mechanism of action for azadirachtin, in *Azadirachta indica* A. Juss., eds. R.P. Singh and R.C. Saxena, Oxford & IBH Publishing Co. Pvt. Ltd., pp. 247-258.

Nasiruddin, M. and Mordue (Luntz), A.J. (1993), The effect of azadirachtin on the midgut histology of the locusts, *Schistocerca gregaria* and *Locusta migratoria*, *Tissue and Cell*, **25**, 875-884.

Nisbet, A.J., Woodford, J.A.T., Strang, R.H.C. and Connolly, J.D. (1993), Systematic antifeedant effects of azadirachtin on the peach-potato aphid *Myzus persicae*, *Entomologia Experimentalis et Applicata*, **68**, 87-98.

Nisbet, A.J., Mordue (Luntz), A.J., and Mordue, W. (1995), Detection of [22, 23-$^3H_2$] dihydroazadirachtin binding sites on membranes from *Schistocerca gregaria* (Forskol) testes, *Insect Biochemistry and Molecular Biology,* **25**, 551-558.

Nisbet, A.J., Mordue (Luntz), A.J., Mordue, W., Williams, L.M. and Hannah, L. (1996), Autoradiographic localisation of [22,23-$^3H_2$] dihydroazadirachtin binding sites in desert locust testes, *Tissue and Cell,* **28**, 725-729.

Nisbet, A.J., Mordue (Luntz), A.J., Grossman, R.B., Jennens, L., Ley, S.V. and Mordue, W. (1997), Characterisation of azadirachtin binding to Sf9 nuclei *in vitro*, *Archives of Insect Biochemistry and Physiology*, **34**, 461-473.

Razada, R.B., Srivastava, M.K., Kaushal, R.A. and Singh, R.P. (2001), Azadirachtin, a neem biopesticide: subchronic toxicity assessment in rats, *Food and Chemical Toxicology,* **39**, 477-483.

Reed, E. and Majumdar, S.K. (1998), Differential cytotoxic effects of azadirachtin on *Spodoptera frugiperda* and mouse cultured cells, *Entomologia experimentalis et Applicata*, **89**, 215-221.

Rembold, H and Annadurai, R.S. (1993), Azadirachtin inhibits proliferation of Sf9 cells in monolayer culture, *Zeitschrift fur Naturforschung*,,**48C**, 495-499.

Rembold, H and Seiber K.P. (1981), Inhibition of oogenesis and ovarian ecdysteroid synthesis by azadirachtin in *Locusta migratotia migratoriodes* (R&F), *Zeitschrift fur Naturforschung*, **36C**, 466-469.

Salehzadeh, A., Jabbar, A., Jennens, L., Ley, S.V., Annadurai, R.S., Adams, R. and Strang, R.H.C. (2002), The effects of phytochemical pesticides on the growth of cultured invertebrate and vertebrate cells, *Pest Management Science*, **58**, 268-276.

Sayah, F., Fayet, C., Idaomar, M. and Karlinsky, A. (1996), Effect of azadirachtin on vitellogenesis of *Labidura riparia* (Insecta, Dermaptera), *Tissue and Cell*, **28,** 741-749.

Sayah, F., Fayet, C. and Laverdure, A.M. (1998), Allatostatin-like peptide in the brain-retrocerebral complex of the earwig *Labidura riparia:* cyclic variations related to the reproductive cycle, *Invertebrate Reproduction and Development*, **34**, 269-275.

Schlüter, U. (1985), Occurrence of weight gain reduction and inhibition of metamorphosis and storage protein formation in last instars of the Mexican bean bettle, *Epilachna varivestis*, after injection of azadirachtin, *Entomologia Experimentalis et Applicata*, **39**, 191-195.

Schlüter, U. (1987), Effects of azadirachtin on developing tissues of various insect larvae, in *Natural Products from the neem Tree and Other Tropical Plants, Proceedings 3rd International neem Conference*, eds.H. Schmutterer and K.R.S. Ascher, GTZ, Eschborn, Germany, pp.331-348.

Simmonds, M.S.J. and Blaney, W.M. (1984), Some effects of azadirachtin on lepidopterous larvae, in *Natural Products from the neem Tree and Other Tropical Plants Proceedings 2nd International Neem conference,* eds. H. Schmutterer and K.R.S. Ascher, GTZ, Eschborn, Germany, pp. 163-180..

Simmonds, M.S.J. and Blaney, W.M., Ley, S.V., Anderson, J.C. and Toogood, P.L. (1990), Azadirachtin: structural requirements for reducing growth and increasing mortality in lepidopterous larvae, *Entomologia Experimentalis et A pplicata*, **55**, 169-181.

Scott, R.H., O'Brian, K., Roberts, L., Mordue, W. and Mordue (Luntz), A.J. (1999), Extracellular and intracellular actions of azadirachtin on the electrophysiological properties of cultured rat DRG neurons, *Comparative Biochemistry Physiology*, **123**, 85-93.

Schmutterer, H. (1985), Which insect pests can be controlled by application of neem seed kernel extracts under field conditions? *Zeitschrift furAangewandte Entomologie*, **100**, 468-475.

Subrahmanyam, B. and Rembold, H. (1989), Effect of azadirachtin A on neuroendocrine activity in *Locusta migratoria*, *Cell Tissue Research*, **256**, 513-517.

Subrahmanyam, B., Muller, T. and Rembold, H. (1989), Inhibition of turnover of neurosecretion by azadirachtin in *Locusta migratoria*, *Journal of Insect Physiology*, **35,** 493-500.

Tanzubil, P.B. and McCaffrey A.R. (1990), Effects of azadirachtin in the African armyworm (*Spodoptera exempta*), *Entomologia Experimentalis et Applicata,* **57**, 115-121.

Timmins, W.A. and Reynolds, S.F. (1992), Azadirachtin inhibits secretion of trypsin in midgut of *Manduca sexta* caterpillars: reduced growth due to impaired protein digestion, *Entomologia Experimentalis et Applicata*, **63**, 47-54.

Turner, C.J., Tempesta, M.S., Taylor, R.B., Zagorski, M.G., Termini, J.C., Schroeder, D.R. and Nakanishi, K. (1987), An NMR spectro-scopic study of azadirachtin and its trimethylether, *Tetrahedron*, **43**, 2789-2803.

Van der Esch, S.A., Gianacovo, G., Maccioni, O. and Vitali, F. (1993), Preliminary results on the production of azadirachtin by plant tissue culture of *Azadirachta indica*, *Giornale Botanico Italiana*, **127**, 927-928.

Chapter 12

NEEM BIOTECHNOLOGY – A SYNTHESIS

OPENDER KOUL[1] and SEEMA WAHAB[2]
[1]*Insect Biopesticide Research Centre, 30 Parkash Nagar, Jalandhar- 144003, India;*
[2]*Department of Biotechnology, CGO Complex, New Delhi- 110 003, India*

1. INTRODUCTION

If we read the positively rhapsodic words of neem tree proponents, one would think there is almost nothing this plant cannot do. That is why it has been dubbed "the miracle or wonder tree". Neem bark, leaves, oil, sap, twigs, seeds, roots, flowers, and fruits all have medicinal properties. On top of all this, neem is used to feed cattle and sheep, build houses, supply firewood, provide mulch, improve soil, make soap, and flavor certain dishes (Koul, 1990). No wonder the United Nations declared neem the tree of the 21st century! Dr. Leonard Smith, in a speech transcribed for a neem site (*www.neemorganics.com*) discussed an *in vitro* study on white blood cells infected with HIV. According to this study, the group exposed to neem bark experienced a reduction in the secretion of the P-24 viral protein that did not occur in the control group. Dr. Smith thinks this indicates that lymphocytes with the HIV virus do not release viral protein when neem is present. He also spoke of a study where cells were placed in a petri dish with neem extract and HIV and the lymphocytes did not become infected. In this case, neem seems to have been able to stimulate immunity in the lymphocytes. Dr. Smith did not explain exactly how neem works to stimulate the immune system. Neem's ability to ward off the numerous and commonplace infections that are associated with AIDS also makes it very attractive for people living with HIV. The studies currently available through medical journals were all conducted on rats and mice and were not specifically testing neem and HIV. What these studies do indicate is that with certain groups of viruses, neem activates macrophages and lymphocytes. Mice given neem leaves show immuno-potentiating effects, meaning the oil of neem stimulates their immune systems in laboratory tests. No doubt, this is why villagers in India call neem tree a complete pharmacy that has materials to cure a number of ailments. Another aspect of neem is its use as a biopesticide. Farm chemical use is a rural economic welfare issue globally and more

O. Koul and S. Wahab (eds.), Neem: Today and in the New Millennium, 243–259.

so in India, where cotton farmers currently spend 16 billion rupees annually on insecticide sprays. Vegetable producers in India currently suffer a $2.5 billion loss annually to insect damage, even while spending (on tomatoes, for example) $100-$200 per hectare on insecticides (Padmanabhan, 2000). Approximately 500,000 kg of pesticides are applied each year in US agriculture, and many non-target species beneficial to the environment are negatively affected. Similarly, it is estimated that genetic engineering targeted for pest control could diminish the need for pesticides (Pimentel *et al.*, 1993). Therefore, insecticide derived from neem is especially important for resource-poor farmers, since it requires lower inputs, and causes no health hazards. Yet, in most developing countries the adaptation rate of neem as a method of pest control is still low because of a variety of reasons, such as availability and the cost of raw materials. Neem grows abundantly in Africa and in Asia, but not in Latin America. Also, neem grows mostly in dry regions, and is not found in the areas most suitable for neem insecticide applications, the humid vegetable growing regions. Therefore, wide-spread application is dependent on marketing structures, which are still underdeveloped in most countries.

Secondly, efficiency of neem-based pesticides is based on the ingredient azadirachtin A, which can vary significantly in trees grown under different conditions (Rangasamy *et al.*, 1996). In addition, if the seeds are not dried and stored properly, it can lead to an easy degradation of azadirachtin A. Although the technology needed to prepare a neem seed extract is simple, the labour intensity of collecting and processing is high. For example, an African farmer spends about 32 hours on extraction for 1 ha crop. Many farmers are not willing to spend so many hours on this, and will therefore rely on the market if the prices are low enough. Other limiting factors are the fact that the neem harvest time can compete with the time needed for other harvests; that the land needed for neem is also needed for other crops; and that the harvesting time of neem does not usually coincide with the need for the insecticide. In general, the potential use of neem will occur in a situation were farmers are environmentally conscious, are aware of health problems of other insecticides, and possess processing knowledge (Hellpap and Leupolz, 1999).

There is a growing demand for quality planting material for plantations of neem. However, individual neem trees vary in their chemical make-up as the oil content and limonoid content of neem tree is governed by genetic and environmental factors. Efforts are lacking for the selection of neem trees based on azadirachtin content and the oil content. Very few studies have been carried out so far, in India and abroad to find the existing variability of azadirachtin content in neem trees and even fewer on oil content variability. Ermel *et al.* (1987, 1995) assessed the wide variability of azadirachtin contents in neem seeds of different countries and found that the highest yield of azadirachtin content per seed kernel is not restricted to a specific country but it is distributed in single trees of different origin. A study carried out with neem ecotypes of India showed varying azadirachtin content of 0.14 to 1.66 per cent (Rengasamy *et al.* 1996). Climate, soil type and altitude affected the azadirachtin content. The azadirachtin levels in seeds from plants grown in six ecotypes of Northen Australia ranged from 0.35 to 0.89 per cent of the dried kernel (Bally *et al.* 1996). Seasonal and annual variations were also observed. Efforts are under way at CRIDA, Hyderabad in India to correlate different factors such as soil type, rainfall, and relative humidity with azadirachtin content in the seeds for trees growing in different parts of Andhra Pradesh. These studies suggest that there is a need to do the chemical evaluation of the seeds from diverse zones to identify neem trees having high amounts of oil and azadirachtin to make the plant economically more attractive.

In this direction in India, an admirable initiative has been taken up by launching a "National Network on Integrated Development of Neem" in 1999 by involving `9' government institutions and 2 NGOs. State-wise systematic collection/survey of neem seeds is one of the objectives of the network. Similarly some institutions have undertaken analysis of these seed collection for chemical parameters such as oil, fatty acid and azadirachtin to identify trees having high azadirachtin and oil content. The analysis will help in biochemical characterization of the selections. As the overall objective of the network is to propagate elite planting material, the biochemical characterization will help in identification of elite trees from different zones, which will inturn produce high yields of neem oil and azadirachtun. The elite trees can then be mass propagated through tissue culture and utilized for plantation programmes. This will ultimately help in creating a choice for superior plantation material for Neem growers (http//www.teriin.org/division/bbdiv/pmb/docs/ft02.htm). In fact, in a very recent study an efficient protocol for the production of triploid plants from endosperm callus of *Azadirachta indica* has been achieved (Chaturvedi *et al.*, 2003)

2. NEEM AND GENETIC ENGINEERING

One of the desirable areas of development for genetic engineering technologies that have the potential to benefit agricultural sustainability, the integrity of the natural environment, and the health and safety of society is enhancing crop resistance to pests. Resistance factors and toxins that exist in nature can be used for insect pest and plant pathogen control.. Although some resistance characteristics have been reduced or eliminated in commercial crops, they still can be found in related wild varieties, which provide an enormous gene pool for the development of host plant resistance (Boulter *et al.*, 1990). For example, a wild relative of tobacco that produces a single acetylated derivative of nicotine is reported to be 1000 times more toxic to the tobacco hornworm than is cultivated tobacco (Jones *et al.*, 1987). Transferring this toxic gene to nonfood crops, such as ornamental shrubs and trees, would protect them from some insect pests. In addition, thionins, proteases, lectins, and chitin binding proteins, which are often present in plants, especially in the seeds, help control some pathogens and pest insects in wild plants (Boulter *et al.*, 1990; Czapla and Lang, 1990; Garcia-Olmedo *et al.*, 1992; Pimentel, 1989; Raikhel *et al.*, 1993). Only limited quantities of botanical pesticides are now used in developed countries in place of some synthetic pesticides. However, in some developing countries, including China and India, botanical pesticides such as neem are effectively used. Increasing the effectiveness of neem and other available botanical pesticides by genetic engineering would be an asset to farmers because they are relatively effective and safe. In order to examine the potential of genetic engineering with respect to neem, it is imperative to synthesize some pertinent questions that would help in developing some specific biotechnological approaches for its widespread use.

It is important to know who grows neem trees and where? Neem trees are grown much more widely in semi-arid and light soil regions. Though these trees grow well in Indo-gangetic plains with good irrigation systems, they cannot compete with more payable trees such as mangoes, guavas or even eucalyptus, etc. Neem grows fast and with in a few years starts to bear fruits. It has not become a crop for agro-forestry systems as of yet, but with increasing demand for its seeds, it might well become so providing an attractive economic option to disadvantaged dry farmers. During drought when most crops fail, neem tree survives well and in fact, its leaves are used as stress fodder for livestock. Thus increase in demand for neem seeds from within country or abroad will not only be of benefit to the environment, but for the economy of dry farmers. There is no other crop that is so sturdy, requires so little inputs and yet offers so much economic gains provided of course that so called well wishers of the Third World in USA do not succeed in killing the market for neem seeds from India and Africa.

Another question is regarding the erosion of neem diversity. Neem trees produce seeds profusely and only a small fraction of these seeds are required for its reproduction needs. Given very low dormancy of neem seeds, these quickly germinate after the rains and die if not replanted or allowed to grow. Thus collection of neem seeds poses no danger to continuance of the diversity. There is one danger though that once commercial interests become dominant, few selected germ plasm sources may be widely grown and may lead to genetic uniformity. This danger is not exclusive to neem alone but to all commercial crops grown on large scale and requires generic solution. However, studies at Central Research Institute of Dry Land Agriculture (CRIDA) have convincingly shown that there exists a great genetic diversity. Not only that, scientists at CRIDA has cloned neem trees successfully. This is an important breakthrough because it was seen that propagation by seeds was not always successful. Cross-pollination of neem also made the task of getting true seed of a particular kind more difficult. Clonal propagation of course accentuates the danger of genetic uniformity even more with all the attendant dangers. But this trade off is inherent, as said earlier, in any strategy to maximize returns per unit of land and other scarce inputs.

The point is that excessive collection of neem seeds is unlikely to lead to genetic erosion on a scale more than caused by environmental degradation. Since increase in demand of neem might lead to its cultivation on marginal and so called waste lands, if any thing, it might lead to an increase in the diversity of neem by its cultivation in diverse ecological conditions. In some recent developments regarding neem studies in India a national database is being generated. For instance, National Oilseeds and Vegetable Oils Development Board, Ministry of Agriculture, Government of India, New Delhi is conducting chemical evaluation of neem germplasm for cataloguing and value addition. The protocols for such studies are evaluation of the total oil content of the seeds obtained from zonal selections, study of the fatty acid composition of the seeds obtained from different centers, evaluation of the diversity of the azadirachtin content in different selections, and information on neem resources, utilization, user industries, etc. (http://www.teriin.org/division/bbdiv/pmb/pmb.htm)

One of the first papers, which renewed the interest of the global community in neem, was by Dr Pradhan *et al.* in 1962. It led to isolation and identification of the active principle, azadirachtin (Butterworth and Morgan, 1968; Butterworth *et al.*, 1972; Zanno *et al.* 1975). The properties of this compound as a growth regulator and reproduction-inhibitor in a number of insects were reviewed (Schumutterer, 2002). Koul (1993) lists products made of neem and highlights the pitfalls of relying on a narrow base of compounds as well. The list includes number of products available today (see chapter 1, this volume). However, still the products have not succeeded in the market. A major reason why many neem-based products have not succeeded in the market place is because they are fast biodegradable, stability is definitely a constraint and standardization is still a problem to be solved. Obviously, if a firm invests in research and development of this material, would seek answers to these problems that cannot be summarily ignored.

3. WHAT ARE THE EXPECTATIONS?

While farmers have used neem products for pest control for thousand of years, scientists and companies have also been looking for ways of making products out of neem for at least three decades or more. Indian or western concerns are free to make any product based on available scientific information. True, none of them so far have paid any special price to those from whom they collect neem seed, but that may change. If the demand for leaves or bark or seeds of neem increases in future that will be a great contribution to both economic development and conservation. When the Khadi and Village Industries Commission began using non-edible oilseeds for making soaps, employment opportunities for poor people increased. The question is: will value addition take neem out of the reach of the poor? Not necessarily; the product if made by decentralized but competitive small sectors will remain within the affordable reach of the poor and nobody should bother about who makes it so long as it reaches poor and serves its purpose. The approach should be to recognize farmer's choice; prevent the onslaught of pests; emphasize on value addition in plant products and good and cheap neem based products for the market. An attempt should be made to get neem growers attractive prices and own factories that process the seeds rather than to depend on the multinationals, whose interests lie mostly on conventional chemical pesticides.

There are no farmers' agitations for continued deprivation of poor people in biodiversity-rich and economically poor areas like hill areas, forest regions, drought prone regions etc. Why then are we making an issue of neem? Is it because an increase in the income of growers and seed collectors of neem and other such trees and herbs in the dry regions will deprive urban and rich people of cheap labour? The social and intellectual inertia has generated a good market in India for half-baked theories and populist slogans even if these are based on inaccurate information. Why make a fuss about `patenting the neem tree' itself when this has not been so, cannot be and will perhaps never be possible?

First of all, it must be stated unequivocally that every farmer in any part of the world is free to use neem in whichever way he/she wants. Any claim to the contrary is a case of misrepresentation. Inventions which are patentable include: any new product that is developed afresh; new methods to produce a known old product; and new use suggested for an old product made by new method or old known method. But the patent rights are restricted only to the new, non-obvious, and inventive step.

The use of neem extract, its seed or leaves or any other part of the plant as pesticide cannot be patented since such uses have been known for hundreds of years and are also acknowledged in many patent applications. Also, the seed itself being a product of nature is not patentable unless considerably modified (In many countries, instead of (or in addition to) patenting, plant variety acts are provided to protect plant varieties. There can be no patent on azadirachtin. However, any synthetic analogue of a naturally occurring product is patentable because, it does not exist in nature in that form. About 90 patents have been granted for neem based matetials and some of them, for instance, are:

W R Grace and Company - USA - patent No 4946681, 1990 for improving the storage stability of neem seed extracts containing azadirachtin.

W R Grace and Company - USA - patent No 5001146, 1991 for storage stable azadirachtin formulation.

The National Institute of Immunology, New Delhi (India) patent No. 5196197, 1993 for developing a reversible contraceptive based on neem extract.

Godrej Soaps (India) patent No.5298247, 1994 for water-soluble, storage stable, and environmentally safe pesticide.

W R Grace & Company (USA) patent No 5124349, 1994 for storage stable insecticidal composition comprising neem seed extract. The major contribution was increasing the shelf-life stability of azadirachtin solution.

W. R. Grace and USDA (USA) patent No. 5356628. 1994 for hydrophobic extracted neem oil. A novel fungicide

Trumo Corporation - Japan - patent No 4537774, 1995 for another method using hot water extract of neem bark for controlling tumors.

To obtain such patents has become possible, apparently due to changing interpretations of the laws. For instance, the concept of prior use and novelty has been defined under Section 35 U.S.C. 101 (inventions patentable) in very broad terms, i.e. " Whoever invents or discovers any new and useful process, machine, manufacture, or composition of matter, or any new and useful improvement thereof, may obtain a patent therefore, subject to the conditions and requirements of this title" (United States Patent Trademark Office, 2003). Section 35 U.S.C. 102 of US patent law (conditions for patentability; novelty and loss of right to patents) elaborates the concept further by saying that "A person shall be entitled to patent unless (i) the invention was known or used by others in this country, or patented or described in a printed publication in this or a foreign country, before the invention thereof by the applicant for patent, or (ii) the invention was patented or described in a printed publication in this or a foreign country or in public use or on sale in this country, more than one year prior to the date of the application for patent in the United States (United State Patents and Trademark Office, 2003). Therefore, despite the fact that prior art includes knowledge that was available at the time of patenting, for instance in case of W. R. Grace's process of extracting stable compounds were widely used prior to the issue of the patent, neem was still patentable product because these requirements had not taken place within US territory.

Should we argue here that any National Institute should let any MNC or national corporation commercialize the technology developed by it without any return and then when it needs funds, it should knock at the doors of insensitive bureaucrats in the corridors of power? Or any company should not develop a product, which may compete with the chemical pesticides and thus save the environment a bit? Incidentally, Indonesia had banned fifty percent pesticide use in rice paddies in 1987 and since then its production has been increasing and consumption of pesticides decreasing.

It is obvious that if various users of neem had to identify the plant without drawing upon any clue from traditional knowledge they would have had to spend lot of time in resources. Availability of traditional knowledge certainly reduces their transaction cost. There is a strong case for national and international patent holders to share part of the profits accruing from the commercialization of their products with the providers of the knowledge. Article 8J of Convention on Biological Diversity clearly requires such a reciprocity. However, in this case the neem tree grows in many countries including India and knowledge about its uses also is widely shared. Therefore, no contribution can accrue to any one community or country. Local communities and entrepreneurs for developing various products can only put it in an international fund to support conservation of neem germplasm as well as research and development. It must however be understood that this contribution can only occur if profits are made. And profits can only be made if an entrepreneur has efficient technology, consumer demand and some protection from others not imitating or copying his/her formulation. Therefore, patents on products per se do not preclude the possibilities of communities benefiting from the international fund. It may however be added that situation is far more serious in the case of human drugs and unfortunately there is no hue and cry on that issue. Studies have shown that as many as 74 per cent of the plant derived human drugs are used for the same purpose for which native people discovered their use. And yet, not a penny has ever gone to those communities. A similar situation exists with improved varieties of fruits, vegetables and other food crops of which the large corporations have developed hybrids or other varieties.

The argument that improved varieties or neem products will be available to farmers and should, therefore, be considered, as sufficient reciprocity is not tenable.

Those who benefit will be either commercial farmer in the west or green revolution farmers in developing countries who grow crops that consume much of the pesticides. However, those who grow neem trees or collect its seeds and provide knowledge about its use are generally the farmers in rain fed regions who will not benefit very much from this.

4. THE NEEM PATENTS

Several IPR actions taken outside of India were also represented as threats to national genetic sovereignty. As mentioned above in 1992 the W.R. Grace company secured a patent in the United States for a distinctive chemical formulation of a naturally occurring pesticide from neem trees. Anti-corporate activists in India began asserting that international companies such as Grace were planning to use such patent protections to appropriate for themselves the locally developed folk-knowledge that had long been available to India's farmers and rural communities for free. Grace argued that this charge was unfounded, since the patent was on a new chemical formulation implying that it did not prevent traditional farmers all over the world from continuing to use neem extract as they always had. Nonetheless the neem case enflamed popular anxieties regarding the foreign appropriation of local knowledge, and led to an international struggle by environmental NGOs to remove the patent. In May 2000, the European Patent Office did revoke Grace's 1995 patent in Europe for the neem oil extraction process, but litigation against the patent in the United States was sure to prove more difficult. Some of India's most respected leaders in the area of agricultural research have shared the concern that patent protections for plants, or even a national move toward a conventional PBR system, might leave the nation's poor farmers at a disadvantage. Rural communities in India have for thousands of years employed their own on-farm seed selection practices to breed a highly diverse stock of plant varieties nicely attuned to local conditions. Under a PBR system, why should IPR protection go only to the professional breeders (working either within international companies or national institutes) who routinely use these already improved local varieties as the basis for their breeding programs? On one hand patents on neem were justified on the ground that companies have to spend a huge amount of money on research and development for such products, and have to protect their interests by taking patents on their discoveries, though the general knowledge becomes a private commodity (Anonymous, 1995). Crespi (1995) suggested ways by which the patents on products developed by biotechnology can be defended. However, at the same time the argument is that patents for natural products should be granted on their genuine originality and not to the extent they conflict with traditional knowledge systems. Large-scale purchase of raw materials by multinationals, having huge resources, would take the price of unprocessed neem beyond the reach of farmers who may be forced to rely on the commercial product rather than on traditional recipes (Balasubramanian, 1995).

Let us go a bit deeper in the story. For a long time, Grace's aggressive interest in Indian neem production had provoked serious objections from Indian scientists, farmers and political activists. Grace had earlier been granted another patent in 1992 on a process for extracting and stabilising azadirachtin. According to Grace, azadirachtin was being destroyed during traditional processing. This is highly inaccurate. The extracts were indeed subject to degradation but this did not amount to any wastage since farmers put such extracts to use as and when required. The problem of stabilization arose only when it needed to be commercially packaged for a long time. Moreover, stabilization and other advances, attributed to modern laboratory technology, had already been developed by Indian scientists in the 1960s and 1970s, well before the US and Japanese companies expressed interest in them. The 1992 patent application was put forward by Grace on the principle that the process supposedly invented by them paved the way for additional extraction in the form of water soluble neem extract and hence is an add-on rather than a substitute to the current neem industry in India. In short, the processes are supposedly novel and an advance on the Indian techniques. However, this novelty exists mainly due to ignorance in the West. In 1995, a coalition of NGOs from 40 countries was established to protest Grace's 1992 patent. The petition was filed primarily on the following grounds:

- Biological resources are common heritage and are not to be patented.
- The patent will restrict the availability of living material to local people, whose ancestors have developed the material through centuries.
- The patent may block economic growth in developing countries.

The coalition of NGOs fear that if this neem patent is allowed to stand it would mean that the indigenous population around the world will not be able to freely use many of the biological resources that have been developed and nurtured by them over hundreds of years. At the time the Neem patent challenge was filed, only four patents had been granted on Neem products by the European Patent Office. Today one can find 40 neem patent applications at various stages in the European Patent Office, and 90 have been granted worldwide on various aspects. These include claims for insecticides, fungicidal effects, methods of extraction, storage stable formulations of one of the active ingredients, azadirachtin, contraceptives, and medical uses. The majority of neem "proprietors" are transnational corporations, such as the pharmaceutical company Rohm and Haas, and the agrochemical giant W.R. Grace. It should be noted that none of the neem patents involve a genetically engineered product; neither has the tree itself been patented, nor any of its parts.

5. THE BIOPIRACY VIS-À-VIS THE NEEM TREE

The neem patents will result in major financial gains for their so-called owners, but the communities, which first understood the neem's uses and shared this knowledge with the rest of the world, will not be compensated at all. The neem patents are just one in a large catalogue of genetic resources originating in the South over which intellectual property rights are being asserted by a few multinational corporations originating, for the most part, in the North. The Northern patent system was not intended to recognise or reward as inventive the products of community innovation processes such as those that created the various uses of the neem today. It is only when these uses are described in the terms of Western science and technology that an "invention" is deemed to have taken place and an individual "inventor" or a set of individual "inventors" is allowed to be rewarded with the monopoly property rights that make a patent worth having. This is the mechanism through which a massive transfer of biological and intellectual wealth is taking place--from the Third World to the North.

The fungicide claimed in the USDA/W.R. Grace patent cannot be produced without naturally occurring neem seeds. One direct impact of the corporate monopoly on neem made possible by the patent system is a staggering increase in the companies' demand for seed. A processing plant set up by Grace in India can handle 20 tons of seed per day. Almost all the seed collected—which was previously freely available to the farmer and healer—is now purchased by the company, causing the price of neem seed to rise beyond the reach of the ordinary people. Neem oil itself, used for lighting lamps, is now practically unavailable, as the local oil millers are not able to access the seed. Poor people have lost access to a resource vital for their survival—a resource that was once widely and cheaply available to them. In an effort to deal with the problems of biopiracy (Johnston, 1995) there were attempts to introduce a mechanism for "prior informed consent" into the EU Directive on "Legal Protection of Biotechnological Inventions." However, this controversial legislation was enacted in July 1998 without building in any of the proposed protective measures. Now efforts are being focused on the Biodiversity Convention as an international legal instrument to require that patent applications involving biological resources identify the source of the material.

5.1 Facts of the Case

On December 12, 1990 the multinational agribusiness corporation W.R. Grace of New York and the United States Department of Agriculture, Washington DC, filed a European Patent application with the European Patent Office (EPO) on the basis of a U.S. priority application of December 26, 1989, covering a method for controlling fungi on plants by the aid of a hydrophobic extracted neem oil. After a very difficult and highly controversial examination procedure, the grant of a European patent for this application was published on September 14, 1994, the main claim having been restricted by the EPO to:

"A method for controlling fungi on plants comprising contacting the fungi with a neem oil formulation containing 0.1 to 10% of a hydrophoobic extracted neem oil which is substantially free of azadirachtin, 0.005 to 5.0% of emulsifying surfactant, and 0 to 99% water."

In June of 1995 a legal opposition against the grant of this patent was filed by Magda Aelvoet, MEP, on behalf of the Green Group in the European Parliament, Brussels, Dr. Vandana Shiva, on behalf of the Research Foundation for Science,

Technology, and Natural Resource Policy, New Delhi, and the International Federation of Organic Agriculture Movements, based in Germany. The Opponents submitted evidence to the EPO that the fungicidal effect of hydrophobic extracts of neem seeds was known and used for centuries on a broad scale in India, both in Ayurvedic medicine to cure dermatological diseases, and in traditional Indian agricultural practice to protect crops from being destroyed by fungal infections. Since this traditional Indian knowledge was in public use for centuries, it would seem that the patent application in question lacked two basic statutory requirements for the grant of a European patent, namely novelty and inventive step (in the U.S. non-obviousness). In addition, the Opponents charged that the fungicidal method claimed in the patent was based on one single plant variety (*Azadirachta indica)* and hence resulted in at least partially monopolising this single plant variety. Since the European Patent Convention (EPC) explicitly prohibits the patenting of plant varieties, the patent should therefore be revoked (www.teriin.org/division/bbdw/pmb/docs/H02.htm).

In a first preliminary statement of September 30, 1997, the Opposition Board of EPO held that in summary, it appeared that "*the present patent cannot be maintained*" in view of the evidence supplied by the Opponents for lack of novelty and inventive step. Moreover, the content of additional evidence filed by the Opponents could "*possibly form a very relevant prior art with regard to the inventive step.*" In a second preliminary statement of June 15, 1999, the Opposition Board of EPO held that according to evidence supplied by the Opponents it appeared that "*all features of the present claim* (of the patent) *have been disclosed to the public prior to the patent application during field trials in the two Indian districts Pune and Sangli* of Maharashtra, Western India, in summer 1985 and 1986. Furthermore, the Opposition Board held that on the basis of other evidence supplied by the Opponents, it appeared to be "*mere routine work for a skilled person to add an emulsifier in an appropriate amount*" and that therefore, "*the present subject-matter was considered not to involve an inventive step.*"

But citizens and scientists around the world have registered strong and growing ethical and scientific concerns about reducing the fundamental building blocks of life to commodities bought and sold in the market place. In May, 2000, Indian and EU activists scored an important and precedent-setting victory against the forces of biopiracy. They persuaded the European Patent Office to remove a patent that had been registered by the pharmaceutical giant W. R. Grace for chemical formulations derived from the Neem tree. The company planned to privatize and profit from the bio-pesticidal and medicinal properties of the Neem that have been known and used for generations by indigenous villagers and farmers in India.

The Neem Patent challenge was initiated in solidarity with the Neem Campaign, which was launched in 1993 by farmers in India who feared that their genetic resources and traditional knowledge were coming increasingly under foreign control through the legal mechanism of patents. They likened what they were experiencing to a modern form of "enclosure of the commons"—but in this case it was not public land that is being privatized, it was public knowledge. A delegation of Indian farmers and scientists is bringing to Munich 500,000 signatures of Indian citizens demanding that the patents on Neem be withdrawn. (www.teriin.org/division/bbdw/pmb/docs/H02.htm).

6. BALANCING ACT

On the whole, we think that the prospect that Indian farmers would experience negative economic consequences because they have to buy seeds will depend on the future availability of the seed. This availability is currently not a problem in India. However, it is hard to estimate if and how this would change. There is, therefore, a need to research these kinds of consequences in its specific socio-economic context, which not only differs from region to region, but also from country to country. These differences will make it hard to extrapolate the Indian experiences to other countries. In addition, this potential availability problem seems not to be the result of the Grace patent, but more of commercialization of neem-related products. In addition, prohibition of the use of traditional extractions seems unlikely. There are, however, some serious reasons for concern.

Firstly, if there was prior knowledge of this process outside the USA, then the commercial value of that knowledge is at least annulled in the USA. For India, it will be more difficult to penetrate the markets of developed countries once specific neem-related products and processes have been patented there.

Secondly, the rules of the *General Agreement on Tariffs and Trade/World Trade Organization* (GATT/WTO) on intellectual property can induce damage to the Indian economy. Patents are always national in character. Nevertheless, under the rules of the GATT/WTO, India has to eliminate the exclusions in its patent law. However, it is unclear whether India can be forced to acknowledge the existing neem-related patent. If so, India may find 'its' knowledge, assuming that the coalition's claim is correct, turned against it in the form of a patent held by a foreign company.

Thirdly, an additional question concerns the control over biological resources (neem is originally native to southeast and southern Asia). Issues such as how and who compensates developing countries or farmers for the use of 'their' biological resources remains an important issue. This deserves the attention that the neem patent debate is now attracting (Kocken and Roozendaal, 1997)

6.1 Neem and TRIPS

It is well known now that biodiversity has been further threatened as US patent laws have become globalized through the WTO and TRIPS (Trade related Intellectual Property Rights). The protection of intellectual propery rights (IPRs) in various forms is considered to provide minimum standard of protection on a global scale and have been enforced through TRIPS agreement, adopted in 1995. It has been found that failure to provide adequate and effective IPR protection is legitimate barrier to trade. Developing countries have opposed this for a number of reasons, including biopiracy. However, Maskus (2000) described some advantages of it, like encouragement of new ideas and innovation, technological progress due to the money incentive for R&D, and the transmission of these new advances on a global scale. At the same time strong patent laws enforced through IPR have been rejected in developing countries due to their inherently monopolistic, exclusionary and a tool of intellectual property law (Waterman, 2003). Developing countries have a genuine concern as they are not adequately compensated when foreign researchers develop products that are based on existing materials or knowledge once taken away from them (Braga, 2000); such as the case with neem based products where Indian biodiversity is patented creating a negative impact on the country. Currently, Third World Network, which represents developing countries to meet these challenges and issues, categorically states that patents cannot be granted for trivial changes to known products and processes. Indian villagers have been extracting the neem tree's chemical for pesticidal use through water and alcohol solvents years ahead of the patented processes of today (Third World Network, 2003). Therefore, lack of inclusion of indigenous knowledge into TRIPS reveals that , "IPRs for corporate interest is strong, to say the least, and are non-existent for the owners of knowledge that firms base their knowledge on" (Perelman, 2002). According to Waterman (2003) it is ironic that TRIPS was established to eradicate the expropriation of knowledge when it successfully does just that to traditional bio-diversity.

7. IMPACT ON ECONOMY

In the present scenario of development of neem based products it is apparent that potential and future economic impact of neem could be significant financially, though any specific numbers are not available. As mentioned above and earlier in chapter 1 of this volume almost every part of this tree has potential to give us a commercial product, besides being an excellent source of biopesticide. According to the estimates of Neem Foundation (2003) environmental service rendered by neem tree at $ 10 per month, would yield a value of US $ 24,000 to 36,000 in its 200 to 300 year life time. Further, the alternative uses in organic agriculture (like fertilizer) and medicinal products, commercialized neem has huge economic potential. It may seem a bit exaggerated at this stage that "demand for neem products, especially the seed as the basic raw material is going to increase by leaps and bounds and is also going to provide solution for creating income and job opportunities" (Neem Foundation, 2003); the statement cannot be summarily ignored. However, it is to be seen that how are exclusive rights to produce and export neem is going to be globally handled, which has tremendous legal implications. So it is to be seen what happens in 2005 when real effects of TRIPS would be felt.

8. ENVIRONMENTAL IMPACT

It is well known that neem provides an environmental friendly alternative for synthetic pesticides, which degrade environment. The neem tree has been recognized as a renewable source for organic agrochemicals and nutrients, which are biodegradable, non-toxic and potentially efficacious. Neem also provides refuge for many beneficial organisms like bats, birds, honey bees, spiders, etc. Several species of birds and fruit eating bats survive on the ripe fruits (Neem Foundation, 2003). Neem is an ideal tree for reforestation as it can rehabilitate degraded, semiarid and arid lands. It may also contribute to the decline of carbon dioxide and subsequently help to slow global warming (Schmutterer, 2002).

According to Waterman (2003), it is possible that in future foreign patents may limit and monopolize the use, production and exportation possibilities of neem. Given the beneficial effects neem has on the ecosystem and environment, Waterman feels that any restriction or limitation would be ecologically and environmentally unbeneficial and perhaps harmful if people seek more environmental harmful means of replacing neem. In fact, patents on biodiversity that is definitely going to effect environment (though indirectly), needs international debate to provide appropriate solutions. Neem has great ecological, environmental and economic potential, at least for India, but they may never be realized if the policies existing and regulation implemented today are not streamlined.

9. CONCLUSIONS

Agricultural sustainability requires a focus on the long run, on intergenerational equity. It must be capable of meeting the needs of the present while leaving equal or better opportunities for the future. It must be ecologically sound and socially responsible as well as economically viable. It must also include, as much as possible, the element of local or regional production, and aim for a reasonable level of regional food security. It encourages a shortening of the distance between producers and consumers, to the benefit of both. In a local economy consumers have influence over the kind and quality of their food, and they contribute to the preservation and enhancement of the local landscape. It gives everybody in the local community a direct, long-term interest in the prosperity, health, and beauty of their homeland.

Organic farming falls under this broader classification of "sustainable agriculture." It is commonly thought of as farming without chemicals, and that is usually the case, but it is much more than that. Organic farmers try to farm holistically - that is, they design production systems that capitalize on the positive synergies among crops, soils, seeds, and animals, in such away that each element of the system promotes the productivity and health of other elements. The rapid growth of organic and sustainable agriculture in Canada is occurring with almost no support from the federal government, whose policies are almost entirely devoted to encouragement of industrial agriculture. Other countries are heading in the opposite direction. Germany's goal, for instance, is to see a 20 percent market share for organic products within ten years. The cornerstone of Germany's new agricultural policy will be sustainability.

ACKNOWLEDGEMENTS

Authors are thankful to Prof. John D. Stark, Washington State University for his helpful comments and suggestions on earlier draft of this chapter.

REFERENCES

Anonymous (1995), Patenting nature now, *Nature London*, **377**, 89-90.

Balasubramanian, D. (1995), Patents and indigenous lore, *Nature London*, **378**, 352.

Bally Ian, S. E., Ruddle, L. and Simpson, B. (1996), azadirachtin level in neem seed grown in northern Australia, *Abstracts International Conference on Neem, Queensland, Australia*, p. 30.

Boulter, D., Edwards, G. A., Gatehouse, A. M. R., Gatehouse, J. A. and Hilder, V. A. (1990) Additive protective effects of incorporating two different higher plant derived insect resistance genes in transgenic tobacco plants. *Crop Protection*, **9**, 351–354.

Braga, P. C. A., Carsten, F. and Claudia, P.S. (2000), *Intellectual Property Rights and Economic Development*, World Bank Working Paper No. 412, Washington DC.

Butterworth, J.H. and Morgan, E. D. (1968), Isolation of a substance that suppresses feeding in locusts, *Journal of Chemical Society Chemical Communications*, 23-24.

Butterworth, J. H., Morgan, E.D. and Percy, G. R. (1972), The structure of azadirachtin; the functional groups, *Journal of Chemical Society Perkin Trans I*, 2445-2450.

Chaturvedi, R., Razdan, M. K. and Bhojwani, S. S. (2003) An efficient protocol for the production of triploid plants from endosperm callus of neem, *Azadirachta indica* A. Juss, *Journal of Plant Physiology*, **160**, 557-564.

Crepsi, S. (1995), Biotechnology patenting: the wicked animal must defend itself, *European Intellectual Property Review*, **17**, 431-444.

Czapla, T. H. and Lang, B. A. (1990), Effect of plant lectins on the larval development of European corn borer (Lepidoptera: Pyralidae) and Southern corn rootworm (Coleoptera: Chrysomelidae). *Journal of Economic Entomology*, **83**, 2480–2485.

Ermel, K., Pahlich, E. and Schmutterer, H. (1987), Azadirachtin content of neem kernels from different geographical locations, and its dependence on temperature. Relative humidity, and light, in *Natural Pesticides from Neem Tree and Other Tropical Plants*, eds. H. Schmutterer and K.R.S. Ascher, GTZ, Eschborn, Germany, pp. 171-184.

Ermel, K. (1995) Azadirachtin content of neem seed kernels from different regions of the world, in *The Neem Tree Azadirachta indica A. Juss and Other Meliaceous Plants- Source of Unique Products for Pest Management and Other Purposes*, ed. H. Schmutterer, VCH Verlagsgesellschaft gmbH, Weinheim, Germany, pp. 89-92.

Garcia-Olmedo, F., Carmona, M.J., Lopez-Fando, J.J., Fernandez, J.A., Castagnaro, A., Molina, A., Hernandez-Lucas, C. and Carbonero, P. (1992), Characterization and analysis of thionin genes, in Genes Involved in Plant Defense, eds. T. Boller and F. Meins, Springer-Verlag, New York, pp. 283-302.

Hellpap, C. and Leupolz, W. (1999), Influence of socioeconomic factors on the use of neem insecticides by farmers, in *Azadirachta indica* A. Juss., eds. R.P. Singh and R.C. Saxena, Oxford & IBH Publishing Co., New Delhi, pp. 317-322.

Johnston, B. (1995), The patenting of neem: modern piracy or political correctness? *Herbalgram*, **35,** 32-33.

Jones, D., Huesing, J., Zador, E. and Heim, C. (1987), The tobacco-insect model system for genetically engineering plants for non-protein insect resistance factors, in *UCLA Symposia on Molecular Cell Biology*, Alan R. Liss, New York, pp. 469-478.

Kocken, J. and Roozendaal, G. Van (1997), The neem tree debate, *Biotechnology and Development Monitor*, **30**, 8-11.

Koul, O. (1993), Plant allelochemicals and insect control : An antifeedant approach, in *Chemical Ecology of Phytophagous Insects*, eds. T. N. Ananthakrishnan and A. Raman , Oxford & IBH Publishing Co. Pvt. Ltd., New Delhi, pp. 51-80.

Koul, O. (1996), Neem research and development : present and future scenario. In S. S. Handa and M. K. Koul (eds.), *Supplement to Cultivation and Utilization of Medicinal Plants*, PID, CSIR, New Delhi, pp. 583-611.

Koul, O., Isman, M. B. and Ketkar, C. M. (1990), Properties and uses of neem, *Azadirachta indica* A. Juss. *Canadian Journal of Botany*, **68**, 1-11.

Maskus, K.E. (2000), *Intellectual Property Rights in the Global Economy*, Institute for International Economics, Washington, DC.

Neem Foundation (2003), www.neemfoundation.org

Padmanabhan, A. (2000), Transgenic Seeds: Key to Green Revolution, Version 2.0, *India Abroad* August 4.

Perelman, M. (2002), *Steal This Idea*, Palgrave, New York.

Pimentel, D., Hunter, M.S., LaGro, J.A., Efroymson, R.A., Landers, J.C., Mervis, F.T., McCarthy, C.A. and Boyd, A.E. (1989), Benefits and risks of genetic engineering in agriculture, *BioScience*, **39**, 606-614.

Pimentel, D., McLaughlin, L., Zepp, A., Lakitan, B., Kraus, T., Kleinman, P., Vancini, F., Roach, W. J., Graap, E., Keeton, W. S. and Selig, G. (1993), Environmental and economic effects of reducing pesticide use in agriculture, *Agriculture, Ecosystem and Environment*, **46**, 273-288.

Pradhan, S., Jotwani, M.G. and Rai, B.K. (1962), The neem seed deterrent to locusts, *Indian Farming*, **12**, 7-11.

Raikhel, N.V., Lee, H.I. and Broekaert, W.F. (1993), Structure and function of chitin-binding proteins, *Annual Review of Plant Physiology and Plant Molecular Biology*, **44**, 591-615.

Rangasamy, S., Kaushik, N., Kumar, J., Koul, O. and Parmar, B. S. (1996), Azadirachtin content and bioactivity of some ecotypes of India. In R. P. Singh, Chari, M. S. Chari, A. K. Raheja and W. Kraus (eds.), *Neem and Environment*, Vol. I, Oxford & IBH Publishing Co. Pvt. Ltd., New Delhi, pp. 207-217.

Schumutterer, H. (2002), *The Neem Tree*, 2nd edition, Neem Foundation, Mumbai, India.

Third World Network (2003), www.twnside.org.sg/title/neem-ch.htm

United States Patent and Trademark Office (2003), www.uspto.gov/web/office/pac/mpep/consolidated_laws.pdf

Waterman, J. (2003), Neem, US patent laws and TRIPS: Expropriation or legitimate innovation? www.econ.ucsb.edu

Zanno, P.R., Miura, I., Nakanishi, K. and Elder, D.L. (1975), Structure of the phagorepellent azadirachtin. Application of PRFT CWD carbon-13 nuclear magnetic resonance, *J. Am. Chem. Soc.*, **97**, 1975-1977.